T0176649

Composite Structures of Steel and Concrete

Composite Structures of Steel and Concrete

Beams, Slabs, Columns and Frames for Buildings

Fourth Edition

Roger P. Johnson
University of Warwick, UK

with Fire Resistance chapter contributed by

Yong C. Wang
University of Manchester, UK

This edition first published 2019
© 2019 John Wiley & Sons Ltd

Edition History
Crosby Lockwood Staples (1e 1975); Blackwell Scientific Publications (2e 1994); Blackwell Publishing
(3e 2004)

Registered Offices
John Wiley & Sons, Inc., 111 River Street, Hoboken, NJ 07030, USA
John Wiley & Sons Ltd, The Atrium, Southern Gate, Chichester, West Sussex, PO19 8SQ, UK

Editorial Office
9600 Garsington Road, Oxford, OX4 2DQ, UK

For details of our global editorial offices, customer services, and more information about Wiley products
visit us at www.wiley.com.

Wiley also publishes its books in a variety of electronic formats and by print-on-demand. Some content that
appears in standard print versions of this book may not be available in other formats.

Library of Congress Cataloging-in-Publication Data

Name: Johnson, R. P. (Roger Paul), author.
Title: Composite structures of steel and concrete : beams, slabs, columns and
    frames for buildings / Roger P. Johnson, University of Warwick, UK.
Description: 4th edition. | Hoboken, NJ, USA : Wiley [2019] | Includes
    bibliographical references and index. |
Identifiers: LCCN 2018016256 (print) | LCCN 2018016962 (ebook) | ISBN
    9781119401414 (Adobe PDF) | ISBN 9781119401384 (ePub) | ISBN 9781119401438
    (hardcover)
Subjects: LCSH: Composite construction. | Building, Iron and steel. |
    Concrete construction.
Classification: LCC TA664 (ebook) | LCC TA664 .J63 2018 (print) | DDC
    624.1/821–dc23
LC record available at https://lccn.loc.gov/2018016256

Cover Design: Wiley
Cover Image provided by Severfield plc

Set in 10/12pt WarnockPro by SPi Global, Chennai, India

Printed in Singapore by C.O.S. Printers Pte Ltd

10 9 8 7 6 5 4 3 2 1

Contents

Preface

This volume provides an introduction to the theory and design of composite structures of steel and concrete. Readers are assumed to be familiar with the elastic and plastic theories for the analysis for bending and shear of cross-sections of beams and columns of a single material, such as structural steel, and to have some knowledge of reinforced concrete. No previous knowledge is assumed of the concept of shear connection within a member composed of concrete and structural steel, nor of the use of profiled steel sheeting in composite slabs. Shear connection is covered in depth in Chapter 2 and Appendix A, and the principal types of composite member in Chapters 3, 4 and 5.

Limit state design philosophy has been used in British codes for structural design for over 40 years. Some familiarity is assumed with ultimate limit states (ULS) and serviceability limit states (SLS), which are the main subject here. The accidental limit state of exposure to fire, important in buildings, is the subject of Chapter 6.

All material of a fundamental nature that is applicable to structures for both buildings and bridges is included, plus more detailed information and a fully worked example relating to buildings. Subjects mainly applicable to bridges, such as box girders and design for fatigue, are not included. The design methods are illustrated by calculations. For this purpose a single structure, or variants of it, has been used throughout the volume. The reader will find that its dimensions, its loadings, and the properties of its materials soon remain in the memory. Its foundations are not included. The design is not optimal, because one object here has been to encounter a wide range of design problems, whereas in practice one seeks to avoid them.

This volume is intended for undergraduate and graduate students, for university teachers, and for engineers in professional practice who seek familiarity with composite structures. Most readers will seek to develop the skills needed both to design new structures and to predict the behaviour of existing ones. This is now always done using guidance from a code of practice. The Eurocodes replaced the former British codes in 2010. Their use is required for building and bridge structures that are publicly funded. Use of the former codes continues for some smaller projects, for private clients, but their methods are increasingly out-of-date.

All the design methods explained and used in this volume are those of the current Eurocodes, except where expected revisions are used. In the worked examples, both tensile and compressive forces and strengths of materials are given as positive numbers, distinguished in symbols by subscripts 't' or 'c' or by words such as 'tensile'. A rigorous tension-positive convention has been used only in Appendix A. Elsewhere, the presentation is in the style normally used for hand calculations in practice.

The British versions of the Eurocodes are numbered BS EN 1990 to 1999, subdivided into 58 Parts. The national versions of the Eurocodes were published by each national standards organization in its chosen language, between 2002 and 2010. Those most relevant to this book are in the list of References, beginning 'BSI'. Similarly, the German standards organization (for example) has published DIN EN 1990, and so forth. Each code or part includes a National Annex, for use for design of structures to be built in the country concerned. Apart from these annexes and the language used, the codes are identical and are applicable in all countries that are members of the European Committee for Standardization, CEN.

The withdrawal of the UK from the European Union is not expected to alter the link between the British Standards Institution and CEN, or the status of the Eurocodes in the UK. CEN already includes non-EU countries, such as Switzerland and Norway.

The Eurocode for composite structures, EN 1994, is based on recent research and current practice. It has much in common with the earlier national codes in Western Europe, but its scope is far wider. Two of its three Parts are used here, with the shortened names: EC4-1-1, General rules and rules for buildings, and EC4-1-2, Structural fire design. The third Part, EC4-2, is for bridges. Eurocode 4 has many cross-references to other Eurocodes, particularly:

- EN 1990, Basis of Structural Design;
- EN 1991, Actions on Structures;
- EN 1992, Design of Concrete Structures; and
- EN 1993, Design of Steel Structures.

The Eurocodes refer to other European (EN) and International (ISO) standards, for subjects such as products made from steel, and execution. 'Execution' is an example of a word used in Eurocodes with a particular meaning that has replaced the word previously used: construction (BSI, 2011). Other examples will be explained as they occur.

The cost of purchasing Eurocodes is quite high. Employers provide them for their staff, and students have access via university libraries. Designers' guides or handbooks to the Eurocodes have been published in the UK by the Institution of Civil Engineers (e.g. Beeby and Narayanan, 2005; Gardner and Nethercot, 2004; Johnson, 2012), the Institution of Structural Engineers, and by associations such as the Steel Construction Institute and the Concrete Society. They start from a higher level of prior knowledge than is assumed here.

The purpose of this book is to present, explain and use the theories, structural models and assumptions used in Eurocode 4, and those needed from Eurocodes 1, 2 and 3. There are few cross-references to Eurocode clauses, and access to them, although helpful, is not assumed.

Readers should not assume that the worked examples are fully in accordance with the Eurocodes as implemented in any particular country. The examples are not comprehensive, and Eurocodes give only 'recommended' values for some numerical values, especially the γ and ψ factors. The recommended values, which are used here, were subject to revision in National Annexes, but few of them were changed in the UK.

The first major revision of the Eurocodes began in 2015, with publication expected in the early 2020s. There will be important additions to the scope of Eurocode 4. Examples are: beams with large web openings, slim-floor construction, use of precast floor slabs, concrete dowel shear connectors, and partial shear connection for beams supporting composite slabs. These subjects are included here, and their consequences for the design examples are explained.

The principal author of the book is R.P. Johnson, who has for many decades shared the challenge of work on Eurocode 4 with other members of committees of BSI and of CEN. The substantial contributions made by these colleagues and friends to his understanding of the subject are gratefully acknowledged. Chapter 6 is contributed by Yong C. Wang who has greatly benefited from interactions with other members of CEN working groups and the project team who are responsible for revising EN 1994-1-2. Both authors are contributing to the revisions of Eurocode 4. These are under discussion at present, and will have to be consistent with the revisions of other Eurocodes. Hence, the references given here to expected changes should be considered as the authors' best understanding of the current state of development of Eurocode 4. Responsibility for what is presented here rests with the writers, who would be glad to be informed of any errors that may be found.

November 2017   *Roger P. Johnson and Yong C. Wang*

Symbols, Terminology and Units

The symbols used in this volume are, wherever possible, the same as those in EN 1994 and in the Designers' Guide to EN 1994-1-1 (Johnson, 2012). They are based on ISO 3898:1987, *Bases for design of structures – Notation – General symbols*. They are more consistent that those used in the British codes, and more informative. For example, in design one often compares an applied ultimate bending moment (an 'action effect' or 'effect of action') with a bending resistance, since the former must not exceed the latter. This is written

$$M_{Ed} \leq M_{Rd}$$

where the subscripts E, d, and R mean 'effect of action', 'design', and 'resistance', respectively. For clarity, multiple subscripts are often separated by commas ($M_{R,d}$ would be an example); but there are many exceptions, as the examples above show.

For longitudinal shear, the following should be noted:

- v, a shear stress (shear force per unit area), but τ is used for a vertical shear stress;
- v_L, a shear force per unit length of member, known as 'shear flow';
- V, a shear force (used also for a vertical shear force).

For subscripts, the presence of three types of steel leads to the use of 's' for reinforcement, 'a' (from the French 'acier') for structural steel, and 'p' or 'ap' for profiled steel sheeting. Another key subscript is k, as in

$$M_{Ed} = \gamma_F M_{Ek}$$

Here, the partial factor γ_F is applied to a *characteristic* bending action effect to obtain a *design* value, for use in a verification for an ultimate limit state. Thus 'k' implies that a partial factor (γ) has not been applied, and 'd' implies that it has been. This distinction is made for actions and resistances, as well as for the action effect shown here.

Other important subscripts are:

- c or C for 'concrete';
- v or V, meaning 'related to vertical or longitudinal shear'.

Terminology

The word 'resistance' replaces the widely-used 'strength', which is reserved for a property of a material or component, such as a bolt.

A useful distinction is made in most Eurocodes, and in this volume, between 'resistance' and 'capacity'. The words correspond respectively to two of the three fundamental concepts of the theory of structures, equilibrium and compatibility (the third being the properties of the material). The definition of a resistance includes a unit of force, such as kN, while that of a 'capacity' does not. A capacity is typically a displacement, strain, curvature, or rotation.

Cartesian axes

In the Eurocodes, x is an axis along a member. A major-axis bending moment M_y acts about the y axis, and M_z is a minor-axis moment. This differs from previous practice in the UK, where the major and minor axes were xx and yy, respectively.

Units

The SI system is used. A minor inconsistency is the unit for stress, where both N/mm^2 and MPa (megapascal) are found in the codes. Similarly, kN/mm^2 corresponds to GPa (gigapascal). The unit for a coefficient of thermal expansion may be given as 'per $°C$' or as 'K^{-1}', where K means kelvin, the unit for the absolute temperature scale. The convention of sign is explained in the Preface.

Symbols

The list of symbols in EN 1994-1-1 extends over eight pages, and does not include many symbols in clauses of other Eurocodes to which it refers. The list can be shortened by separation of main symbols from subscripts. In this book, commonly-used symbols are listed here in that format. Rarely-used symbols are defined where they appear. Fire-related symbols from EN 1994-1-2 are listed in Chapter 6.

Latin upper-case letters

A	accidental action; area
B	breadth
C	factor
E	modulus of elasticity; effect of actions
(EI)	stiffness of a composite section (the same whether transformed into 'steel' or 'concrete')
F	action; force; force per unit length
G	permanent action; shear modulus
H	horizontal load or force per frame per storey
I	second moment of area
J	property of an end-plate connection
K	stiffness; coefficient
L	length; span
M	moment in general; bending moment; modal mass
M'	hogging bending moment
N	axial force

P	shear force or resistance for a shear connector
Q	variable action
R	resistance; resistance function (as R); response factor; ratio
S	stiffness; width (of floor)
T	tensile force or resistance; total time
ULS	ultimate limit state
V	shear force; vertical load per frame per storey
W	section modulus; wind load
X	property of a material
Z	shape factor

Latin lower-case letters

a	acceleration; lever arm; dimension
b	width; breadth; dimension
c	outstand; thickness of concrete cover; dimension
d	diameter; depth
e	eccentricity; dimension
f	strength (of a material); natural frequency; coefficient; factor
g	permanent action per unit length or area; gravitational acceleration
h	depth of member; thickness; height
k	coefficient; factor; property of a composite slab; stiffness
l	length
m	property of a composite slab; mass per unit length or area; number
n	modular ratio; number
p	spacing (e.g. of shear connectors)
q	variable action per unit length or area
r	radius; ratio
s	spacing; slip of shear connection
t	thickness
u	perimeter
v	shear stress; shear strength; shear force per unit length; dimension
w	crack width; load per unit length
x	dimension to neutral axis; depth of stress block; co-ordinate along member
y	major axis; co-ordinate; distance from elastic neutral axis to extreme fibre
$\bar{y}$	distance of excluded area from centre of area
z	lever arm; dimension; co-ordinate

Greek letters

α	angle; ratio; factor; coefficient; reduction factor
β	angle; factor; coefficient
γ	partial factor
Δ	difference in … (precedes main symbol)
δ	steel contribution ratio; deflection; slip capacity
ε	strain; coefficient
η	coefficient; degree of shear connection

θ	angle
κ	curvature
λ	(or $\bar{\lambda}$ if non-dimensional) slenderness ratio
μ	coefficient of friction; ratio of bending moments; exponent (as superscript)
υ	Poisson's ratio
ρ	reinforcement ratio; density (unit mass)
σ	normal stress
τ	shear stress
ϕ	diameter of a reinforcing bar; rotation; angle of sidesway
φ	creep coefficient
χ	reduction factor (for buckling)
Ψ	combination factor for variable actions; ratio; exponent

Subscripts

A	accidental; area; structural steel
a	structural steel; spacing
ap	profiled steel sheeting
b	buckling; bolt; beam
bot	bottom
C	concrete
c	compression; concrete; composite; connection; cylinder compressive strength
cf	concrete flange
cr	critical
cs	strain in concrete (e.g. from shrinkage)
cu	concrete cube compressive strength
d	design; diameter
E	effect of action
eff	effective
e	effective (with further subscript); elastic
el	elastic
eq	equivalent
F	action
f	flange; full shear connection; surface finish (in h_f); full interaction
fl	flange
G	permanent (referring to actions)
g	centroid; permanent load
H or h	horizontal
hog	hogging bending
i	index (replacing a numeral)
imp	imperfection
ini	initial
j	joint
k	characteristic
L	longitudinal (e.g. in v_L, shear flow)
LT	lateral-torsional

l or ℓ	longitudinal; lightweight-aggregate
M	material
m	relating to bending moment; mean; mass; measured; number (of columns)
max	maximum
min	minimum
N	(allowing for) axial force
n	number; neutral axis
o	particular value; concrete in tension neglected; overhang
P	profiled steel sheeting
p	profiled steel sheeting; perimeter; plastic
pa, pr	properties of profiled sheeting (Section 3.3.1)
pe	effective, for profiled sheeting
pl	plastic
Q	variable (referring to actions)
R	resistance
r	reduced; rib of profiled sheeting; reinforcement
red	reduced
rms	root mean square
S	reinforcing steel; action effect (from the French, *sollicitation*)
s	reinforcing steel; shear span; slab; slip at interface; spacing; stiffness
sag	sagging bending
sc	shear connector
T	tensile force; steel T-section in a connection
t	tension; torsion; time; transverse; top; total
u	ultimate; maximum
V	shear; vertical
Vs	shear (composite slab)
v	vertical; shear; shear connection
w	web; weighted
wp	web post
x	axis along member
y	major axis of cross-section; yield
z	minor axis of cross-section
0, 1, 2, etc.	particular values
0	combination value (in Ψ_0); fundamental (in f_0); overall (in γ_0); mean; short-term (in n_0)
1	frequent value (in Ψ_1); uncracked; end moment for a column length
2	quasi-permanent value (in Ψ_2); cracked reinforced; end moment for a column length
0.05, 0.95	fractiles

1

Introduction

1.1 Composite beams and slabs

The design of structures for buildings and bridges is mainly concerned with the provision and support of load-bearing horizontal surfaces. Except in some long-span structures, these floors or decks are usually made of reinforced concrete, for no other material provides a better combination of low cost, high strength, and resistance to corrosion, abrasion and fire.

The economic span for a uniform reinforced concrete slab is little more than that at which its thickness becomes sufficient to resist the point loads to which it may be subjected or, in buildings, to provide the sound insulation required. For spans of more than a few metres it is cheaper to support the slab on beams, ribs or walls than to thicken it. Where the beams or ribs are also of concrete, the monolithic nature of the construction makes it possible for a substantial breadth of slab to act as the top flange of the beam that supports it.

At spans of more than about 10 m, and especially where the susceptibility of steel to loss of strength from fire is not a problem, as in most bridges, steel beams often become cheaper than concrete beams. It was at first customary to design the steelwork to carry the whole weight of the concrete slab and its loading; but by about 1950 the development of shear connectors had made it practicable to connect the slab to the beam, and so to obtain the T-beam action that had long been used in concrete construction. The term 'composite beam' as used in this book refers to this type of structure.

The same term is in use for beams in which prestressed and *in situ* concrete act together; and there are many other examples of composite action in structures, such as between brick walls and beams supporting them, or between a steel-framed shed and its cladding; but these are outside the scope of this book.

No income is received from money invested in construction of a multistorey building such as a large office block until the building is occupied. The construction time is strongly influenced by the time taken to construct a typical floor of the building, and here structural steel has an advantage over *in situ* concrete.

Even more time can be saved if the floor slabs are cast on permanent steel formwork that acts first as a working platform, and then as bottom reinforcement for the slab. The use of this formwork, known as *profiled steel sheeting*, began in North America (Fisher, 1970) and is now standard practice in Europe and elsewhere. These floors span in one direction only, and are known as *composite slabs*. Where the steel sheet is flat, so that

Composite Structures of Steel and Concrete: Beams, Slabs, Columns and Frames for Buildings, Fourth Edition. Roger P. Johnson.
© 2019 John Wiley & Sons Ltd. Published 2019 by John Wiley & Sons Ltd.

two-way spanning occurs, the structure is known as a *composite plate*. These occur in box-girder bridges.

Steel profiled sheeting and partial-thickness precast concrete slabs are known as *structurally participating* formwork. Cement or plastic profiled sheeting reinforced by fibres is sometimes used. Its contribution to the strength of the finished slab is normally ignored in design.

The degree of fire protection that must be provided is another factor that influences the choice between concrete, composite and steel structures, and here concrete has an advantage. Little or no fire protection is required for open multistorey car parks, a moderate amount for office blocks, and most of all for public buildings and warehouses. Many methods have been developed for providing steelwork with fire protection.

Design against fire and the prediction of fire resistance is known as fire engineering (Wang et al., 2012). Several of the Eurocodes have a Part 1.2 devoted to it. Full encasement of steel beams, once common, is now more expensive than the use of lightweight non-structural materials. Concrete encasement of the web only, done before the beam is erected, is more common in continental Europe than in the UK, and is covered in EN 1994-1-1 (BSI, 2004). It enhances the buckling resistance of the member (Section 4.2.4) as well as providing fire protection.

The choice between steel, concrete and composite construction for a particular structure thus depends on many factors that are outside the scope of this book. Composite construction is particularly competitive for medium- or long-span structures where a concrete slab or deck is needed for other reasons, where there is a premium for rapid construction, and where a low or medium level of fire protection to steelwork is sufficient.

1.2 Composite columns and frames

Composite columns may be constructed by encasing steel sections in concrete or by infilling steel tubes with concrete. When the columns in steel frames were first encased in concrete to protect them from fire, they were designed for the applied load as if uncased. Tests by Faber and others (Faber, 1956) then showed that savings could be made by using better-quality concrete and designing the column as a composite member. Full or partial encasement is economical for steel columns because the casing makes the column much stronger, although the need for formwork for the concrete increases costs and may lengthen the construction time.

Composite columns made of concrete-filled steel tubes (CFSTs) are structurally efficient, can often be used without external fire protection, and can be constructed rapidly (Wang, 2014). A notable early use of filled tubes was in a four-level motorway interchange (Kerensky and Dallard, 1968). CFST columns are widely used in building construction in many parts of the world, such as China and Australia, and are becoming more common in the UK. Their design is covered in Chapter 5.

In framed structures, there may be steel members, composite beams, composite columns, or all of these, and there are many types of beam-to-column connection. Their behaviour can range from 'nominally pinned' to 'rigid', and may influence bending moments throughout the frame. Two buildings with rigid-jointed composite frames were built in England in the early 1960s, in Cambridge (Johnson et al., 1965) and

London (Cassell et al., 1966). These were trials of new methods and, with other work, found that in buildings, the cost of making joints stiff enough to be treated as rigid could outweigh the saving from the use of shallower beams.

Research (e.g. Couchman and Way, 1998; Anderson et al., 2000; Brown and Anderson, 2001) enabled design rules for joints in steel and composite frames to be given in EN 1993 (BSI, 2005a) and EN 1994 (BSI, 2004) for beam-to-column joints with nominally pinned, rigid and semi-rigid behaviour. Some of them lead to extensive calculations, as shown in Section 5.10. In British design practice, joints are usually treated as nominally pinned, even though many have sufficient stiffness to reduce significantly the deflections of beams.

1.3 Design philosophy and the Eurocodes

1.3.1 Background

In design, account must be taken of the random nature of loading, the variability of materials, and the defects that occur in construction, to reduce the probability of unserviceability or failure of the structure during its design life to an acceptably low level. Extensive study of this subject since about 1950 has led to the incorporation of the older 'safety factor' and 'load factor' design methods into a comprehensive 'limit state' design philosophy. Its first important application in British standards was in 1972, in CP 110, *The structural use of concrete*. It is used in current British and European codes for the design of structures.

Work on international codes began after the Second World War, first on concrete structures and then on steel structures. A committee for composite structures, set up in 1971, prepared the Model Code (European Convention for Constructional Steelwork, 1981). The Commission of the European Communities has supported work on Eurocodes since 1982, and has delegated its management to the Comité Europeén Normalisation (CEN), based in Brussels. This is an association of the national standards institutions (NSIs) of the countries of the European Union, the European Free Trade Area, and a growing number of other countries from central and eastern Europe. Further information will be found in the Preface.

The Eurocodes provide a coherent system, in which duplication of information has been minimized. For example, EN 1994 refers to EN 1990, *Basis of structural design* (BSI, 2002c), for design philosophy, limit state requirements and most definitions.

Values for loads and other actions that do not depend on the material used for the structure (the great majority) are given in EN 1991, *Actions on structures* (BSI, 2002b,d). All provisions for structural steel that apply to both steel and composite structures are in EN 1993, *Design of steel structures* (BSI, 2005a, 2006a,b, 2014b). Similarly, for concrete, EN 1994 (BSI, 2004) refers to but does not repeat material from EN 1992, *Design of concrete structures* (BSI, 2014a).

Within Eurocode 4, material is divided between that which applies to both buildings and bridges, to buildings only, and to bridges only. The first is found in the 'General' clauses of EN 1994-1-1, the second in clauses in EN 1994-1-1 marked 'for buildings', and the third in EN 1994-2, 'Rules for bridges', which also repeats the 'General' clauses. Structural fire design is found in EN 1994-1-2 (BSI, 2014c), which cross-refers for the high-temperature properties of materials to the 'Fire' parts of EN 1992 and EN 1993, as appropriate.

Design of foundations is covered in EN 1997, *Geotechnical design*, and seismic design in EN 1998, *Design of structures for earthquake resistance*.

The British codes for composite structures that preceded Eurocode 4 have not been revised or updated since 2010, and their scope is narrower. For example, columns, web-encased beams and box girders are not covered.

1.3.2 Limit state design philosophy

1.3.2.1 Basis of design, and actions

Parts 1.1 of ENs 1992, 1993 and 1994 each have a Section 2, 'Basis of design', that refers to EN 1990 for the presentation of limit state design as used in the Eurocodes. Its Section 4, 'Basic variables', classifies these as actions, environmental influences, properties of materials and products, and geometric data (e.g. initial out-of-plumb of a column). Actions are either:

- direct actions (forces or loads applied to the structure), or
- indirect actions, such as deformations imposed on the structure, for example by settlement of foundations, change of temperature, or shrinkage of concrete.

'Actions' thus has a wider meaning than 'loads'. Similarly, the Eurocode term 'effect of actions' has a wider meaning than 'stress resultant', because it includes stresses, strains, deformations and crack widths, as well as bending moments, shear forces, and so on. The Eurocode term for 'stress resultant' is 'internal force or moment'.

The scope of the following introduction to limit state design is limited to that of the design examples in this book. There are two classes of *limit states*:

- ultimate (denoted ULS), which are associated with structural failure, whether by rupture, crushing, buckling, fatigue, or overturning; and
- serviceability (SLS), such as excessive deformation, vibration, or width of cracks in concrete.

Either type of limit state may be reached as a consequence of poor design, construction or maintenance, or from overloading, insufficient durability, fire, and so forth.

There are three types of *design situation*:

- persistent, corresponding to normal use;
- transient, for example, during construction, refurbishment or repair;
- accidental, such as fire, explosion or earthquake.

There are three main types of action:

- permanent (G or g), such as self-weight of a structure (formerly 'dead load'), and including shrinkage of concrete;
- variable (Q or q), such as imposed, wind, or snow load (formerly 'live load'), and including expected changes of temperature;
- accidental (A), such as impact from a vehicle, or high temperature from a fire.

The spatial variation of an action is either:

- fixed (typical of permanent actions); or

- free (typical of other actions), and meaning that the action may occur over only a part of the area or length concerned.

Permanent actions are represented (and specified) by a *characteristic value*, G_k. 'Characteristic' implies a defined fractile of an assumed statistical distribution of the action, modelled as a random variable. For permanent loads, it is usually the mean value (50% fractile).

Variable actions have four *representative values*:

- characteristic (Q_k), normally the upper 5% fractile;
- combination ($\psi_0 Q_k$), for use where the action is assumed to accompany the design ultimate value of another variable action, which is the 'leading action';
- frequent ($\psi_1 Q_k$), for example, occurring at least once a week; and
- quasi-permanent ($\psi_2 Q_k$).

Recommended values for the combination factors ψ_0, ψ_1 and ψ_2 (all less than 1.0) are given in EN 1990. Definitive values, usually those recommended, are given in national annexes. For example, for imposed loads on the floors of offices, the recommended values are $\psi_0 = 0.7$, $\psi_1 = 0.5$, and $\psi_2 = 0.3$.

Design values of actions are, in general, $F_d = \gamma_F F_k$, and in particular:

$$G_d = \gamma_G G_k \tag{1.1}$$

$$Q_d = \gamma_Q Q_k \text{ or } Q_d = \gamma_Q \psi_i Q_k \tag{1.2}$$

where γ_G and γ_Q are partial factors for actions, recommended in EN 1990 and given in national annexes. They depend on the limit state considered, and on whether the action is unfavourable or favourable (i.e. tends to increase or decrease the action effect considered). The values used in this book are given in Table 1.1 and, for fire, in Chapter 6.

The *effects of actions* are the responses of the structure to the actions, for example:

$$E_d = E(F_d) \tag{1.3}$$

where the function E represents the process of structural analysis. Where linear-elastic or plastic global analysis are used and the action effect is an internal force or moment, *verification for an ultimate limit state* consists of checking that

$$E_d \leq R_d \tag{1.4}$$

where R_d is the relevant design resistance of the system or member or cross-section considered. These methods of analysis are explained in Sections 1.4 and 1.6.

Table 1.1 Values of γ_G and γ_Q for persistent design situations

Type of action	Permanent unfavourable	Permanent favourable	Variable unfavourable	Variable favourable
Ultimate limit states	1.35*	1.35*	1.5	0
Serviceability limit states	1.0	1.0	1.0	0

*Except for checking loss of equilibrium, or where the coefficient of variation is large.

1.3.2.2 Resistances

Resistances, R_d, are calculated using design values of properties of materials, X_d, given by

$$X_d = X_k/\gamma_M \tag{1.5}$$

where X_k is a characteristic value of the property and γ_M is the partial factor for that property.

The characteristic value is typically a 5% lower fractile (e.g. for compressive strength of concrete). Where the statistical distribution is not well established, it is replaced by a *nominal* value (e.g. the yield strength of structural steel), so chosen that it can be used in design in place of X_k.

The subscript M in γ_M is often replaced by a letter that indicates the material concerned, as shown in Table 1.2, which gives the values of γ_M used in this book. A welded stud shear connector is treated like a single material, even though its design resistance to shear, P_{Rk}/γ_V, is influenced by the properties of both steel and concrete. In Eurocode 3, where the resistance depends on properties of a single material, γ_M is applied to the resistance, not to the property.

For resistance to fracture of a steel cross-section in tension, $\gamma_A = 1.25$. Subscripts A or a are used in Eurocode 4 for structural steel (French, 'acier') because s is used for reinforcement.

1.3.2.3 Combinations of actions

The Eurocodes treat systematically a subject for which many empirical procedures have been used in the past. For ultimate limit states, the principles are:

- permanent actions are present in all combinations;
- each variable action is chosen in turn to be the 'leading' action (i.e. to have its full design value), and is combined with lower 'combination' values of other variable actions that may co-exist with it;
- the design action effect for a cross-section or member is the most unfavourable of those found by this process.

The use of combination values allows for the limited correlation between time-dependent variable actions.

As an example, let us assume that a bending moment M_{Ed} in a member is influenced by its own weight (G), by an imposed vertical load (Q_1) and by wind loading (Q_2). The

Table 1.2 Recommended values for γ_M for strengths of materials and for resistances

Material	Structural steel	Profiled sheeting	Reinforcing steel	Concrete	Shear connection
Property	f_y	f_{yp}	f_{sk}	f_{ck} or f_{cu}	P_{Rk}
Symbol for γ_M	γ_A	γ_P	γ_S	γ_C	γ_V or γ_{Vs}
Ultimate limit states	1.0	1.0	1.15	1.5	1.25
Serviceability limit states	1.0	1.0	1.0	1.0	1.0

Notation: For concrete, f_{ck} and f_{cu} are respectively characteristic cylinder and cube strengths; symbol γ_{Vs} is for shear resistance of a composite slab.

two fundamental combinations for verification for persistent design situations are:

$$\gamma_G G_k + \gamma_{Q1} Q_{k,1} + \gamma_{Q2} \psi_{0,2} Q_{k,2} \tag{1.6}$$

$$\gamma_G G_k + \gamma_{Q1} \psi_{0,1} Q_{k,1} + \gamma_{Q2} Q_{k,2} \tag{1.7}$$

Each term in these expressions gives the value of the action for which a bending moment is calculated, and the + symbols apply to the bending moments, not to the values of the actions. This is sometimes indicated by placing each term between quotation marks.

In practice, it is usually obvious which combination will govern. For low-rise buildings, wind is rarely critical for floors, so Expression 1.6 with imposed load leading would be used; but for a long-span lightweight roof, Expression 1.7 would govern, and both positive and negative wind pressure would be considered, with the negative pressure combined with $Q_{k,1} = 0$.

For serviceability limit states, three combinations are defined. The most onerous, the 'characteristic' combination, corresponds to the fundamental combination (above) with the γ factors reduced to 1.0. For the example in Expressions 1.6 and 1.7, it is

$$G_k + Q_{k,1} + \psi_{0,2} Q_{k,2} \tag{1.8}$$

$$G_k + \psi_{0,1} Q_{k,1} + Q_{k,2} \tag{1.9}$$

It is normally used for verifying irreversible limit states, for example, deformations that result from the yielding of steel.

Assuming that Q_1 is the leading variable action, the others are:

- frequent combination,

$$G_k + \psi_{1,1} Q_{k,1} + \psi_{2,2} Q_{k,2} \tag{1.10}$$

- quasi-permanent combination,

$$G_k + \psi_{2,1} Q_{k,1} + \psi_{2,2} Q_{k,2} \tag{1.11}$$

The frequent combination is used for reversible limit states, for example, the elastic deflection of a floor under imposed loading. However, if that deformation causes cracking of a brittle floor finish or damage to fragile partitions, then the limit state is not reversible, and the check should be done for the higher (less probable) loading of the characteristic combination.

The quasi-permanent combination is used for long-term effects (e.g. deformations from creep of concrete) and for the appearance of the structure.

Some combination factors used in this book are given in Table 1.3.

Table 1.3 Combination factors

Factor	ψ_0	ψ_1	ψ_2
Imposed floor loading in an office area of a building	0.7	0.5	0.3
Wind loading on a building	0.6	0.2	0

1.3.2.4   Comments on limit state design philosophy

The use of limit states has superseded earlier methods partly because limit states provide identifiable criteria for satisfactory performance. Stresses cannot be calculated with the same confidence as resistances of members, and high values may or may not be significant.

An apparent disadvantage of limit states design is that several sets of design calculations may be needed, whereas with some older methods, one was sufficient. This is only partly true, for it has been found possible to identify many situations in which design for, say, an ultimate limit state will ensure that certain types of serviceability will not occur, and vice versa. In the rules of EN 1994 for buildings it has generally been possible to avoid specifying limiting stresses for serviceability limit states, by using other methods to check deflections and crack widths.

1.4   Properties of materials

Information on the properties of structural steel, profiled sheeting, concrete and reinforcement is readily available. Only that which has particular relevance to composite structures is given here.

The determination of the bending moments and shear forces in a beam or framed structure is known as *global analysis*. The most widely used method is elastic analysis, in which all the materials are assumed to behave in a linear-elastic manner, although an effective modulus is used for the concrete to allow for its creep under sustained compressive stress.

Global analyses are often done before reinforcement has been designed. Concrete is then assumed to be as stiff in tension as in compression, to compensate for the omission of the reinforcement. Where reinforcement is included, the tensile stiffness and strength of concrete are usually taken as zero. Calculations of crack widths take account of the stiffness of the uncracked concrete between the cracks. The effects of its shrinkage are rarely significant in buildings.

Rigid-plastic global analysis can sometimes be used (Section 4.3.3), despite the profound difference between a typical stress–strain curve for concrete in compression and those for structural steel or reinforcement, in tension or compression, that are illustrated in Figure 1.1.

Concrete reaches its maximum compressive stress at a strain of between 0.002 and 0.003, and at higher strains it crushes, losing almost all of its compressive strength. It is very brittle in tension, having a strain capacity of only about 0.0001 (i.e. 0.1 mm per metre) before it cracks. Figure 1.1 also shows that the maximum stress reached by concrete in a beam or column is well below its cube strength.

Steel yields at a strain similar to that given for the maximum stress in concrete, but on further straining the stress continues to increase slowly, until (for a typical structural steel) the total strain is at least 30 times the yield strain. Its subsequent necking and fracture is of little significance for composite members, because the useful resistance of a cross-section is reached when all of the steel has yielded, when steel in compression buckles, or when concrete crushes.

Resistances of cross-sections are determined ('section analysis') using plastic analysis wherever possible, because results of elastic analyses are unreliable, unless careful

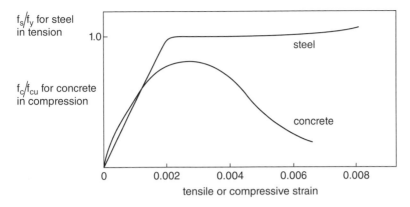

Figure 1.1 Stress–strain curves for concrete and structural steel

account is taken of the cracking, shrinkage, and creep of concrete (which is difficult), and also because plastic analysis is simpler and leads to more economical design.

The use of a higher value of γ_M for concrete than for steel (Table 1.2) includes allowance for the higher variability of the strength of test specimens, and the variation in the strength of concrete over the depth of a member, due to migration of water before setting. It also allows for the larger errors in the dimensions of cross-sections, particularly in the positions of reinforcing bars.

Global analysis is the subject of Section 1.6, with non-linear global analysis treated in Section 1.6.2. Brief comments are now given on individual materials.

1.4.1 Concrete

In EN 1992, a typical strength class for normal-density concrete is denoted C30/37, where the specified characteristic compressive strengths at age 28 days are $f_{ck} = 30\,\text{N/mm}^2$ (cylinder test) and $f_{cu} = 37\,\text{N/mm}^2$ (cube test). In strict SI units, the stress $1\,\text{N/mm}^2$ becomes 1 MPa. Both variants are used in the Eurocodes, and also in this book. The design formulae in EN 1994 use f_{cd}, which is f_{ck}/γ_C. The definition of f_{cd} in Eurocode 2 is different: $\alpha_{cc}f_{ck}/\gamma_C$, where α_{cc} is a nationally determined parameter in the range 0.8 to 1.0. This anomaly can cause confusion. There are proposals to remove it from the next generation of Eurocodes.

The normal-density concrete used in worked examples here is C25/30, which is 'Grade 30' in British terminology, with f_{ck} taken as $25\,\text{N/mm}^2$. It is used for the columns and for encasement of beam webs. Its weight density is about $24\,\text{kN/m}^3$, but allowance is made for the much denser reinforcement by using $25\,\text{kN/m}^3$ for reinforced concrete.

The floor slabs are constructed with lightweight-aggregate concrete of density class 1.6 (density between 1401 and $1800\,\text{kg/m}^3$). The strength class is LC25/28 (BSI, 2016). Properties for these two concretes are given in Table 1.4. The highest strength class included in EN 1994-1-1 is C60/75 generally, but the limit for composite columns in clause 6.7 is C50/60.

The strength classes for concrete defined in EN 1992-1-1 extend to class C90/105. Class C80/95 concrete was used in columns for the Shard building in London in 2012, and concrete-filled steel tube (CFST) columns with $f_{ck} > 100\,\text{MPa}$ have been used in

Table 1.4 Properties of concretes used in the examples, at age 28 days

Concrete grade		C25/30	LC25/28
Characteristic cylinder strength, N/mm^2	f_{ck}	25	25
Mean cylinder strength, N/mm^2	f_{cm}	33	33
Lower tensile strength, N/mm^2	$f_{ct,0.05}$	1.80	1.60
Mean tensile strength, N/mm^2	f_{ctm}	2.60	2.32
Upper tensile strength, N/mm^2	$f_{ct,0.95}$	3.30	2.94
Mean elastic modulus, kN/mm^2	E_{cm}	31.0	20.7
Weight density, reinforced concrete, kN/m^3		25.0	19.5

buildings in the US and Japan (Liew and Xiang, 2015). A worked example of a CFST column using high-strength materials is given in Section 5.11.

1.4.2 Reinforcing steel

Strength grades for reinforcing steel are given in EN 10080 (BSI, 2005b) in terms of a characteristic yield strength f_{sk}. The value used here is 500 N/mm^2, for both ribbed bars and welded steel fabric. The modulus of elasticity for reinforcement, E_s, is normally taken as 200 kN/mm^2; but in a composite section it may be assumed to have the value for structural steel, $E_a = 210$ kN/mm^2, as the error is negligible.

1.4.3 Structural steel

Strength grades for structural steel are given in EN 10025 (BSI, 1990b) in terms of a nominal yield strength, f_y and ultimate tensile strength f_u. The grade used in worked examples here is S355, for which $f_y = 355$ N/mm^2, $f_u = 510$ N/mm^2 for elements of all thicknesses up to 40 mm. For columns, clause 6.7 of EN 1994-1-1 is applicable for grades up to S460. Some concrete-filled steel tube columns using steel with a yield strength of 780 MPa were built in Japan in 2014 (Liew and Xiang, 2015), and that Report extends the scope of the 'Simplified method' of clause 6.7.3 of Eurocode 4 to S550 steel used with C90/105 concrete.

The density of structural steel is assumed to be 7850 kg/m^3. Its coefficient of linear thermal expansion is 12×10^{-6} per °C. The difference between this value and that for normal-density concrete, 10×10^{-6} per °C, can usually be ignored.

1.4.4 Profiled steel sheeting

This product is available with yield strengths ranging from 235 N/mm^2 to at least 460 N/mm^2, in profiles with depths ranging from 45 mm to over 200 mm, and with a wide range of shapes. These include both re-entrant ribs, and trapezoidal troughs as in Figure 3.8. There are various methods of achieving composite action with a concrete slab, discussed in Section 2.4.3.

Sheets are normally between 0.8 mm and 1.5 mm thick, and are protected from corrosion by a zinc coating about 0.02 mm thick on each face. Elastic properties of the material may be assumed to be as for structural steel.

1.4.5   Shear connectors

In the early years of composite construction, many types of connector were in use. This market is now dominated by automatically-welded headed studs. Details of these and other types of connector, and of the measurement of their resistance to shear, are given in Chapter 2.

1.5   Direct actions (loading)

The characteristic loadings to be used in worked examples are now given. They are taken from EN 1991.

The *permanent loads* (dead load) are the weight of the structure and its finishes. The structural steel or profiled sheeting component of a composite member is invariably built first, so a distinction must be made between load resisted by the steel component only, and load applied to the member after the concrete has developed sufficient strength for composite action to become effective. The division of the permanent load between these categories depends on the method of construction.

Composite beams and slabs are classified as *propped* ('shored' in North America) or *unpropped*. In propped construction, the steel member is supported at intervals along its length until the concrete has reached a certain proportion, usually three-quarters, of its design strength. When the props are removed, the whole of the dead load is assumed to be carried by the composite member. Unpropped composite slabs can be built with propped beams, but if propped slabs are built with unpropped beams, the deflection of the beams can affect the sheeting. Where no props are used, it is assumed in elastic analysis that the steel member alone carries its own weight and that of the formwork and the concrete slab. Other dead loads such as floor finishes and internal walls are added later, and so are assumed to be carried by the composite member. In ultimate-strength methods of analysis (Section 3.5.3), inelastic behaviour causes extensive redistribution of stress before failure, and it can be assumed that the whole load is applied to the composite member, whatever the method of construction.

The principal vertical *variable action* in a building is a uniformly distributed load on each floor. EN 1991-1-1 gives ranges of loads, depending on the use to be made of the area, with a recommended value. For 'office areas', this value is

$$q_k = 3.0 \, \text{kN/m}^2 \tag{1.12}$$

Account is taken of point loads (e.g. a safe being moved on a trolley with small wheels) by defining an alternative point load, to be applied anywhere on the floor, on an area about 50 mm square. For the type of area defined above, this is

$$Q_k = 4.5 \, \text{kN} \tag{1.13}$$

The ability of floor slabs of uniform thickness to resist point or line loads is greater than that of composite slabs. Details are given in Section 3.5.5.

Where a member such as a column is carrying loads q_k from n storeys ($n > 2$), the total of these loads may be multiplied by a reduction factor α_n. The recommended value is

$$\alpha_n = [2 + (n - 2)\psi_0]/n \tag{1.14}$$

where ψ_0 is the combination factor (e.g. as in Table 1.3). This allows for the low probability that all n floors will be fully loaded at once.

In an office block, the location of partitions can be unknown at the design stage. Their weight is usually allowed for by increasing the imposed loading q_k by an amount that depends on the expected weight per unit length of the partitions. The increases given in EN 1991-1-1 range from 0.5 to $1.2\,\text{kN/m}^2$.

The principal horizontal variable load for a building is wind. These loads are given in EN 1991-1-4. They usually consist of pressure or suction on each external surface. On large flat areas, frictional drag may also be significant. Wind loads rarely influence the design of composite beams, but can be important in framed structures not braced against sidesway and in tall buildings.

Methods of calculation for distributed and point loads are sufficient for all types of direct action. Indirect actions such as subsidence or differential changes of temperature, which occasionally influence the design of structures for buildings, are not considered in this book.

1.6 Methods of analysis and design

The purpose of this section is to provide a preview of the principal methods of analysis used for composite members and frames, and to show that most of them are straightforward applications of methods in common use for steel or reinforced concrete structures. Non-linear global analysis is treated in Section 1.6.2.

The steel designer will be familiar with the elementary elastic theory of bending, and the simple plastic theory in which the whole cross-section of a member is assumed to be at yield, in either tension or compression. Both theories are used for composite members, the main differences being as follows:

- in the elastic theory, concrete in compression is 'transformed' into an equivalent area of steel by dividing the width of the cross-section by the modular ratio E_a/E_c;
- in the plastic theory, the design 'yield stress' of concrete in compression is taken as $0.85f_{cd}$, where $f_{cd} = f_{ck}/\gamma_C$. Transformed sections are not used. Examples of this method are given in Sections 3.4.2 and 3.11.1.

In the strength classes for concrete in EN 1992, the ratios f_{ck}/f_{cu} range from 0.78 to 0.83, so for $\gamma_C = 1.5$, the stress $0.85f_{cd}$ corresponds to a value between $0.44f_{cu}$ and $0.47f_{cu}$. This agrees closely with BS 5950 (BSI, 1990a), which uses $0.45f_{cu}$ for the plastic resistance of cross-sections.

The factor 0.85 takes account of several differences between a standard cylinder test and what concrete experiences in a structural member. These include the longer duration of loading in the structure, the presence of a stress gradient across the section considered, and the differences in the boundary conditions for the concrete. It is the equivalent of α_{cc} in Eurocode 2.

The concrete designer will be familiar with the method of transformed sections, and with the rectangular-stress-block theory outlined above. The basic difference from the elastic behaviour of a reinforced concrete beam is that the steel section in a composite beam is more than tension reinforcement. It has a significant bending stiffness of its own, and resists most of the vertical shear.

The formulae for the elastic properties of composite sections are more complex than those for steel or reinforced concrete sections. The chief reason is that the neutral axis for bending may lie in the web, the steel flange, or the concrete flange of the member. The theory is not in principle any more complex than that for a steel I-beam.

1.6.1 Typical analyses

1.6.1.1 Longitudinal shear

Students may find this subject troublesome, even though the formula

$$\tau = VA\bar{y}/Ib \tag{1.15}$$

is familiar from their study of vertical shear in elastic beams, so a note on its use may be helpful. The proof can be found in any undergraduate-level textbook on strength of materials.

First, let us consider the shear stresses τ in the elastic I-beam shown in Figure 1.2, due to a vertical shear force V. For a cross-section 1–2 through the web, the 'excluded area', A in the formula, is the top flange, of area A_f. The distance $\bar{y}$ of its centroid from the neutral axis (line X–X in Figure 1.2) is $(h - t_f)/2$. The shear stress τ_{12} on plane 1–2, of breadth t_w, is therefore

$$\tau_{12} = VA_f(h - t_f)/(2I t_w) \tag{1.16}$$

where I is the second moment of area of the whole cross-section about the axis X–X through its centre of area.

This result may be recognized as the *vertical* shear stress at this cross-section. However, a shear stress is always associated with a complementary shear stress at right angles

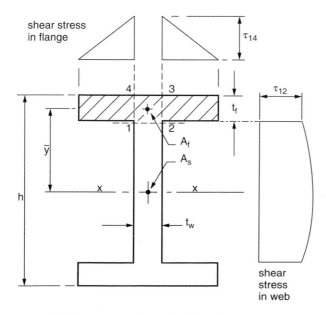

Figure 1.2 Shear stresses in an elastic I-section

to it, of equal value; in this case, the *longitudinal* shear stress. This will now be denoted v, rather than τ, as v is used in EN 1994.

If the cross-section in Figure 1.2 is a composite beam, with the cross-hatched area representing the transformed area of a concrete flange, shear connection is required on plane 1–2. It has to resist this longitudinal shear stress over a width t_w, so the shear force per unit length is vt_w. This is named the *shear flow*, and has the symbol v_L. From these results,

$$v_{L,12} = VA_f(h - t_f)/(2I) \tag{1.17}$$

Consideration of the longitudinal equilibrium of the small element 1234 in Figure 1.2 shows that if its area $t_w t_f$ is much less than A_f, the shear flows on planes 1–4 and 2–3 are each approximately $v_{L,12}/2$, and the mean shear stress on these planes is given approximately by

$$\tau_{14} t_f = \tau_{12} t_w / 2$$

This stress is needed for checking the resistance of the concrete slab to longitudinal shear.

Repeated use of Equation 1.15 for various cross-sections shows that the variation of longitudinal shear stress is parabolic in the web and linear in the flanges, as shown in Figure 1.2.

The second example is the elastic beam shown in section in Figure 1.3. This represents a composite beam in sagging bending, with the neutral axis at depth x, a concrete slab of thickness h_c, and the interface between the slab and the structural steel (which is assumed to have no top flange) at level 6–5. The concrete has been transformed to steel, so the cross-hatched area is the equivalent steel section. The concrete in area ABCD is assumed to be cracked. As in the theory for reinforced concrete beams, it resists no longitudinal stress, but is capable of transferring shear stress.

Equation 1.15 is based on rate of change along the beam of bending stress, so in applying it here, area ABCD is omitted when the 'excluded area' is calculated. Let the

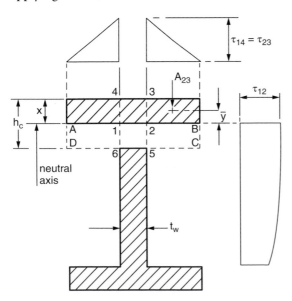

Figure 1.3 Shear stresses in a composite section with the neutral axis in the concrete slab

cross-hatched area of the flange be A_f, as before. The longitudinal shear stress on plane 6–5 is given by

$$v_{65} = VA_f\bar{y}/It_w \tag{1.18}$$

where $\bar{y}$ is the distance from the centroid of the excluded area to the neutral axis, *not to plane 6–5*. If A and $\bar{y}$ are calculated for the cross-hatched area below plane 6–5, the same value v_{65} is obtained, because it is the equality of the two products '$A\,\bar{y}$' that determines the value x.

The preceding theory relies on the assumption that the flexibility of shear connectors is negligible, and is used in bridge design and for fatigue generally. Ultimate-strength theory (Sections 3.3.2 and 3.6.1) provides an alternative that takes advantage of the plastic behaviour of stud connectors, and is widely used in design for buildings.

For a plane such as 2–3 in Figure 1.3, the shear flow is

$$v_{L,23} = VA_{23}\bar{y}/I \tag{1.19}$$

The design shear stress for the concrete on this plane is $v_{L,23}/h_c$. It is not equal to $v_{L,23}/x$ because the cracked concrete can resist shear. The depth h_c does not have to be divided by the modular ratio, even though the transformed section is of steel, because the transformation is of widths, not depths. This is consistent with the assumption in the simplified model used for elastic analysis that the shear strain in the plane of a T-beam flange is uniform through its thickness.

1.6.1.2 Longitudinal slip

Shear connectors are not rigid, so that a small longitudinal slip occurs between the steel and concrete components of a composite beam. The problem does not arise in other types of structure, and relevant analyses are quite complex (Section 2.6 and Appendix A). They are not needed in design, for which simplified methods have been developed.

1.6.1.3 Deflections

The effects of creep and shrinkage make calculation of deflections in reinforced concrete beams more complex than for steel beams. The subject is extensively treated in the worked examples in Sections 3.4, 3.11 and 4.6. Eurocodes do not specify limits to deflections or to span/depth ratios for beams, because so much depends on the circumstances; for example, on whether there is a risk of cracking of brittle finishes. In some situations, long-term deflection under total loading is a potential problem; in others, it may be the change of deflection due to imposed loading, or the amplitude of load-induced vibration.

1.6.1.4 Vertical shear

The stiffness in vertical shear of the concrete slab of a composite beam is often much less than that of the steel component, and is usually neglected in design. For vertical shear, the methods used for steel beams are then applicable also to composite beams.

1.6.1.5 Buckling of flanges and webs of beams

This will be a new problem to many designers of reinforced concrete. It leads to restrictions on the breadth/thickness ratios of unstiffened steel webs and flanges in compression (Section 3.5.2). These do not apply to the steel part of the top flange of a composite T-beam at mid-span, because local buckling is prevented by its attachment to the concrete slab.

1.6.1.6 Crack-width control

The maximum spacings for reinforcing bars given in codes for reinforced concrete are intended to limit the widths of cracks in the concrete, for reasons of appearance and to avoid corrosion of reinforcement. In composite structures for buildings, cracking is likely to be a problem only where the top surfaces of continuous beams support brittle finishes or are exposed to corrosion. The principles of crack-width control are the same as for reinforced concrete. The calculations are different, but can normally be avoided by using the simplified methods given in EN 1994.

1.6.1.7 Continuous beams

The Eurocode design methods for continuous beams (Chapter 4) make use of both simple plastic theory (as for steel beams) and redistribution of moments (as for concrete beams).

1.6.1.8 Columns

The only British code that gave a design method for composite columns was BS 5400:Part 5, *Composite bridges*. EN 1994 gives a simpler method, which is described in Section 5.6.

1.6.1.9 Framed structures for buildings

Composite members normally form part of a frame that is essentially steel, rather than concrete, so the design methods of EN 1994 are based on those of EN 1993 for steel structures. Beam-to-column joints are classified in the same way; the same assumptions are made about geometrical imperfections, such as out-of-plumb columns; and similar allowance is made for second-order effects (increases in bending moments and reduction in stability, caused by interaction between vertical loads and lateral deflections). Joints are introduced in Section 4.1 and continued in Section 5.3. Analysis of frames to Eurocodes (Section 5.4.3) can be more complex than in previous practice, but includes methods for unbraced frames.

1.6.1.10 Structural fire design

The high thermal conductivity and slenderness of structural steel members and profiled sheeting cause them to lose strength in fire more quickly than concrete members. Structures for buildings are required to have fire resistance of minimum duration (typically 30 minutes to 2 hours) to enable the structure to survive burn-out under realistic fire conditions. This has led to the provision of minimum thicknesses of concrete and areas of reinforcement and of thermal insulation of steelwork.

Extensive research and the recent subject of fire engineering (Wang et al., 2012) have enabled the Eurocode rules for resistance to fire to be less onerous than older rules. Advantage is taken in design of membrane effects associated with large deformations, and of the provisions for accidental design situations. These allow for over-strength of members and the use of 'frequent' rather than 'characteristic' load levels. Explanations and worked examples are given in Chapter 6.

1.6.2   Non-linear global analysis

In laboratory testing, load on a member is usually increased gradually. In a slender column, the maximum bending moment increases faster than the loading; but in a tension member, it increases more slowly. These behaviours are respectively *over-proportional* and *under-proportional*. Accurate analysis of structures with members of these types requires non-linear global analysis. It is also needed where simplified stress–strain properties of materials (Section 1.4) are not accurate enough. The development of finite-element methods has made its use practicable.

Second-order analysis, sometimes required for columns, is introduced in Sections 5.1 and 5.4.3. Its method can be linear or non-linear elastic. 'Second-order' means that account is taken of the deformations of the structure caused by the loading. Conditions for its use are given in clause 5.2 of Eurocode 4.

Non-linear global analysis is treated in Eurocode 4 simply by referring in clause 5.4.3 to Eurocodes 2 and 3. This is unhelpful, as those methods differ in their details. A simpler presentation of the method in clause 5.7 of EN 1992-2 is being considered for the next Eurocode 4. An outline of it is now given.

It is assumed that a finite-element analysis is being developed to replicate a set of test results, and then for use in parametric studies with ranges of the variables wider than those used in the tests. The properties of the materials are mean values from test measurements, assumed to be the best estimate of the actual values.

If a slender column is being tested, failure will occur at a cross-section where the resistance function is $R_{pl,m}(N, M)$, with N the axial compression and M a bending moment. The N-M resistance curve for a column cross-section, shown in Figure 5.12, is reproduced as the inner curve in Figure 1.4. It is obtained by calculations explained in Section 5.6.4, using design (factored) properties of the materials. The outer curve is similar, using mean properties of the materials. The computed function for action effects from increasing load, $E(N,M)$, follows a curve such as OAB, shown in the figure with M over-proportional.

The method uses an overall partial factor γ_0, defined as $R_{pl,m}/R_{pl,d}$, and assumes that curve OAB is linear. In practice, a similar value would be obtained from the dashed line shown. This factor allows for the use of mean properties of the materials.

Figure 1.4 Interaction curves for a column cross-section, for finding factor γ_0

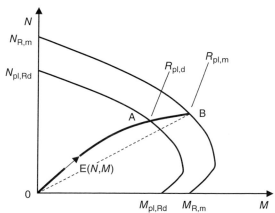

For characteristic permanent and variable loads G_k and Q_k and their usual partial factors γ_G and γ_Q, the expression to be verified is

$$\gamma_{Sd}\gamma_{Rd}E(\gamma_G G_k + \gamma_Q Q_k) \leq R_{pl,m}/\gamma_0$$

The factors γ_{Sd} and γ_{Rd} allow for model uncertainties in the action effects and the resistances, respectively. Recommended values for these factors are: for γ_{Sd}, from 1.0 to 1.15, depending on details of the analysis method; and for γ_{Rd}, 1.0 where the finite element model is calibrated with test results, and 1.1 otherwise.

2

Shear Connection

2.1 Introduction

Eurocodes 2 (for reinforced concrete) and 3 (for structural steel) do not include design of connections between the concrete and structural steel parts of a single member. In composite members, the force applied to this connection is mainly, but not entirely, shear on the longitudinal interface between the materials. As with bolted and welded joints, the connection is a region of severe and complex stress that defies accurate analysis, and so methods of connection have been developed empirically and verified by tests. They are described in Section 2.4. The next Eurocode 2 will include a new Part 4 on connections to concrete. Shear connectors for composite members will remain in Eurocode 4.

The simplest type of composite member used in practice occurs in floor structures of the type shown in Figure 3.1. The concrete floor slab is continuous over the steel I-sections, and is supported by them. It is designed to span in the y-direction in the same way as when supported by walls or the ribs of reinforced concrete T-beams. When shear connection is provided between the steel member and the concrete slab, the two together span in the x-direction as a composite beam. The steel member has not been described as a 'beam', because its main function at mid-span is to resist tension, as does the reinforcement in a T-beam. The compression is assumed to be resisted by an 'effective' width of slab, as explained in Section 3.5.1.

In buildings, but not in bridges, these concrete slabs are often composite with profiled steel sheeting (Figure 2.9), which rests on the top or bottom flange of the steel beam. Some of the many types of beam cross-section in use are shown in Figure 2.1.

The ultimate-strength design methods used for shear connection in beams and columns in buildings are described in sections 3.6 and 5.6.6, respectively.

The subjects of the present chapter are: the effects of shear connection on the behaviour of very simple beams, current methods of shear connection, standard tests on shear connectors, and shear connection in composite slabs.

The clauses on shear connection in the current Eurocode 4 (2004) relate only to members of the types covered in earlier codes. They do not include shear connection for slabs and beams where deep decking, shallow floor construction or concrete encasement is used (sections 3.4.6, 3.12 and 3.13). Since 2008 the Technical Committee TC11 of the European Convention for Constructional Steelwork has developed a deeper understanding of the behaviour and design of shear connection for flexural members, covering

Composite Structures of Steel and Concrete: Beams, Slabs, Columns and Frames for Buildings, Fourth Edition. Roger P. Johnson.
© 2019 John Wiley & Sons Ltd. Published 2019 by John Wiley & Sons Ltd.

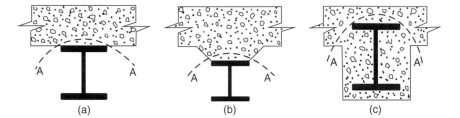

Figure 2.1 Composite beams: (a) typical cross-section; (b) haunched beam; (c) fully-encased beam

both static and repeated loading. Its 118-page publication of this work (Leskela, 2017) cites its sources, over 100 references, and should make possible wider coverage of shear connection in the next Eurocode 4.

2.2 Simply-supported beam of rectangular cross-section

Flitched beams, whose strength depended on shear connection between parallel timbers, were used in medieval times, and survive today in the form of glued-laminated construction. Such a beam, made from two members of the same cross-section (Figure 2.2b), will now be studied. Its span/depth ratio is typical, 15 to 25 (not shown to scale in Figure 2.2a), so that shear deformation is negligible compared with flexural deformation. The loading w per unit length is much greater than the weight of the beam, which is now neglected. Its components are made of an elastic material with Young's modulus E.

2.2.1 No shear connection

It is assumed first that there is no shear connection or friction on the interface AB. The upper beam cannot deflect more than the lower one, so each carries load $w/2$ per unit length as if it were an isolated beam of second moment of area $bh^3/12$, and the vertical compressive stress across the interface is $w/2b$. The mid-span bending moment in each beam is $wL^2/16$. By elementary beam theory, the longitudinal stress distribution at mid-span is given by the dashed line in Figure 2.2c, and the maximum bending stress in each component, σ, is given by

$$\sigma = \frac{My_{max}}{I} = \frac{wL^2}{16}\frac{12}{bh^3}\frac{h}{2} = \frac{3wL^2}{8bh^2} \tag{2.1}$$

The maximum longitudinal and vertical shear stress, τ, occurs near a support. The two parabolic distributions given by simple elastic theory are shown in Figure 2.2d; and at the horizontal centre-line of each beam:

$$\tau = \frac{3}{2}\frac{wL}{4}\frac{1}{bh} = \frac{3wL}{8bh} \tag{2.2}$$

The maximum deflection, δ, is given by the usual formula

$$\delta = \frac{5(w/2)L^4}{384EI} = \frac{5wL^4}{64Ebh^3} \tag{2.3}$$

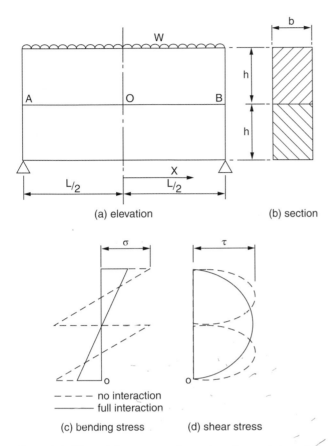

Figure 2.2 Effect of shear connection on bending and shear stresses

The bending moment in each beam at a section distant x from mid-span is $M_x = w(L^2 - 4x^2)/16$, so that the longitudinal strain ε_x at the bottom fibre of the upper beam is

$$\varepsilon_x = \frac{M y_{\max}}{EI} = \frac{3w}{8Ebh^2}(L^2 - 4x^2) \tag{2.4}$$

There is an equal and opposite strain in the top fibre of the lower beam, so that the difference between the strains in these adjacent fibres, known as the *slip strain*, is $2\varepsilon_x$.

It is easy to show by experiment with two or more flexible wooden laths or rulers that, under load, the end faces of the two-component beam have the shape shown in Figure 2.3a. The slip at the interface, s, is zero at $x = 0$ (from symmetry) and a maximum at $x = \pm L/2$. The cross-section at $x = 0$ is the only one where a plane cross-section through both members remains plane. The slip strain, defined above, is not the same as slip. In the same way that strain is rate of change of displacement, slip strain is the rate of change of slip along the beam. Thus from Equation 2.4:

$$\frac{ds}{dx} = 2\varepsilon_x = \frac{3w}{4Ebh^2}(L^2 - 4x^2) \tag{2.5}$$

Integration gives:

$$s = \frac{w}{4Ebh^2}(3L^2x - 4x^3) \tag{2.6}$$

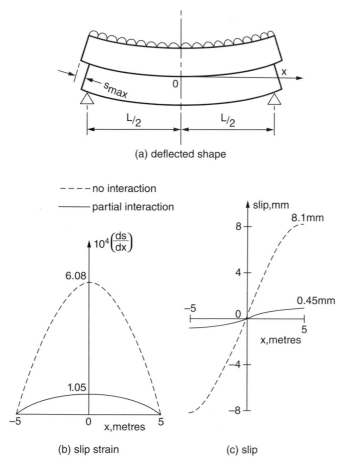

(a) deflected shape

(b) slip strain

(c) slip

Figure 2.3 Deflections, slip strain and slip

The constant of integration is zero, since $s = 0$ when $x = 0$, so that Equation 2.6 gives the distribution of slip along the beam.

Results (2.5) and (2.6) for the beam studied in Section 2.7 are plotted in Figure 2.3. The curves 'no interaction' show that, at mid-span, slip strain is a maximum and slip is zero and, at the ends of the beam, slip is a maximum and slip strain is zero. From Equation 2.6, the maximum slip (when $x = L/2$) is $wL^3/4Ebh^2$. Some idea of the magnitude of this slip is given by relating it to the maximum deflection of the two beams. From Equation 2.3, the ratio of slip to deflection is $3.2h/L$. For a beam with a typical span/depth ratio ($L/2h$ here) of 20, the end slip is less than a tenth of the deflection. This shows that *shear connection must be very stiff if it is to be effective*.

2.2.2 Full interaction

It is now assumed that the two halves of the beam shown in Figure 2.2 are joined together by an infinitely stiff shear connection. The two members then behave as one. Slip and slip strain are everywhere zero, and it can be assumed that plane sections remain plane.

This situation is known as *full interaction*. Except in design with partial shear connection (sections 3.5.3 and 3.7.1), all design of composite beams and columns in practice is based on the assumption that full interaction is achieved.

For the composite beam of breadth b and depth $2h$, $I = 2bh^3/3$, and elementary theory gives the mid-span bending moment as $wL^2/8$. The extreme fibre bending stress is

$$\sigma = \frac{My_{\max}}{I} = \frac{wL^2}{8}\frac{3}{2bh^3}h = \frac{3wL^2}{16bh^2} \tag{2.7}$$

The vertical shear at section x is:

$$V_x = wx \tag{2.8}$$

so the shear stress at the neutral axis is

$$\tau_x = \frac{3}{2}wx\frac{1}{2bh} = \frac{3wx}{4bh} \tag{2.9}$$

and the maximum shear stress is

$$\tau = \frac{3wL}{8bh} \tag{2.10}$$

The stresses are compared in Figure 2.2 (c) and (d) with those for the non-composite beam. The provision of the shear connection does not change the maximum shear stress, but the maximum bending stress is halved.

The mid-span deflection is

$$\delta = \frac{5wL^4}{384EI} = \frac{5wL^4}{256Ebh^3} \tag{2.11}$$

which is a quarter of the previous deflection (Equation 2.3). Thus the provision of shear connection increases both the bending strength and stiffness of a beam of given size, and in practice leads to a reduction in the size of the beam required for a given loading, and usually to a reduction in its cost.

In this example – but not always – the interface AOB coincides with the neutral axis of the composite member, so that the maximum longitudinal shear stress at the interface is equal to the maximum vertical shear stress, which occurs at $x = \pm L/2$ and is $3wL/8bh$, from Equation 2.10.

The shear connection must be designed for the longitudinal shear per unit length, v_L, which is known as the *shear flow*. In this example it is given by

$$v_{L,x} = \tau_x b = \frac{3wx}{4h} \tag{2.12}$$

The total shear flow in a half span is found, by integration of Equation 2.12, to be $3wL^2/(32h)$. Typically, $L/2h = 20$, so the shear connection in the whole span has to resist a total shear force

$$2 \times \frac{3}{32}\frac{wL^2}{h} = 7.5wL$$

Thus, this shear force is almost eight times the total load carried by the beam. A useful rule of thumb is that the resistance of the shear connection for a beam should be an order of magnitude greater than the load to be carried; it shows that *shear connection must be very strong*.

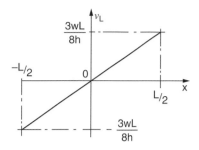

Figure 2.4 Shear flow for 'triangular' spacing of connectors

In design based on elastic behaviour, the shear connectors are spaced in accordance with the shear flow. Thus, if the design shear resistance of a connector is P_{Rd}, the pitch or spacing at which single connectors should be provided, p, is given by $pv_{L,x} \le P_{Rd}$. From Equation 2.12 this is

$$p \le \frac{4P_{Rd}h}{3wx} \tag{2.13}$$

This is known as 'triangular' spacing, from the shape of the graph of v_L against x (Figure 2.4).

2.3 Uplift

In the preceding example, the stress normal to the interface AOB (Figure 2.2) was everywhere compressive, and equal to $w/2b$ except above the supports. The stress would have been tensile if the load w had been applied to the lower member. Such loading is unlikely, except when travelling cranes are suspended from the steelwork of a composite floor above; but there are other situations in which stresses tending to cause vertical separation of the two members (uplift) can occur at the interface. These arise from complex effects such as the torsional stiffness of reinforced concrete slabs forming flanges of composite beams, the triaxial stresses in the vicinity of shear connectors and, in box-girder bridges, the torsional stiffness of the steel box.

Tension across the interface can also occur in beams of non-uniform section or with partially completed flanges. Two members without shear connection, as shown in Figure 2.5, provide a simple example. Beam AB is supported on CD and carries distributed loading. It can easily be shown by elastic theory that if the flexural rigidity of AB exceeds about a tenth of that of CD, then the whole of the load on AB is transferred to CD at points A and B, with separation of the beams between these points. If AB were connected to CD, there would be uplift forces at mid-span.

Almost all connectors used in practice are therefore so shaped that they provide resistance to uplift as well as to slip. Uplift forces are so much less than shear forces that it is not normally necessary to calculate or estimate them for design purposes, provided that connectors with some uplift resistance are used.

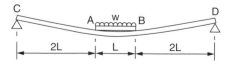

Figure 2.5 Uplift forces

2.4   Methods of shear connection

2.4.1   Bond

Until the use of deformed bars became common, most of the reinforcement for concrete consisted of smooth mild-steel bars. The transfer of shear from steel to concrete
was assumed to occur by bond or adhesion at the concrete–steel interface. Where the
steel component of a composite member is surrounded by reinforced concrete, as in
an encased beam (Figure 2.1c) or a concrete-encased column (Figure 5.14), the analogy
with reinforced concrete suggests that no shear connectors need be provided. Tests have
shown that this can be true for encased columns and concrete-filled steel tubes, where
bond stresses are usually low, and also for encased beams in the elastic range. In design
it is necessary to restrict bond stress to a low value, to provide a margin for the complex
effects of shrinkage of concrete, poor adhesion to the underside of steel surfaces, and
stresses due to variations of temperature.

Research on the ultimate strength of encased beams has shown that, at high loads,
calculated bond stresses have little meaning, due to the development of cracking and
local bond failures. If longitudinal shear failure occurs, it is usually on a surface such as
AA in Figure 2.1c, and not around the perimeter of the steel section. For these reasons,
codes of practice do not generally allow ultimate-strength design methods to be used
for composite beams without shear connectors.

Other types of composite beam are described in sections 3.10 and 3.12, but most
beams have cross-sections of types (a) or (b) in Figure 2.1. Tests on such beams show
that, at low loads, most of the longitudinal shear is transferred by bond at the interface,
that bond breaks down at higher loads, and that once broken it cannot be restored. So
in design calculations, bond strength is taken as zero and, in research on shear connection, the bond is deliberately destroyed by greasing the steel flange before the concrete
is cast. For uncased beams, the most practicable form of shear connection is some form
of dowel welded to the top flange of the steel member and subsequently surrounded by
in situ concrete when the floor or deck slab is cast.

2.4.2   Shear connectors

2.4.2.1   Headed stud connectors

The most widely used type of connector is the headed stud (Figure 2.6). These range
in diameter from 13 to 25 mm, and in length, h_{sc}, from 65 to 150 mm, although longer
and thicker studs are sometimes used. Studs should have an ultimate tensile strength
of at least 450 N/mm^2 and be made from steel with an elongation of at least 15%. The

Figure 2.6 Headed stud shear connector

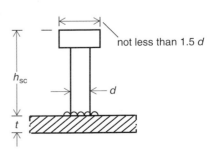

advantages of stud connectors are that the welding process is rapid, they provide little obstruction to reinforcement in the concrete slab, and they are equally strong and stiff in shear in all directions normal to the axis of the stud.

There are two factors that influence the diameter of studs. One is the welding process, which becomes increasingly difficult and expensive at diameters exceeding 20 mm, and the other is the thickness, t (Figure 2.6), of the plate or flange to which the stud is welded. A study in the USA (Goble, 1968) found that the full static strength of the stud can be developed if d/t is less than about 2.7, and a limit of 2.5 is given in EN 1994-1-1. Tests using repeated loading have led to the rule that where the flange plate is subjected to fluctuating tensile stress, d/t may not exceed 1.5. These rules prevent the welding of studs to the thin sheeting used in composite slabs. The maximum shear force that can be resisted by a 25 mm stud is relatively low, at about 130 kN.

2.4.2.2 Other types of connector

The preliminary version of Eurocode 4 (CEN, 1992) gave design rules for some other types of connector with higher strength, developed primarily for use in bridges. These are bars with hoops (Figure 2.7a), tees with hoops, channels (Figure 2.7b), angles and high-strength friction-grip bolts. Bars with hoops are the strongest of these, with ultimate shear strengths up to 1000 kN, but are not ductile. All these connectors were omitted from EN 1994-1-1 to shorten the code. There has been little recent research on

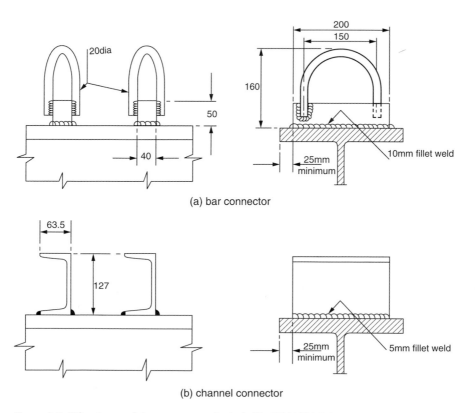

(a) bar connector

(b) channel connector

Figure 2.7 Other types of shear connector included in ENV 1994-1-1

them. It may be difficult to establish that they are ductile, but no other reason is known why the design rules given should not be used.

There is a connector made from thin-gauge steel and fixed by shot-fired pins. Its supplier, Hilti, provides extensive information to assist designers. It finds application mainly where longitudinal shear forces are relatively low.

Epoxy adhesives have been tried, but it is not clear how resistance to uplift can reliably be provided where the slab is in contact with the steel member only at its bottom surface.

In the 1980s, a bridge was being designed for a remote location in Venezuela, where use of welded studs was expected to be uneconomic. The consultants developed a new type of connector, known as Perfobond. It consisted of lengths of steel plate that formed an upwards continuation of the steel web and projected into the concrete, as in Figure 2.8a. Reinforcing bars could be threaded through holes in the plate, and the concrete in the holes acted as dowels to resist shear. This concept was the basis for several subsequent types of connector. The research on them has been summarized (Costa-Neves et al., 2016).

The design of a new connector has to reach a compromise between stiffness in shear at serviceability loads, flexibility (slip capacity) at ultimate loads, avoidance of stress concentrations that crack, split or spall surrounding concrete, ability to interact with reinforcement in the slab, resistance to forces tending to cause uplift, simplicity in construction, and, of course, cost. The connector should be suitable for a wide range of shear force per unit length of beam ('shear flow'), and for bridges it needs good fatigue performance.

Design rules with wide scope need to be based on mechanical models of behaviour, especially as failure is approached. The lack of these models for some types of connector has been one reason why studs are often preferred.

2.4.2.3 Perforated strips with concrete dowels and closed holes

The rules proposed for Perfobond connectors are based on curve-fitting to results from sets of tests. The parameters are: the properties of the three materials involved, the thickness, length and height of the strip, the size, level and spacing of the holes, the amount of reinforcement passing through them, and the thickness of the slab.

The shear resistance is a combination of:

- bearing of the end of a strip on the concrete (not ductile)

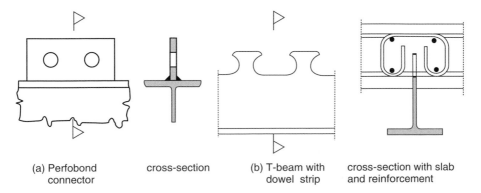

(a) Perfobond   cross-section   (b) T-beam with   cross-section with slab
    connector   dowel strip   and reinforcement

Figure 2.8 Concrete dowel shear connectors with (a) closed and (b) open holes for reinforcement

- shear in reinforcing bars passing through holes (ductile)
- dowel action from shear in the concrete on each side of a hole (fairly ductile)
- bond or friction on the sides of the strip (a relatively small part of the total resistance).

Most of the design rules proposed in the literature give the shear resistance as a sum of terms, one for each of the four contributions above. Many are not dimensionally correct, all have limited scope, and none of them provides a good fit to results from all of the test series reported by Costa-Neves et al. (2016). The relative importance of the four terms and the ductility depend on the details of the test specimens.

2.4.2.4 Perforated strips with open-top holes

The threading of reinforcing bars through holes in the Perfobond strips is an awkward process, which can be avoided by replacing holes by recesses in the upper side of the strip (Figure 2.8b). These were known as puzzle-strips, through their resemblance to an edge of a piece of a jigsaw puzzle. Many shapes for their profile have been tried. The design rules now proposed for Eurocode 4 name them *composite dowel connectors*. They use a standard shape based on the clothoid curve, with 'teeth' from 60 to 160 mm high, spaced at 2.5 times their height. Even the smallest size can be stronger in shear than a 19 mm stud.

Extensive research and development have led to their use in several bridges in central and eastern Europe. There have been many publications, including a complete issue of the journal *Steel Construction* (e.g. Feldmann et al., 2016; Lorenc et al., 2014). A report from the European Convention for Constructional Steelwork, *Shear connectors that use concrete dowels*, expected in 2018, gives prominence to the Perfobond type.

The clothoid type of connector has the same resistance to shear in both longitudinal directions. Its performance does not rely on an adjacent steel flange, so it can be made by gas cutting the top edge of a steel web to the standard profile. The steel section is then an inverted T. This top edge can be embedded in a slab of uniform thickness, or in a haunched slab. The design rules even apply where the inverted T forms the bottom of the web of a reinforced concrete T-beam, although the design for vertical shear would then be outside the scope of Eurocode 4. On a wide steel top flange, dowel strips can be welded side by side, so the method is well suited to bridge beams with high longitudinal shear force per unit length.

The detailing rules make the dowel connectors unsuitable for slabs less than about 180 mm thick, and transverse profiled sheeting cannot be continuous across a beam, so they are less suitable for buildings.

In a concrete flange of uniform thickness, more transverse reinforcement is needed than that required where studs are used. Stirrups are required on both sides of each tooth, as shown in the cross-section in Figure 2.8b, to prevent splitting of the slab. The local reinforcement is thus more complex than for studs, and helps to provide the high ductility that is observed in tests.

As with studs, the shear resistance is given by the lesser of two expressions, one based on yielding of the steel strip, and the other on the failure of a cone of concrete between each tooth, including the effect of the transverse reinforcement. The strips are usually continuous for long lengths, so bearing pressure on the end of a strip is unimportant. Bond or friction on the sides of a strip could be significant if there is pressure from transverse hogging bending; but it could be negligible where transverse tension is present, so

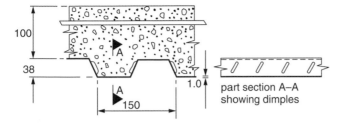

Figure 2.9 Composite slab

it does not appear in the resistance expressions. Rules for fatigue resistance have been developed.

Tests find that these connectors have high initial stiffness in shear, and at service loads are slightly stiffer than studs. They then have good plastic behaviour and reach maximum shear force at a slip usually exceeding 6 mm, with little loss of resistance at 10 mm, so they are 'ductile' as defined in Eurocode 4.

2.4.3 Shear connection for profiled steel sheeting

This material is commonly used as permanent formwork for floor slabs in buildings, then known as *composite slabs*. The usual profile shape is *trapezoidal*. Some profiles have flat top surfaces (Figure 2.9); others have a small projecting rib (Figures 2.16 and 3.11) that prevents buckling of the top flange, while the sheeting is supporting wet concrete, and not composite. As it is impracticable to weld shear connectors to material that may be less than 1 mm thick, shear connection is provided by pressed or rolled dimples that project into the concrete. Trapezoidal profiles can be over 200 mm deep, as shown in the example in Section 3.13.

Another type of profile is *re-entrant* (on the right in Figure 2.15). This shape prevents separation of the sheeting from the concrete slab, provides strong shear connection that can be brittle, and is convenient for attaching supports for services or false ceilings.

Additional shear connection can be provided where sheeting is supported on a steel flange, by studs welded through the sheeting. The example in Section 3.4.3 includes this. At a support, there is also shear connection from friction, proportional to the vertical shear in the slab. This can be allowed for in design (Section 3.3.2).

2.5 Properties of shear connectors

The property of a shear connector most relevant to design is the relationship between the shear force transmitted, P, and the slip at the interface, s. This load–slip curve should ideally be found from tests on composite beams, but in practice a simpler specimen is necessary. Most of the data on connectors have been obtained from various types of 'push-out' or 'push' test. The flanges of a short length of steel I-section are connected to two small concrete slabs. The details of the standard push test of EN 1994-1-1 are shown in Figure 2.10. The slabs are bedded onto the lower platen of a compression-testing machine or frame, and load is applied to the upper end of the steel section. Slip between the steel member and the two slabs is measured at several points, and the average slip is

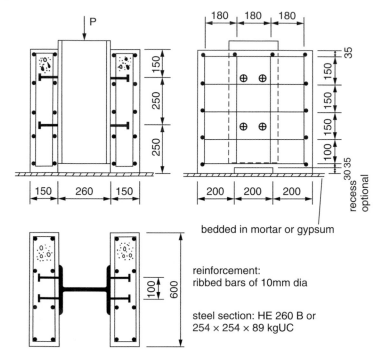

Figure 2.10 Standard push test

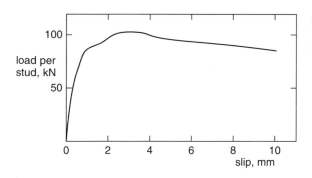

Figure 2.11 Typical load–slip curve for 19 mm stud connectors in a composite slab

plotted against the load per connector. A typical load–slip curve is shown in Figure 2.11, from a test using composite slabs (Mottram and Johnson, 1990).

In practice, designers normally specify shear connectors for which strengths have already been established, for it is an expensive matter to carry out sufficient tests to determine design strengths for a new type of connector. If reliable results are to be obtained, the test must be specified in detail, as the load–slip relationship is influenced by many variables, including:

(1) number of connectors in the test specimen;
(2) mean longitudinal stress in the concrete slab surrounding the connectors;
(3) size, arrangement and strength of slab reinforcement in the vicinity of the connectors;

(4) thickness of concrete surrounding the connectors;
(5) freedom of the base of each slab to move laterally, and so to impose uplift-type forces on the connectors;
(6) bond at the steel–concrete interface;
(7) strength of the concrete slab;
(8) degree of compaction of the concrete surrounding the base of each connector;
(9) significant external tension, which reduces shear strength (Shen and Chung, 2017).

The details shown in Figure 2.10 include requirements relevant to items 1 to 6. The amount of reinforcement specified and the size of the slabs were increased in Eurocode 4, as compared with the former British test, which had barely changed since it was introduced in 1965. The Eurocode test gives results that are less influenced by splitting of the slabs, and so give better predictions of the behaviour of connectors in beams (Johnson, 2012).

The 'standard' test is intended for slabs of uniform thickness. The alternative is a 'specific' push test. This has been found to be necessary for composite slabs, where other failure mechanisms can occur, such as a stud pulling out from the slab, with much deformation of the trough of sheeting that surrounds it.

The tensile force that causes pull-out is higher in a push test than it would be in a loaded beam. To allow for this, a new push test for studs in open-trough profiled sheeting may be added to the next Eurocode 4. The main difference from the standard test is that distributed horizontal forces are applied to the outer surfaces of the slabs, to reduce this tension. They represent the vertical loading on a floor, and are limited to $P/10$, where P is shown in Figure 2.10.

At present, the length of time that an increment of deformation must be maintained in a push test, before the load is recorded, is not specified. Practice has varied. It is being proposed that the increment should be 1 mm slip, and the time five minutes. This, if adopted, may lead to slightly lower maximum loads.

Tests have to be done for a range of concrete strengths, because the strength of the concrete influences the mode of failure, as well as the failure load. Studs may reach their maximum load when the concrete surrounding them fails, but in stronger concrete, they shear off. This is why the design shear resistance of studs with $h_{sc}/d \geq 4$ is given in Eurocode 4 as the lesser of two values:

$$P_{Rd} = \frac{0.8 f_u (\pi d^2/4)}{\gamma_V} \tag{2.14}$$

and

$$P_{Rd} = \frac{0.29 d^2 (f_{ck} E_{cm})^{1/2}}{\gamma_V} \tag{2.15}$$

where f_u is the ultimate tensile strength of the steel (≤ 500 N/mm^2), and f_{ck} and E_{cm} are the cylinder strength and mean secant (elastic) modulus of the concrete, respectively. Dimensions h_{sc} and d are shown in Figure 2.6. The value recommended for the partial safety factor γ_V is 1.25, based on statistical calibration studies. When $f_u = 450$ N/mm^2, Equation 2.14 governs when f_{ck} exceeds about 30 N/mm^2.

Ignoring γ_V, it is evident that Equation 2.14 represents shear failure in the shank of the stud at a mean stress of $0.8 f_u$. To comment on Equation 2.15, it is assumed that the force P_R is distributed over a length of connector equal to twice the shank diameter,

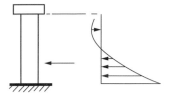

Figure 2.12 Bearing stress on the shank of a stud connector

because research has shown that the bearing stress on a shank is concentrated near the base, as sketched in Figure 2.12. An approximate mean stress is then $0.145\,(f_{ck}E_{cm})^{1/2}$. Its value, using data from EN 1992, ranges from $110\,\text{N/mm}^2$ for class C20/25 concrete to $171\,\text{N/mm}^2$ for class C40/50 concrete, so for these concretes the mean bearing stress at concrete failure ranges from $5.5f_{ck}$ to $4.3f_{ck}$. This estimate ignores the enlarged diameter at the weld collar at the base of the stud, shown in Figure 2.6; but it is clear that the effective compressive strength is several times the cylinder strength of the concrete.

This very high strength is possible only because the concrete bearing on the connector is restrained laterally by the surrounding concrete, its reinforcement, and the steel flange. The results of push tests are likely to be influenced by the degree of compaction of the concrete, and even by the arrangement of particles of aggregate, in this small but critical region. This is thought to be the main reason for the scatter of the results obtained.

The extent of the scatter can only be found by repeating 'nominally identical' tests. Eurocode 4 gives a method for estimating the characteristic resistance P_{Rk} from only three results, provided that they are all within 10% of their mean. It is to take P_{Rk} as 10% below the lowest of the three. If the scatter of the results is too great, more tests are required. The strength of the connector material is often found to be above the specified minimum value, which leads to a downwards correction of P_{Rk}.

The other property that can be deduced from a push-test result is the *slip capacity*, δ_u. This is defined in EN 1994-1-1 as the maximum slip at the load level P_{Rk}, normally on the falling branch of the load–slip curve. The characteristic slip capacity, δ_{uk}, is the minimum value of δ_u from a set of tests, reduced by 10%, unless there are sufficient test results for the 5% lower fractile to be determined.

Where connectors have sufficient slip capacity ('ductility'), design of shear connection can be simplified. In the current Eurocode 4, a connector may be taken as 'ductile' if $\delta_{uk} \geq 6\,\text{mm}$. A new category, 'super-ductile' ($\delta_{uk} \geq 10\,\text{mm}$), is likely to be added, as explained in Section 3.6.2.

The load–slip curve for a connector in a beam is influenced by the difference between the longitudinal stress in a concrete flange and that in the slabs in a push test. Where the flange is in compression the load/slip ratio (the stiffness) in the elastic range exceeds the push-test value, but the ultimate strength is about the same. For slabs in tension (e.g. in a region of hogging moment), the connection is significantly less stiff (Johnson et al., 1969), but the ultimate shear resistance is only slightly lower. The reduction in stiffness is one reason why partial shear connection (Section 3.6) is allowed in Eurocode 4 only in regions of sagging bending moment. [That statement implies an assumption that is often found in design rules for composite beams: that the concrete slab is above the steel beam. Slabs attached only to bottom flanges are rare, but *double composite action*, where slabs are attached to both flanges, is found in some bridges.]

Figure 2.13 Haunch with connectors too close to a free surface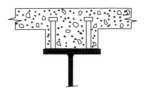

There are two situations in which the resistance of a connector found from push tests may be too high for use in design. One is repeated loading, such as that due to the passage of traffic over a bridge. The other is where the lateral restraint to the concrete in contact with the connector is less than that provided in a push test, as in a haunched beam with connectors too close to a free surface (Figure 2.13). For this reason, the use of the standard equations for resistance of connectors is allowed in haunched beams only where the cross-section of the haunch satisfies certain conditions. In EN 1994-1-1 these are that the concrete cover to the side of the connectors may not be less than 50 mm (line AB in Figure 2.14), and that the free concrete surface may not lie within the line CD, which runs from the base of the connector at an angle of 45° with the steel flange. A haunch that just satisfies these rules is shown as EFG in Figure 2.14.

EN 1994-2 has an annex, 'Headed studs that cause splitting forces in the direction of the slab thickness'. It was first developed for studs with their axes horizontal, and known as 'lying studs'. It covers a wider range of situations than those shown in Figures 2.13 and 2.14. It applies also to buildings, and is in the 'bridge' part of Eurocode 4 only because EN 1994-1-1 was already complete. The current version of these rules is explained in Section 2.5.2, and is likely to be included in the next revision of EN 1994-1-1.

Tests show that the ability of lightweight-aggregate concrete to resist the high local stresses at shear connectors is slightly less than that of normal-density concrete of the same cube strength. This is allowed for in EN 1994-1-1 by the lower value of E_{cm} that is specified for lightweight concrete (Table 1.4).

2.5.1 Stud connectors used with profiled steel sheeting

Where profiled sheeting is used, stud connectors are located within concrete ribs that have the shape of a haunch, which may run in any direction relative to the direction of span of the composite beam. Tests show that the shear resistance of connectors is sometimes lower than it is in a solid slab, for materials of the same strength, because of local failure of the concrete rib. For this reason, EN 1994-1-1 specifies reduction factors, applied to the resistance P_{Rd} found from Equations 2.14 or 2.15.

Figure 2.14 Detailing rules for haunches

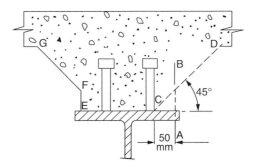

In this section, symbol h for the overall height of a stud is replaced by h_{sc}, as in Eurocode 4. For sheeting with ribs parallel to the beam the factor is:

$$k_\ell = 0.6\frac{b_0}{h_p}\left(\frac{h_{sc}}{h_p} - 1\right) \le 1.0 \tag{2.16}$$

The dimensions b_0 and h_p are illustrated in Figure 2.15, and h_{sc} is taken as not greater than $h_p + 75$ mm. The height h_p does not include small top ribs of the type shown in Figure 2.16.

For sheeting with ribs transverse to the beam, the factor is:

$$k_t = \frac{0.7}{\sqrt{n_r}}\frac{b_0}{h_p}\left(\frac{h_{sc}}{h_p} - 1\right) \tag{2.17}$$

where n_r is the number of connectors in one rib where it crosses a beam, not exceeding two.

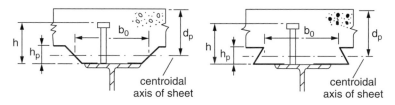

Figure 2.15 Composite beam with composite slab spanning in the same direction. Left: trapezoidal; right: re-entrant

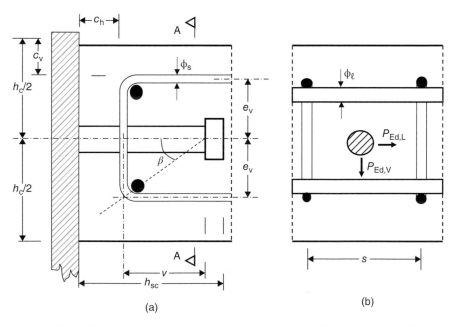

Figure 2.16 (a) Cross-section of a slab with a lying stud supported by a steel plate. (b) Cross-section A–A showing the spacing s of the links

EN 1994-1-1 gives upper limits to k_t that range from 0.6 to 1.0. The limit depends on the thickness of the sheeting, t, the diameter of the studs, and on whether n_r is 1 or 2. For $n_r > 2$, comment is given in Section 3.11.2. The limits also distinguish between studs welded to the steel flange through a hole in the sheeting (the usual practice in some countries) and the British (and North American) practice of through-deck welding.

Since the origin of Equations 2.16 and 2.17 (Grant et al., 1977), tests using newer types of sheeting have found problems with this use of the solid-slab resistances as their basis. These are now explained.

2.5.1.1 Sheeting with ribs transverse to the beam

(1) Troughs in sheeting have stiffening ribs that limit the locations of studs. The 'central' position, shown in Figure 2.17, is best, but central ribs are common. Studs then have to be close to one wall of the trough. These positions are known as 'favourable' and 'unfavourable', or 'strong' and 'weak', depending on the direction of the shear force. Position affects shear resistance, slip capacity and failure mode. Weak positions

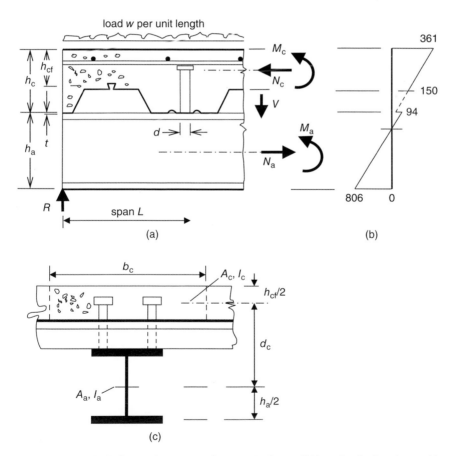

Figure 2.17 (a) End of a simply-supported composite beam. (b) Longitudinal strain at mid-span, $\times 10^6$, for $\eta = 0.46$. (c) Cross-section through top rib of the sheeting

should never be used, but shear from imposed loading can act in either direction. It is now required that non-central studs should be 'staggered': placed in both the strong and weak positions, alternately along the span, even though their average behaviour is not always as good as that of a central stud (Johnson, 2008).

(2) The interaction between studs placed within the same trough of transverse sheeting is not accurately represented by the term $n^{1/2}$ in Equation 2.17, even though more than two studs per trough is no longer permitted.

(3) The shape of a concrete rib is not sufficiently represented by the dimensions b_0 and h_p.

(4) The limits to the scope of Equation 2.17, which are $h_p \leq 85$ mm and $b_0 \geq h_p$, need to be widened to suit current sheeting profiles.

(5) The term h_{sc}/h_p in both equations does not always ensure that the projection of a stud above the sheeting is sufficient to prevent pull-out failure.

(6) Many failure modes have been observed in tests other than the two solid-slab modes upon which Equations 2.16 and 2.17 are based.

Sets of design rules for transverse sheeting that give stud resistances directly have been proposed. One of them (Odenbreit and Nellinger, 2017) may be added to Eurocode 4. Equations 2.16 and 2.17 and their limits already involve the material properties f_u, f_{ck} and E_{cm}, and the geometric properties d, b_0, h_{sc}, h_p, n_r and t, most of which are shown in the Figures. These new rules also use the mean tensile strength of the concrete, f_{ctm}, the maximum width of a trough, b_{max}, a factor β that depends on whether the sheeting has re-entrant ribs or not, a penalty factor k_u for studs in the weak position, and factors α_{c1} and α_{c2} to compensate for the short duration of the loading used in push testing. These factors, new to Eurocode 4, have been controversial, and are nationally determined parameters. Their values are likely to be in the range 0.9 to 1.0.

The existing rules have been so widely used that they will not be omitted. Instead, their scope will be further limited to exclude applications that could be unsafe. So that users do not have to choose which set of rules to use, it seems likely that their respective scopes will be mutually exclusive: for example, $h_p < 75$ mm and $b_0 < 110$ mm for the current rules, and $h_p \geq 75$ mm and $b_0 \geq 110$ mm for the new ones.

To compare the sets of rules, resistances P_{Rk} are calculated below by both methods, using this pair of values for h_p and b_0, which are, of course, marginally outside the reduced scope of the current rules.

The further reductions in the scope of the existing rules are expected to include:

- for the projection of studs above a trough: $h_{sc} - h_p \geq 2dn_r^{1/2}$
- for through-deck welding, the stud diameter d not to exceed 20 mm
- studs to be central within a rib, or alternately in the 'strong' and 'weak' positions.

The new rules have a limit $h_p \leq 140$ mm, and so do not include the deepest decks now in use, which are over 200 mm deep (Section 3.4.6). Their required height of studs is more liberal: $h_{sc} - h_p \geq 2d$.

The following illustration of both methods is limited to trapezoidal decking with two studs per trough ($n_r = 2$) of height that satisfies a condition in the new rules, and in central, favourable or staggered positions.

Equations 2.14 and 2.15 represent failure of the studs, or of a cone of concrete surrounding them, treated as alternatives. The new rules treat these resistances as contributions to be added together. They appear (for the limited scope used here) in reversed sequence as the two terms within the [...] brackets in the resistance equation:

$$P_{\text{Rk}} = C_2 \alpha_{c2} \left[\frac{f_{\text{ctm}} W}{(h_p n_r)} + \frac{n_y M_{\text{pl}}}{(h_s - d/2)} \right] \tag{2.18}$$

There is also an equation for stud failure only, similar to Equation 2.14. Although it gives a resistance over 25% lower, it rarely governs.

The terms in Equation 2.18 are now explained.

- C_2 depends on the depth of the profile: $C_2 = 2.5 - h_p/50$, $0.9 \le C_2 \le 1.3$.
- α_{c2} is explained above,
- W is an effective section modulus for cone failure: $W = (2.4 h_{sc} + e_t) b_{\text{max}}^2/6$, where b_{max} is the width of the top of the trough in the sheeting, and e_t is the spacing of a pair of studs.
- n_y depends on the projection of the studs above the sheeting. For $n_r = 2$:

$$n_y = 1 + (h_{sc} - h_p - 2d)/(0.52d), \le 2$$

- M_{pl} is the bending resistance of a stud, $f_{uk} d^3/6$.
- h_s depends on the height of the studs. For trapezoidal decking: $h_s = 0.45 h_{sc}, \le h_p$.

The new method is based on extensive testing, which appears to have shown that the thickness of the sheeting, t, has little influence, because t does not appear in these equations, although the current method for k_t allows for it.

The resistance P_{Rk} has been calculated by both methods for these values, which satisfy their various conditions:

- for materials: $f_{uk} = 450\,\text{MPa}$, $f_{ck} = 30\,\text{MPa}$, $f_{\text{ctm}} = 2.9\,\text{MPa}$, $E_{cm} = 33\,\text{GPa}$, $\alpha_{c1} = \alpha_{c2} = 0.95$.
- for the studs and sheeting: $d = 19\,\text{mm}$, $h_{sc} = 130\,\text{mm}$, $h_p = 75\,\text{mm}$, $b_0 = 110\,\text{mm}$, $b_{\text{max}} = 138\,\text{mm}$, $n_r = 2$, $t \ge 1\,\text{mm}$, $e_t = 80\,\text{mm}$.

From Equation 2.17, the current Eurocode gives

$$k_t = \frac{0.7}{\sqrt{2}} \left(\frac{110}{75} \right) \left(\frac{130}{75} - 1 \right) = \mathbf{0.532}, \le 0.7$$

and using Equation 2.15, with factor α_{c2} added

$$P_{\text{Rk}} = \frac{0.532 \times 0.95 \times 0.29 \times 19^2 (30 \times 33{,}000)^{1/2}}{1000} = \mathbf{52}\,\text{kN/stud}$$

The new method, as currently understood, would give:

- $C_2 = 2.5 - 75/50 = 1.0$
- $W = (2.4 \times 130 + 80) \times 138^2/6 = 1.24 \times 10^6\,\text{mm}^3$
- $n_y = 1 + (130 - 75 - 38)/(0.52 \times 19) = 2.72 \le \mathbf{2.0}$
- $M_{\text{pl}} = 450 \times 19^3/6000 = 514\,\text{N m}$
- $h_s = 0.45 \times 130 = \mathbf{58.5} \le 75\,\text{mm}$
- $P_{\text{Rk}} = 1 \times 0.95\,[2.9 \times 1.24 \times 10^6/(75 \times 2) + 2 \times 514 \times 1000/(58.5 - 9.5)]/1000 = \mathbf{43}\,\text{kN/stud}$

The new method widens the scope of Eurocode 4, but is much longer, and at the boundary with the current method, it is more conservative. It remains to be seen in what form it will be included in the next Eurocode 4.

2.5.1.2 Sheeting with ribs parallel to the beam

There has been relatively little research on stud resistances in sheetings parallel to the span of the beam, and almost none where the sheeting is neither transverse nor parallel. It is therefore not known whether Equation 2.16 needs revision, and no new method is being proposed. The equation can be applied both where the sheeting is continuous across the beam, and b_0 is as defined for transverse sheeting, and where it is discontinuous (Figure 2.15). The layout of studs is assumed to be governed by the rules on spacing used for solid slabs. In practice, it would be prudent to seek guidance also from the rules for unsheeted haunches (Figure 2.14) and for lying studs (Section 2.5.2).

2.5.2 Stud connectors in a 'lying' position

Figure 2.16a shows a cross-section through a concrete slab that is supported by a steel plate, to which it is attached by a row of studs at longitudinal spacing p. Figure 2.16b shows the forces applied by the stud to the plate: longitudinal shear $P_{Ed,L}$ and vertical shear $P_{Ed,V}$. To prevent the longitudinal shear from splitting the slab, links of diameter ϕ_s are provided at spacing s, and enclose two longitudinal bars of diameter ϕ_ℓ.

This layout is known as an 'edge position'. It can occur, for example, in a box girder with double composite action, where a slab rests on the bottom steel plate, so $P_{Ed,V} \approx 0$ and the steel web extends above the slab.

The shanks of the studs are parallel to the plane of the slab, so they are known as 'lying studs'. Research on them led to Annex C of EN 1994-2, 'Headed studs that cause splitting forces in the direction of the slab thickness'. Clause 6.6.4 of EN 1994-2 gives minimum requirements for the dimensions h_{sc}, v and e_v, and refers to Annex C for more complex but less restrictive rules.

In buildings, a more likely layout is where the steel beam is a T-section with no top flange, and the slab extends above it, with studs on both sides of the steel web. This is a 'middle position', for which a revised Annex C is expected to be included in the next EN 1994-1-1. This account of its design method is limited to a single row of studs at mid-depth of the slab, in the edge position shown in Figure 2.16. The main differences from a typical shear connection are the use of links and much longer studs.

The method was developed by fitting equations to the results of tests (Kuhlmann and Kürschner, 2004; Kuhlmann and Raichle, 2008). Their scope is limited and some are dimensionally inconsistent, so that the units for the symbols have to be specified. All dimensions are in mm, strengths of materials are in N/mm^2 and shear resistances are in kN.

The main limitations to scope are:

- effective edge distance, $e_v \geq 50$ mm
- stud diameter, $19 \leq d \leq 25$ mm
- stud spacing, $110 \leq p \leq 440$ mm and $p \leq 250$ mm for vertical shear
- link spacing, $p/2 \leq s \leq p$ and $s/e_v \leq 3$

- link diameter, $\phi_s \geq 8\,\text{mm}$ and $\phi_s \leq 12\,\text{mm}$ for vertical shear
- longitudinal bar diameter, $\phi_\ell \geq 10\,\text{mm}$ and $\phi_\ell \leq 16\,\text{mm}$ for vertical shear.

In the absence of vertical shear the design resistance of a stud to longitudinal shear is

$$P_{\text{Rd,L}} = \frac{1.4(f_{ck}de_v)^{0.4}(p/s)^{0.3}}{\gamma_V} \quad \text{kN} \tag{a}$$

Pull-out of a stud is prevented by requiring a minimum ratio e_v/v, that depends on whether the slab is cracked or uncracked. Where this is not known, the 'cracked' limit, which requires longer studs, should be chosen. Referring to Figure 2.16a, the limits are:

$$\text{for cracked concrete: } \beta \leq 23° \text{ or } v \geq \max{(160\,\text{mm},\ 2.4\,e_v,\ 1.2s)} \tag{b}$$

$$\text{for uncracked concrete, } \beta \leq 30° \text{ or } v \geq \max{(110\,\text{mm},\ 1.7\,e_v,\ 0.85s)} \tag{c}$$

The resistance of a stud in an edge position to vertical shear alone is

$$P_{\text{Rd,V}} = \frac{0.012(f_{ck}\phi_\ell)^{0.5}(dp/s)^{0.4}\phi_s^{0.3}e_v^{0.7}}{\gamma_V} \quad \text{kN} \tag{d}$$

Where both types of shear are present, the interaction limit is

$$\left(\frac{P_{\text{Fd,L}}}{P_{\text{Rd,L}}}\right)^{1.2} + \left(\frac{P_{\text{Ed,V}}}{P_{\text{Rd,V}}}\right)^{1.2} \leq 1 \tag{e}$$

The tensile resistance per unit length of beam of the links provided to resist splitting is

$$T_{\text{Rd}} = \frac{f_{sd}(\pi\phi_s^2/4)}{s} \tag{f}$$

This must be not less than the splitting force per unit length, which is given by

$$T_{\text{Ed}} = \frac{(0.3P_{\text{Ed,L}} + P_{\text{Ed,V}})}{p} \tag{g}$$

The preceding results are for edge positions. For middle positions, the resistances $P_{\text{Rd,L}}$ and $P_{\text{Rd,V}}$ are increased by 14% and 25%, respectively. Rules are also given for studs placed off-centre in the slab, for two rows of studs, and for resistance to fatigue.

Lying studs are at present rarely used in structures for buildings. The design method is quite complex and needs unusually long studs.

2.5.3 Example: stud connectors in a 'lying' position

The purpose is to find shear resistances for a particular design of shear connection. For longitudinal shear, this example uses the minimum values given above: $e_v = 50\,\text{mm}$, $d = 19\,\text{mm}$, $\phi_s = 8\,\text{mm}$ and $\phi_\ell = 10\,\text{mm}$, with the partial factor $\gamma_V = 1.25$. The concrete covers are assumed to be $c_v = c_h = 30\,\text{mm}$. The thickness of the slab must satisfy

$$h_c \geq 2e_v + \phi_s + 2\,c = 100\ +\ 8\ +\ 60 = 168\ \ \text{mm, so } h_c = 170\ \ \text{mm is assumed.}$$

The spacing of the studs is assumed to be $p = 200\,\text{mm}$. An upper limit to the spacing of links is $3e_v$, or 150 mm here, and a lower limit is $p/2$, which is 100 mm. Both values will be tried.

Considering cracked concrete first, from Expression (b):

$$e_v/v \leq \tan 23° = 0.424, \text{ so } v \geq 50/0.424 = 118 \text{ mm}$$

This is less than an alternative in Expression (b), 160 mm, and so is used. Assuming that the head of the stud is 10 mm thick, this gives

$$h_{sc} \geq 118 + 10 + 4 + 30 = 162 \text{mm, so } h_{sc} = 170 \text{mm is assumed.}$$

For uncracked concrete, with $\beta \leq 30°$, a similar calculation gives $h_{sc} \geq 131$ mm, so 140 mm studs are used. Both values exceed the typical lengths of 19 mm studs: 100 mm and 125 mm.

The shear resistance of these studs in a solid slab is given by the lesser of the values from Equations 2.14 and 2.15. It is $P_{Rd} = 83.3$ kN, from Equation 2.15. For lying studs with no vertical shear, Equation (a) gives, for $s = 150$ mm:

$$P_{Rd,L} = \frac{1.4(30 \times 19 \times 50)^{0.4}(1.33)^{0.3}}{1.25} = \mathbf{73.8 \text{ kN}}$$

With $s = 100$ mm, from Equation (a), $P_{Rd,L} = 73.8 \, (1.5)^{0.3} = 83.4$ kN, but $\leq \mathbf{83.3}$ kN, so the usual resistance can be reached with stirrups at 100 mm spacing. Assuming the slab to be cracked does not alter these resistances; it leads to the use of longer studs. The results are shown in Table 2.1.

Equations (f) and (g) are now used to check the 8-mm links, assuming the wider spacing, $s = 150$ mm, $P_{Ed,L} = P_{Rd,L} = 73.8$ kN with no vertical shear, and $f_{sd} = 500/1.15 = 435$ MPa for the links. From Equation (f),

$$T_{Rd} = \frac{0.435(\pi \times 8^2/4)}{0.15} = \mathbf{146 \text{ kN/m}}$$

Equation (g) with $P_{Ed,V} = 0$ shows that the links have sufficient resistance, because

$$T_{Ed} = \frac{(0.3 \times 73.8)}{0.2} = \mathbf{111 \text{ kN/m}}$$

Vertical shear in a beam is normally less than longitudinal shear per unit length. In this example, it is now assumed that $P_{Ed,V} = 0.2P_{Ed,L}$. From Equation (d) with links at 100 mm spacing,

$$P_{Rd,V} = \frac{0.012(30 \times 10)^{0.5}(19 \times 2)^{0.4} \times 8^{0.3} \times 50^{0.7}}{1.25} = \mathbf{20.5 \text{ kN}}$$

Expression (e), taken as an equality, is used to find the maximum $P_{Ed,L}$:

$$\left(\frac{P_{Ed,L}}{83.3}\right)^{1.2} + \left(\frac{0.2P_{Ed,L}}{20.5}\right)^{1.2} = 1$$

Table 2.1 Shear resistances for a lying stud in an edge position, with $P_{Rd} = 83.3$ kN

Link spacing	$P_{Rd,L}$, kN	$P_{Rd,V}$, kN	Maximum $P_{Ed,L}$, kN with $P_{Ed,v} = 0.2\,P_{Ed,L}$
$s/p = 0.5$	83.3	20.5	51.5
$s/p = 0.75$	73.8	17.4	44.8

This gives $P_{Ed,L} \leq 51.5$ kN. Similar calculations with links spaced at 150 mm give $P_{Rd,V} = 17.4$ kN and $P_{Ed,L} \leq 44.8$ kN.

The resistances of the links are checked using Equations (f) and (g). For 100-mm spacing, from Equation (g):

$$T_{Ed} = \frac{(0.3 \times 51.5 + 0.2 \times 51.5)}{0.2} = 129 \text{ kN/m}$$

From Equation (f),

$$T_{Rd} = \frac{0.435(\pi \times 8^2/4)}{0.10} = 219 \text{ kN/m}$$

This is sufficient. From Equation (a), a wider spacing reduces the limit to $P_{Ed,L}$ to below 51.5 kN. These results show that the presence of vertical shear causes a disproportionate reduction in resistance to longitudinal shear.

2.6 Partial interaction

In studying the simple composite beam with full interaction (Section 2.2.2), it was assumed that slip was everywhere zero. However, the results of push tests show (e.g. Figure 2.11) that even at the smallest loads, slip is not zero. It is therefore necessary to know how the behaviour of a beam is modified by the presence of slip. This is best illustrated by an analysis based on elastic theory. It leads to a differential equation that has to be solved afresh for each type of loading, and is therefore too complex for use in design offices. Even so, partial-interaction theory is useful, for it provides a starting point for the development of simpler methods for predicting the behaviour of beams at working load, and finds application in the calculation of interface shear forces due to shrinkage and differential thermal expansion.

The problem to be studied and the relevant variables are defined below. The details of the theory, and of its application to a composite beam, are given in Appendix A. A consistent convention of sign is used. Positive quantities are tensile force, stress and strain, sagging bending moment and anticlockwise vertical shear, as shown in Figure A.1. The free shrinkage strain of concrete, ε_{cs}, is therefore a negative number. The results and comments on them are given below and in Section 2.7.

Elastic analysis is relevant to situations in which the loads on connectors do not exceed about half their ultimate strength. For slip up to about 1 mm, the load–slip curve (Figure 2.11) can be replaced with little error by a straight line. The ratio of load to slip given by this line is known as the *connector modulus*, k. Shear connectors of modulus k are provided at uniform spacing p along the length of the beam. The stiffness of the shear connection is $k_c = k/p$.

For simplicity, the scope of the analysis is restricted here to a simply-supported composite beam of span L, supporting a composite slab of thickness h_c, with propped construction. One end of the beam is shown in Figure 2.17a. The total distributed load is w per unit length. Some profiled sheetings have a small top rib, as shown here, so the effective thickness of the concrete flange is h_{cf}. Any concrete between the ends of lengths of sheeting is ignored, so the centre of area of the concrete in compression is at depth $h_{cf}/2$ (Figure 2.17c). The effective flange width (not to scale in the Figure) is b_c. There is a symmetrical steel I-section of depth h_a, cross-sectional area A_a and second moment of area

I_a. The distance between the centres of area of the concrete and steel cross-sections, d_c, is given by

$$d_c = \frac{(h_a + 2h_c - h_{cf})}{2}$$

The elastic modulus of the steel is E_a, and that of the concrete for short-term loading is E_c. Allowance is made for creep of concrete by using an effective modulus $E_{c,eff}$ in the analysis, where

$$E_{c,eff} = \frac{E_c}{1 + \varphi} \tag{2.19}$$

Symbol φ is the creep coefficient, which is the ratio of the creep strain to the elastic strain. The short-term modular ratio is $n_0 = E_a/E_c$. The modular ratio n is defined by

$$n = \frac{E_a}{E_{c,eff}} \tag{2.20}$$

The concrete is assumed to be as stiff in tension as in compression, for it is found that in a concrete flange that is mainly in compression, the tensile stresses are low enough for little error to result from this analysis, except where the degree of shear connection is very low.

In a composite beam the steel section is thinner than the concrete section, and the steel has a much higher coefficient of thermal conductivity. The steel responds more rapidly than the concrete to changes of temperature. If the two components were free, their lengths would change at different rates; but the shear connection prevents this, and the resulting stresses in both materials can be large enough to influence design. The shrinkage of the concrete slab has a similar effect. A simple way of allowing for shrinkage in this analysis is to assume that, after connection to the steel, the concrete slab extends uniformly, by an amount ε_{cs} per unit length, relative to the steel.

The governing equation gives the longitudinal slip, s, as a function of distance x along the beam, with $x = 0$ at mid-span, and is

$$\frac{d^2s}{dx^2} - \alpha^2 s = -\alpha^2 \beta w x \tag{2.21}$$

where α and β are complex functions of the geometric and elastic properties of the cross-section and the shear connection, listed in Appendix A. The theory was first published by Newmark, Siess and Viest (1951). There are other derivations, with applications, in Oehlers and Bradford (1995) and Ranzi et al. (2003).

In the present analysis, there is symmetry at mid-span, and the bending moment at the supports is zero, so the boundary conditions are:

$$\left.\begin{array}{ll} s = 0 & \text{when } x = 0 \\[2mm] \dfrac{ds}{dx} = -\varepsilon_{cs} & \text{when } x = \pm L/2 \end{array}\right\} \tag{2.22}$$

The solution of Equation 2.21 is then

$$s = \beta w x - \left(\frac{\beta w - \varepsilon_{cs}}{\alpha}\right) \operatorname{sech}\left(\frac{\alpha L}{2}\right) \sinh \alpha x \tag{2.23}$$

Expressions for the slip, slip strain, curvature and stresses throughout the beam can be obtained as functions of x. All results are proportional to the loading w. In full-interaction analysis, the slip is zero, and the results depend only on the bending moment and shear force at the cross-section considered. Here, they depend on the loading, boundary conditions and shear connection for the whole beam.

Few algebraic solutions are available, and are usually limited to shear connection of uniform stiffness along the span (Appendix A). In beams for bridges, shear connection is rarely uniform; and neither is the cracking of concrete, which is ignored in these analyses. In design methods simple enough for use in practice, slip is allowed for by empirical corrections to results based on full-interaction theory.

2.7 Effect of degree of shear connection on stresses and deflections

For uniformly spaced shear connectors of given strength, the number needed to provide 'full shear connection', n_f, is defined in terms of the ultimate resistance to bending of the beam considered. It is the number such that adding more would not increase this resistance. The *degree of shear connection* is defined by $\eta = n/n_f$ (≤ 1), where n is the number provided. *Partial shear connection* occurs in any shear span where $n < n_f$. 'Shear span' is defined later; for a simply supported beam it is the length between a support and a cross-section of maximum bending moment.

In the worked example in Section 3.11, η is 0.46. Partial-interaction analyses of this beam, of span 8.6 m, are given in Appendix A. Table 2.2 compares these results with those from analyses for $\eta = 0$ (no connection) and $\eta = 1$ (full connection).

This linear-elastic analysis is appropriate for checks on serviceability using unfactored loading. The loading for this composite member is 30 kN/m. Load previously applied to the steel beam is not included, so its stresses are higher than those shown. The shear connection for $\eta = 0.46$ is 19-mm studs with design shear strength $P_{Rd} = 42$ kN, in pairs at 0.3 m spacing. Their stiffness (shear force/slip) was taken as 150 kN/mm (Appendix A). Allowance was made for creep by using twice the short-term modular ratio (E_a/E_c) for all the loading.

The action effects in the concrete and steel components are shown separately in Figure 2.17a. The bending moments M_c and M_a cause the same curvature, which is

$$\kappa = \frac{M_c}{E_{c,eff}I_c} = \frac{M_a}{E_a I_a}$$

Table 2.2 Elastic analyses of a composite beam with various degrees of shear connection

Degree of shear connection, η	0	0.46	1	Ratio R, %
Ratio $Nd_c/(M_a + M_c + Nd_c)$	0	0.57	0.59	3.4
Compressive force in the slab, N, kN	0	568	593	4.2
Maximum slip, at a support, mm	5.2	0.24	0	4.6
Shear force on stud nearest a support, kN	0	36.5	41.2	11.1
Mean shear force on all studs, kN	0	20.3	20.6	1.4
Mid-span deflection, mm	45.7	17.2	16.0	4.0
Tensile stress at steel bottom surface, MPa	253	169	166	3.4

As there is no external longitudinal force, the axial forces N_c and N_a are equal and opposite, and are replaced here by N. An explanation of this sign convention is given in the Preface.

The external bending moment M_{Ed} $(= M_a + M_c + Nd_c)$ is resisted by flexure of the concrete and steel components and by the couple Nd_c. Table 2.2 shows that the ratio of Nd_c to M_{Ed} varies little with degree of shear connection. It is also almost constant along the span.

The compressive force in the slab at mid-span, N_c in Figure 2.17, is high enough at $\eta = 0.46$ to outweigh the tensile stress from bending. This is shown by the strain distribution in Figure 2.17b, where the slip strain at the steel–concrete interface is 66×10^{-6}.

The final column of Table 2.2 gives for each row, as a percentage, the ratio R of the change when η is reduced from 1 to 0.46 to the change when it is reduced from 1 to 0. It shows that the performance of 46% shear connection in this example is very close to that of the more expensive full connection. Partial shear connection is widely used, both for this reason and because, as in this example, there may not be enough space in troughs of the profiled sheeting for full connection to be provided.

The longitudinal slip at a support is 0.24 mm, from which the shear on the end stud is 36.5 kN. This gives the only ratio R that is above 5%. This maximum shear force is quite high, at 70% of P_{Rk}, and shows that at lower degrees of shear connection or higher load levels, non-linear load/slip behaviour would cause redistribution of shear force from end studs towards those nearer to mid-span.

In the elastic range, partial and uniformly spaced shear connection changes the distribution of longitudinal shear force per unit length (the shear flow) only slightly. Full shear connection is almost rigid shear connection, for which, with uniform loading, both the mean vertical shear and the mean shear flow are half the maximum values. Here, the mean of the shear forces per stud, at 20.3 kN, is close to half of 36.5 kN. This similarity between shear flow and vertical shear does not occur in fixed-ended or continuous beams (Appendix A).

The mid-span deflection with zero shear connection is almost treble the full-interaction value. Without shear connection, it would be excessive. The maximum tensile stress in the unpropped steel beam is 109 MPa at the end of the construction phase. Adding this to the stresses in the bottom row of Table 2.2 gives 362 MPa with $\eta = 0$, which exceeds the yield strength of 355 MPa. Just 46% of full shear connection reduces this stress to 278 MPa.

2.8 Longitudinal shear in composite slabs

There are three types of shear connection between a profiled steel sheet and a concrete slab. At first, reliance was placed on the natural bond between the two. This is unreliable unless separation at the interface ('uplift') is prevented, so sheets with re-entrant profiles were developed. This type of shear connection is known as 'frictional interlock'. Failures can be brittle and are difficult to predict, as there is no satisfactory conceptual model.

The second type is 'mechanical interlock', provided by pressing dimples or ribs (Figure 2.9) into the sheet. The effectiveness of these embossments depends entirely on their depth, which must be accurately controlled during manufacture. Design for longitudinal shear is based on test data, normally obtained by the manufacturer of the sheeting. When a new profile is developed, further testing is needed, in principle, for each thickness of sheeting, each overall depth of slab to be used, and for a range of concrete strengths. Codes allow some simplification, but the testing remains a costly process.

The third type of shear connection is 'end anchorage'. This can be provided where the end of a sheet rests on a steel beam, by means of shot-fired pins, or by welding studs through the sheeting to the steel flange. Its use is explained in Section 3.4.3.

2.8.1 The shear-bond test

The effectiveness of shear connection is studied by means of loading tests on simply-supported composite slabs, as sketched in Figure 2.18. Specifications for such tests are given in EN 1994-1-1. The length of each shear span, L_s, is always $L/4$, where L is the span, and is varied by changing the span. There are three possible modes of failure:

- in flexure, at a cross-section such as 1–1 in Figure 2.18;
- in longitudinal shear, along a length such as 2–2; and
- in vertical shear, at a cross-section such as 3–3.

The expected mode of failure in a test depends on the ratio of L_s to the effective depth d_p of the slab, shown in Figure 2.19. The modes can be distinguished on a diagram plotted with axes V/bd_p and A_p/bL_s, shown in Figure 2.20, where V is the maximum vertical

Figure 2.18 Critical sections for a composite slab

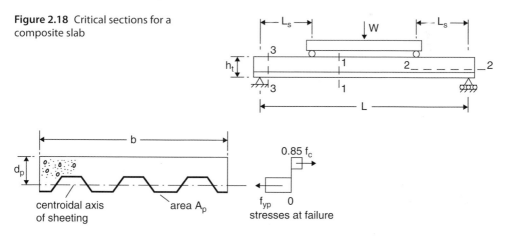

Figure 2.19 Bending resistance of a composite slab

Figure 2.20 Maximum shear force in a test, related to shear span L_s

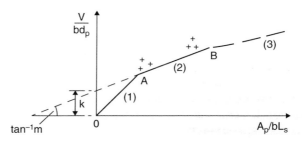

shear found in a test, assumed to be much greater than the self-weight of the slab. Area A_p is the effective cross-section of a width b of the sheeting, which should include several complete wavelengths. The lines in the figure are now explained.

At high L_s/d_p, flexural failure occurs. The maximum bending moment, M_u, is given by

$$M_u = VL_s \tag{2.24}$$

Flexural failure is modelled by simple plastic theory, with all the steel at its yield stress, f_{yp} (Figure 2.19), and sufficient concrete at $0.85f_c$, where f_c is the cylinder strength, for longitudinal equilibrium. The lever arm is a little less than d_p, but approximately,

$$M_u \propto A_p f_{yp} d_p \tag{2.25}$$

From Equation 2.24,

$$\frac{V}{bd_p} = \frac{M_u}{bd_p L_s} \propto \frac{A_p f_{yp}}{bL_s} \tag{2.26}$$

The strength f_{yp} is not varied during a series of tests, and has no influence on longitudinal shear failure. It is therefore omitted from the axes in Figure 2.20, and Equation 2.26 shows that flexural failure should plot as a straight line through the origin, (1) in Figure 2.20.

At low L_s/d_p, vertical shear failure occurs. The mean vertical shear stress on the concrete is roughly equal to V/bd_p. It is assumed in current codes that the ratio A_p/bL_s has little influence on its ultimate value, so vertical shear failures would be represented by a horizontal line. However, Patrick and Bridge (1993) has shown that this should be a rising curve, shown as (3) in Figure 2.20.

Longitudinal shear failures occur at intermediate values of L_s/d_p, and lie near the line

$$\frac{V}{bd_p} = m\left(\frac{A_p}{bL_s}\right) + k \tag{2.27}$$

shown as AB in Figure 2.20, where m and k are constants to be determined by testing. Design based on Equation 2.27 is one of the two methods given in EN 1994-1-1. It modifies an earlier method that had been in use for several decades, which took account of the strength of the concrete, as shown in Equation 2.28:

$$V = bd_p(f_c)^{1/2}\left[m\frac{A_p}{bL_s(f_c)^{1/2}} + k\right] \tag{2.28}$$

where f_c is the measured cylinder or cube strength of the concrete. This equation can give unsatisfactory results for m and k when f_c varies widely within a series of tests, so in Eurocode 4, f_c has been omitted from Equation 2.33. A comparison of the two methods (Johnson, 2012) has shown that this has little effect on m; but the two equations give different values for k, in different units. A value found by the earlier method cannot be used in design to Eurocode 4; but a new value can sometimes be determined from the original test data (Johnson, 2006).

The influence of bond is minimized, in the standard test, by the application of several thousand cycles of repeated loading up to 60% of the expected failure load, before loading to failure.

2.8.2 Design by the *m–k* method

With the safety factor added, Equation 2.27 appears in Eurocode 4 as

$$V_{\ell,\mathrm{Rd}} = bd_{\mathrm{p}}\left[\left(\frac{mA_{\mathrm{p}}}{bL_{\mathrm{s}}}\right) + k\right]\Big/\gamma_{\mathrm{Vs}} \tag{2.29}$$

where m and k are constants with dimensions of stress, and $V_{\ell,\mathrm{Rd}}$ is the design *vertical* shear resistance for a width of slab b. This must exceed the vertical shear at an end support at which longitudinal shear failure could occur in a shear span of length L_{s}, shown by line 2–2 in Figure 2.18.

For uniformly distributed load on a span L, the length L_{s} is taken as $L/4$. The principle that is used when calculating L_{s} for other loadings is now illustrated by an example. The composite slab shown in Figure 2.21a has a distributed load w per unit length and a centre point load wL, so the shear force diagram is as shown in Figure 2.21b. A new shear force diagram is constructed for a span with two point loads only, and the same two end reactions, such that the areas of the positive and negative parts of the diagram equal those of the original diagram. This is shown in Figure 2.21c, in which each shaded area is $3wL^2/8$. The positions of the point loads define the lengths of the shear spans. Here, each one is $3L/8$.

2.8.3 Defects of the *m–k* method

The method proved to be an adequate design tool for profiles with short spans and sometimes brittle behaviour, which have been widely used in North America. However, to exploit fully the ductile behaviour of profiles now available, with good mechanical interlock and longer spans, it is necessary to use a partial-interaction method. This is described in Section 3.3.2, with further details of the shear-bond test.

The defects of the *m–k* method and of profiles with brittle behaviour are given in papers that set out the new methods, by Bode et al. (1996) in Germany, and by Patrick and Bridge (1994) in Australia. They are as follows:

(1) The *m–k* method is not based on a mechanical model, so conservative assumptions have to be made in design when the dimensions, materials or loading differ from those used in the tests.

Figure 2.21 Calculation of L_{s} for a composite slab

(a) composite slab

(b) shear force

(c) shear force

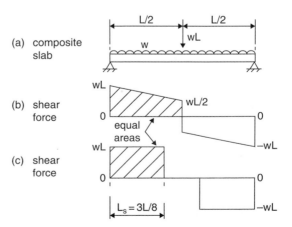

(2) Many additional tests are needed before the range of application can be extended; for example, to include end anchorage or the use of longitudinal reinforcing bars.

(3) The method of evaluation of test data is the same, whether the failure is brittle or ductile. Eurocode 4 adds a penalty factor of 0.8 for brittle behaviour, but this does not adequately represent the advantage of using sheeting with good mechanical interlock, because the advantage increases with span.

(4) The method does not allow correctly for the beneficial effect of friction above supports, which is greater in short shear spans.

It is expected that the partial-interaction method will be the only one given in the next Eurocode 4. The *m–k* method was included in the 2004 Eurocode 4, as an alternative, because it had been so widely used.

3

Simply-supported Composite Slabs and Beams

3.1 Introduction

The subjects of this and subsequent chapters are treated in the sequence in which they developed. Relevant structural behaviour is discovered by experience or research, and is then represented by mathematical models. These make use of standardized properties of materials, such as the yield strength of steel, and enable the behaviour of a member under load to be predicted. The models are developed into design rules, as found in codes of practice, by simplifying them wherever possible, defining their scope and introducing partial safety factors.

Research workers often propose alternative models, and language barriers have been such that the model preferred in one country may be little known elsewhere. The writers of codes try to select the most rational and widely applicable of the available models, but must also consider existing design practices and the need for simplicity. The design rules used in this book are taken from the Eurocodes, which in the UK have since 2010 replaced the corresponding British codes; but the underlying models are usually the same, and significant differences will be explained.

The methods to be described are illustrated by the design calculations for part of a framed structure for a building. To avoid repetition, the results obtained at each stage are used in subsequent work. Some aspects of this worked example were chosen to raise problems, to enable the use here of design processes that a better design concept might avoid. The example cannot cover all the options available for this structure (e.g. type of composite slab or column, method of fire protection), and makes no claim to represent best practice.

The notation used is that explained and listed in the section 'Symbols, terminology and units'.

3.2 Example: layout, materials and loadings

In a framed structure for a wing of a building, the columns are arranged at 4 m centres in two rows 9 m apart. A design is required for a typical floor, which consists of a composite floor slab supported by, and composite with, steel beams that span between the columns as shown in Figure 3.1.

Composite Structures of Steel and Concrete: Beams, Slabs, Columns and Frames for Buildings,
Fourth Edition. Roger P. Johnson.
© 2019 John Wiley & Sons Ltd. Published 2019 by John Wiley & Sons Ltd.

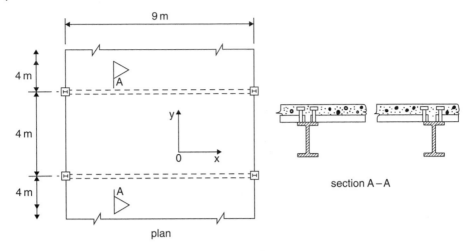

Figure 3.1 Design example – structure for a typical floor

3.2.1 Properties of concrete

Lightweight-aggregate concrete grade LC25/28 will be used for the floor slabs, and normal-density concrete grade C25/30 for the encasement of the columns and, if used for fire protection, the webs of the beams. The properties of these concretes are given in Table 1.4. They both have design compressive strength:

$$f_{cd} = f_{ck}/\gamma_C = 25/1.5 = 16.7 \text{ N/mm}^2$$

The short-term elastic moduli and modular ratios are:

LC25/28: $E_{cm} = 20.7 \text{ kN/mm}^2$; $n_0 = 210/20.7 = 10.1$

C25/30: $E_{cm} = 31.0 \text{ kN/mm}^2$; $n_0 = 210/31 = 6.8$

The long-term compressive strain of these concretes under permanent loads is about three times the initial elastic strain, due to creep. In elastic analysis, this would require separate calculations for permanent and variable loads. For buildings, EN 1994 permits the simplification that all strains may be assumed to be twice their short-term value. This is done by using modular ratios $n = 2n_0$. Hence, for grade LC25/28, $n = 20.2$; for grade C25/30, $n = 13.6$.

3.2.2 Properties of other materials

The partial factors γ_M to be used in ULS verifications are those recommended in the Eurocodes. A national annex may prescribe other values. The following grades of the materials are those widely used:

- structural steel: yield strength $f_y = f_{yd} = 355 \text{ N/mm}^2$ ($\gamma_A = 1.0$)
- profiled steel sheeting: yield strength $f_{yp} = f_{yp,d} = 350 \text{ N/mm}^2$ ($\gamma_A = 1.0$)
- reinforcement: yield strength $f_{sk} = 500 \text{ N/mm}^2$, $f_{sd} = 435 \text{ N/mm}^2$ ($\gamma_S = 1.15$)
- welded fabric: yield strength $f_{sk} = 500 \text{ N/mm}^2$, $f_{sd} = 435 \text{ N/mm}^2$ ($\gamma_S = 1.15$)

- shear connectors: 19-mm headed studs 100 mm high; ultimate strength $f_u = 500\,\text{N/mm}^2, f_{ud} = 400\,\text{N/mm}^2$ ($\gamma_V = 1.25$).

The elastic modulus for structural steel is $E_a = 210\,\text{kN/mm}^2$. In design of beams and slabs, the value for reinforcement is assumed, for simplicity, also to be $210\,\text{kN/mm}^2$; but the more accurate value, $E_s = 200\,\text{kN/mm}^2$, is used in column design.

3.2.3 Resistance of the shear connectors

The design shear resistance is given by Equation 2.15 and is

$$P_{Rd} = \frac{0.29 \times 19^2 (25 \times 20\,700)^{1/2}}{1.25 \times 1000} = 60.2\,\text{kN} \tag{3.1}$$

Equation 2.14 gives the higher value 91 kN, and so does not apply.

3.2.4 Permanent actions

From Table 1.4, the unit weights of the concretes, including reinforcement, are: for LC25/28, $19.5\,\text{kN/m}^3$; for C25/30, $25.0\,\text{kN/m}^3$. For design of formwork, each value is increased by $1\,\text{kN/m}^3$, from EN 1991-1-6 (BSI, 2002b), to allow for the higher moisture content of fresh concrete.

The unit weight of structural steel is taken as $77\,\text{kN/m}^3$. The characteristic weight of floor and ceiling finishes is taken as $1.3\,\text{kN/m}^2$.

3.2.5 Variable actions

The floors to be designed are assumed to be in category C3 of EN 1991-1-1: 'Areas where people may congregate, without obstacles for moving people' (e.g. exhibition rooms, etc.) (BSI, 2002a). The characteristic loadings are:

$$q_k = 5.0\,\text{kN/m}^2 \text{ on the whole floor area or any part of it} \tag{3.2}$$

or

$$Q_k = 4.0\,\text{kN on any area 50 mm square} \tag{3.3}$$

The load q_k is high for a building not intended for storage or industrial use. Its use here enables many aspects of design to be illustrated. For comparison, typical imposed loads for office areas are given by Equations 1.12 and 1.13. An allowance of $1.2\,\text{kN/m}^2$ for non-structural partition walls is treated as an additional imposed load, because their position is unknown.

3.3 Composite floor slabs

Composite slabs have for many decades been the most widely used method of suspended floor construction for steel-framed buildings in North America. Within the last 50 years there have been many advances in design procedures, and a wide range of profiled sheetings has become available in Europe. The first British Standard for the design of composite floors appeared in 1982. There are Eurocodes for design of both the sheeting alone (BSI, 2006a) and the composite slab (BSI, 2004). The steel sheeting has to support not only the wet concrete for the floor slab, but other loads that are imposed during concreting. These may include the heaping of concrete and pipeline or pumping loads. For construction loading, EN 1991-1-6 recommends a distributed loading between 0.75 and 1.5 kN/m². The loading used here is: $q_k = 1.0$ kN/m².

This section is based on the current Eurocode 4. The new EN 1994-1-1 is expected to include these changes:

- Revision of the equations for resistance to bending to include longitudinal reinforcement in the bottom of troughs of sheeting, which improves resistance to fire. The method is otherwise the same.
- Removal of the *m–k* method for resistance to longitudinal shear, leaving only the partial-interaction method.

Profiled steel sheeting The sheeting is very thin, for economic reasons; usually between 0.8 mm and 1.2 mm. It has to be galvanized to resist corrosion, and this adds about 0.04 mm to the overall thickness. It is specified in EN 1993-1-3 that where design is based on the nominal thickness of the steel, the sheet must have at least 95% of that thickness – but it is not a simple matter for the user to check this. The sheets are pressed or cold rolled, and are typically about 1 m wide and up to 6 m long. They are designed to span in the longitudinal direction only. For many years, sheets were typically 50 mm deep, and the limiting span was about 3 m.

The cost of propping the sheets during concreting, to reduce deflections, led to the development of deeper profiles. The current Eurocode 4 covers sheets with overall depth up to 100 mm. This type is used in the main example in Section 3.4. There is also 'deep decking', capable of unpropped spans up to about 6 m. A typical cross-section is shown in Figure 3.12, and is used in the example in Section 3.4.6. Asymmetric beams to support this decking are introduced in Section 3.12, with an example in Section 3.13. As the examples will show, the design of composite slabs is still often governed by a limit on deflection.

There are two main types of profile for sheetings covered by Eurocode 4. Some have re-entrant troughs ('Holorib' is an example), but most have trapezoidal troughs. The latter were at first typically as shown in Figure 3.2, and Eurocode 4 used the words 'concrete above the ribs' (depth h_c in the figure). The top steel flanges then became wider, and stiffening ribs were added, as shown in Figure 2.17. There are now two depths of

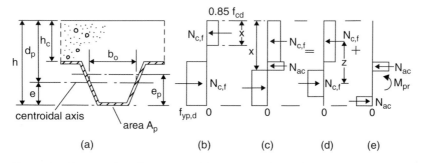

Figure 3.2 Cross-section of composite slab, and stress blocks for sagging bending

concrete slab, named 'gross', as in Figure 3.2, and 'net', labelled h_{cf} in Figure 2.17. Many design rules refer to 'depth of concrete', but in Eurocode 4, none of them state which depth. The next Eurocode 4 will do so, and the correct depth is given with the rules used in this book.

The local buckling stress of a flat panel within sheeting should ideally exceed its yield strength, but this requires breadth/thickness ratios of less than about 35. Modern profiles have local stiffening ribs, but it is difficult to achieve slendernesses less than about 50, so that for flexure, the sections are in Class 4 (i.e. the buckling stress is below the yield stress). Calculation of the resistance to bending then becomes complex, and involves iteration.

The specified or nominal yield strength is that of the flat sheet from which the sheeting is made. In the finished product, the yield strength is higher at every bend and corner, because of work hardening.

To enable it to fulfil its second role, as reinforcement for the concrete slab, dimples are pressed into the surface of the sheeting, to act as shear connectors. These dimpled areas may not be fully effective in resisting longitudinal stress, so both they and the local buckling reduce the effective cross-sectional area (A_{pe}) and second moment of area (I_{pe}) of the sheeting to below the value calculated for the gross steel section.

For these reasons, manufacturers commission tests on prototype sheets, and provide designers either with test-based values of resistance and stiffness, or with load tables calculated from those values.

Design of composite slab The cross-sectional area of steel sheeting that is needed for the construction phase often provides more than enough bottom reinforcement for the composite slab. It is then usual to design the slabs as simply-supported. The concrete is of course continuous over the supporting beams, and the sheets may be as well (e.g. if 6-m sheets are used for a succession of 3-m spans).

These 'simply-supported' slabs require top longitudinal reinforcement at their supports, to control the widths of cracks. The amount is specified in EN 1994-1-1 as 0.2% of the 'cross-sectional area of concrete above the ribs' for unpropped construction, and 0.4% for propped construction. This wording leaves open the question whether a small projection above the main rib (Figure 2.17) is a 'rib' or a 'stiffener'. It is usually taken as a rib, so the net depth of concrete slab should be used.

There are two situations to be checked in design: during construction, when the sheeting is supporting the wet concrete, using Eurocode 3; and after completion, when the composite slab supports the permanent and variable loads, using Eurocode 4.

The stresses and deflections of the sheeting during construction depend on the extent to which it is propped ('shored' in North America), because the spacing of props may not be close enough for the bending moments in the propped member to be negligible. In the following example, props are assumed to be at mid-span only. For both limit states, elastic analysis without redistribution is used, and the deflection of the sheeting at this stage is added to the deflection of the composite slab from the superimposed dead loading and the imposed loading.

When the design ultimate load acts on a simply-supported span, all or nearly all of the sheeting is in longitudinal tension. The bending resistance of the slab is found by plastic theory. Stresses from the construction stage are ignored, and the whole of the loading is assumed to act on the composite slab.

Long-span slabs are sometimes designed as continuous over their supports. Their analysis and resistance to hogging bending are described in Section 4.7. The several resistances needed for the design of composite slabs as simply-supported are now explained. The methods are illustrated by the worked example in Section 3.4.

3.3.1 Resistance of composite slabs to sagging bending

The width of slab considered in calculations, b, is usually taken as one metre, but for clarity only a width of one wavelength is shown in Figure 3.2. The overall thickness h is required by EN 1994-1-1 to be not less than 80 mm; and the thickness of concrete above the 'main flat surface' of the top of the ribs of the sheeting, to be not less than 40 mm. Normally, this thickness is 60 mm or more, to provide sufficient sound or fire insulation, and resistance to concentrated loads.

Except where the sheeting is unusually deep, the neutral axis for bending lies in the concrete where there is full shear connection, but in regions with partial shear connection, there is usually a second neutral axis within the steel section. Local buckling of compressed sheeting then has to be considered. This is done by using effective widths for flat regions of sheeting. These widths are allowed (in EN 1994-1-1) to be up to twice the limits given for Class 1 steel web plates in beams, because the concrete prevents the sheeting from buckling upwards, which shortens the wavelength of the buckles.

For sheeting in tension, the width of embossments should be neglected in calculating the effective area, unless tests have shown that a larger area is effective.

For these reasons, the effective area of width b of sheeting, A_{pe}, and the height of the centre of area above the bottom of the sheet, e, are usually based on tests. These may show that e_p, the height of the plastic neutral axis of the sheeting, is different from e.

For the calculation of the bending resistance of width b of composite slab by simple plastic theory, there are three situations as follows.

(1) *Neutral axis above the sheeting*

The assumed distribution of longitudinal bending stresses is shown in Figure 3.2b. There must be full shear connection, so that the design compressive force in the concrete is denoted $N_{c,f}$ (f for 'full'). It is equal to the yield force for the steel:

$$N_{c,f} = A_{pe} f_{yp,d} \tag{3.4}$$

where $f_{yp,d}$ is the design yield strength of the sheeting. The depth of the stress block in the concrete is given by

$$x = x_{pl} = \frac{N_{c,f}}{0.85 f_{cd} b} \tag{3.5}$$

For simplicity, and consistency with the method for composite beams, the depth to the neutral axis is assumed also to be x_{pl}, even though this is not in accordance with EN 1992. This method is therefore valid when $x_{pl} \le h_c$.

Here, h_c can be taken as the gross depth, assuming that the area enclosed by a small top rib is negligible compared with bh_c. This gives

$$M_{Rd} = N_{c,f}(d_p - 0.5x_{pl}) \tag{3.6}$$

where M_{Rd} is the design resistance to sagging bending.

(2) *Neutral axis within the sheeting, and full shear connection*

The stress distribution is shown in Figure 3.2c. The force $N_{c,f}$ is now less than that given by Equation 3.4, and is

$$N_{c,f} = 0.85 f_{cd} b h_c \tag{3.7}$$

because compression within ribs is neglected, for simplicity. There is a compressive force N_{ac} in the sheeting. There is no simple method of calculating x or the force N_{ac}, because of the complex properties of profiled sheeting, so the following approximate method is used. The tensile force in the sheeting is represented, as shown in Figure 3.2d and e, by a tensile force at the bottom equal to N_{ac} and the remaining tension N_p, where, for equilibrium,

$$N_p = N_{c,f} \tag{3.8}$$

The equal and opposite forces N_{ac} provide a resistance moment M_{pr}. It is less than the resistance moment of the sheeting alone, M_{pa} (provided by the manufacturer) because of the axial force $N_{c,f}$. It should be noted that in EN 1994-1-1, the symbol $N_{c,f}$ is the lesser of the two values from Equations 3.4 and 3.7, and so depends on the location of the neutral axis. For clarity here, a further symbol N_{pa} is introduced. It always has the value from Equation 3.4:

$$N_{pa} = A_{pc} f_{yp,d} \tag{3.9}$$

When Equation 3.7 governs, the relationship between M_{pr}/M_{pa} and $N_{c,f}/N_{pa}$ depends on the profile, but is typically as shown by the dashed curve ABC in Figure 3.3a. This is approximated in Eurocode 4 by the equation

$$M_{pr} = 1.25 M_{pa} \left(1 - \frac{N_{c,f}}{N_{pa}} \right) \le M_{pa} \tag{3.10}$$

which is shown as ADC. The resistance moment is then given by

$$M_{Rd} = N_{c,f} z + M_{pr} \tag{3.11}$$

as shown in Figure 3.2d and e. The lever arm z is found by the approximation shown by line EF in Figure 3.3b. This is clearly correct when $N_{c,f} = N_{pa}$, because N_{ac} is then zero, so M_{pr} is zero. Equation 3.6 with $x_{pl} = h_c$ then gives M_{Rd}. The lever arm is

$$z = d_p - 0.5 h_c = h - e - 0.5 h_c \tag{3.12}$$

as given by point F.

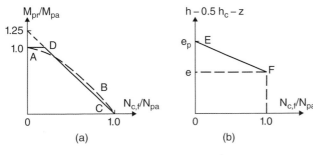

Figure 3.3 Equations 3.10 and 3.14

To check point E, we assume that $N_{c,f}$ is nearly zero (e.g. if the concrete is very weak), so that $M_{pr} \approx M_{pa}$. The neutral axis for M_{pa} alone is at height e_p above the bottom of the sheet, and the lever arm for the force $N_{c,f}$ is

$$z = h - e_p - 0.5h_c \tag{3.13}$$

as given by point E. This method has been validated by tests.

The line EF is given in Eurocode 4 by

$$z = h - 0.5h_c - e_p + (e_p - e)N_{c,f}/N_{pa} \tag{3.14}$$

(3) *Partial shear connection*

The compressive force in the slab, N_c, is now less than $N_{c,f}$ and is determined by the strength of the shear connection. The depth x of the stress block is given by

$$x = \frac{N_c}{0.85f_{cd}b} \leq h_c \tag{3.15}$$

There are two neutral axes: one at depth x and another within the steel sheeting. The stress blocks are as shown in Figure 3.2b, for the slab (with force N_c, not $N_{c,f}$), and Figure 3.2c for the sheeting. The calculation of M_{Rd} is as for method (2), except that $N_{c,f}$ is replaced by N_c, and h_c by x, so that:

$$z = h - 0.5x - e_p + (e_p - e)N_c/N_{pa} \tag{3.16}$$

$$M_{pr} = 1.25M_{pa}\left(1 - \frac{N_c}{N_{pa}}\right) \leq M_{pa} \tag{3.17}$$

$$M_{Rd} = N_c z + M_{pr} \tag{3.18}$$

3.3.2 Resistance of composite slabs to longitudinal shear by the partial-interaction method

In Section 2.8, the three types of shear connection between profiled sheeting and a concrete slab and the shear-bond test are described. The empirical *m–k* method of design is explained. Equation 2.29 predicts the resistance of a slab to *vertical shear*, even though the relevant failure mode is longitudinal shear. This method is expected to be removed from Eurocode 4 for the reasons given in Section 2.8.

The basis of the partial-interaction method is different. It predicts resistance to *bending* at each cross-section, assuming longitudinal shear failure between that section and the adjacent support. The design is verified if the bending moment from the design actions nowhere exceeds this value. The resistance diagram can easily be modified to take advantage of any end anchorage or slab reinforcement, and the loading diagram can be of any shape, so the method is more versatile than the *m–k* method. A worked example using data from shear-bond tests and end anchorage is given in Section 3.4.3.

3.3.2.1 Testing for the partial-interaction method

Further details are given of the shear-bond test, introduced in Section 2.8.1. A separate set of tests is required for each profile of steel sheeting, of thickness *t*, with a slab of a particular overall thickness *h* (in Figure 3.2) and the same mean strengths of the materials, f_{cm} for the concrete and $f_{yp,m}$ for the sheeting. Limits are given in Eurocode 4 for the ranges of values of t, f_{ck} and f_{yp} within which a set of test results can be used in practice. The thickness of the slab can presumably differ from h, and no limits are given.

It should not be assumed, without relevant evidence, that lightweight-aggregate concrete can replace normal-density concrete of the same specified strength. Eurocode 4 does not refer to this.

At least four tests are required for a set. The first is done with a short shear span, to determine whether the behaviour is ductile or brittle, based on the load-deflection curve. It is 'ductile', in Eurocode 4, if the failure load exceeds the load causing an end slip of 0.1 mm by more than 10%. This is a requirement for the use of the partial-interaction method. For the other three tests, Eurocode 4 says 'the shear span (L_s in Figure 2.18) should be as long as possible while still providing failure in longitudinal shear'. The shear span is always $L/4$ and is altered by changing the span, L. The choice of the span may need previous experience, or even preliminary tests.

3.3.2.2 Determination of the mean ultimate shear strength, τ_u

It is assumed that in each test, before maximum load is reached, there is complete redistribution of longitudinal shear stress at the interface from a free end to the nearer load point, a length $L_s + L_0$, where L_0 is the length of the short end overhang, Figure 2.18. Using the measured strengths of the materials, the compressive force in the slab for full shear connection, $N_{c,f}$, is calculated as $A_{pe} f_{yp,m}$.

The full-interaction bending resistance $M_{pl,m}$ is calculated by the relevant method in Section 3.3.1. The resistance of the sheeting alone is $M_{pl,a}$, usually provided by the manufacturer. These resistances correspond to $\eta = 1$ or $\eta = 0$, respectively, where η is the degree of shear connection, as shown in Figure 3.4. By assuming a range of values η (changing the compressive force in the slab), the partial-interaction curve AB is calculated for the test specimen.

For each test, the maximum bending moment at a load point is determined, and entered on Figure 3.4. The path C→D→E gives the degree of shear connection.

The ultimate shear stress is assumed to act on a plane of width b and length $L_s + L_0$, and so is a nominal value, because shear failure occurs at the surface of the sheeting, which has a width greater than the overall width, b. Its value is calculated from

$$\tau b(L_s + L_0) = \eta N_{c,f} \tag{3.19}$$

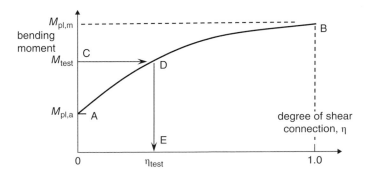

Figure 3.4 Determination of degree of shear connection from a shear-bond test

This is done for a range of shear spans. The characteristic shear strength τ_{Rk} is found from the results as the lower 5% fractile value, by a statistical method.

The preceding process takes account of the longitudinal shear transferred by friction above a support, which can be significant for short shear spans. If it is included directly in design calculations, τ_{Rk} has to be reduced. Eurocode 4 does this by replacing Equation 3.19 by

$$\tau b (L_s + L_0) + \mu V_t = \eta N_{c,f}$$

where V_t is the support reaction at the ultimate test load, and μ is a coefficient of friction, to be taken as 0.5.

3.3.2.3 Partial-interaction design for longitudinal shear

For a given slab, materials, loading and value for τ_{Rd} ($= \tau_{Rk}/\gamma_V$), the length of shear span, L_{sf} say, needed for the bending resistance to reach the full-interaction value, $M_{pl,Rd}$, can be calculated. In a diagram similar to Figure 3.4, and shown as Figure 3.5, an axis for L_x, the distance from an end support, is added, with L_{sf} matched with $\eta = 1$. The bending-moment curve M_{Ed} for the relevant shear span is also shown.

The resistance is $M_{pl,a,Rd}$ at $L_x = 0$. The resistance curve, shown dashed, is found by the partial-interaction method, assuming various values for η. If this curve, M_{Rd}, lies above curve M_{Ed}, the longitudinal shear resistance is sufficient. There may be a mid-span region where full shear connection is achieved and M_{Rd} is independent of x.

If the loading is increased until the curves touch, the position of the point of contact gives the location of the cross-section of flexural failure and, if the interaction is partial, the length of the shear span.

The only type of end anchorage for which design rules are given in Eurocode 4 is the headed stud, welded through the sheeting to the top flange of a steel beam. The resistance of the anchorage is based on local failure of the sheeting, as explained elsewhere (Johnson, 2012).

3.3.3 Resistance of composite slabs to vertical shear

Tests show that resistance to vertical shear is provided mainly by the concrete ribs. For open profiles, their effective width b_0 should be taken as the mean width, although the

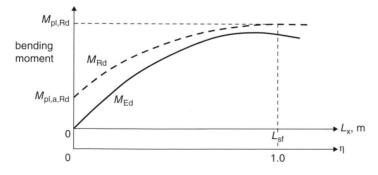

Figure 3.5 Bending moment M_{Ed} and bending resistance, governed by longitudinal shear

width at the centroidal axis (Figure 3.2a) is accurate enough. For re-entrant profiles, the minimum width should be used.

This shear resistance is given by the method of EN 1992-1-1 for concrete beams. Reinforcement contributes to the resistance only where it is fully anchored beyond the cross-section considered. The sheeting is unlikely to satisfy this condition. The resistance per unit width of a composite slab with ribs of effective width b_0 at spacing b is then

$$V_{Rd} = (b_0/b)d_p v_{min} \tag{3.20}$$

where d_p is the depth to the centroidal axis (Figure 3.2a) and v_{min} is the shear strength of the concrete.

The recommended value for v_{min} is

$$v_{min} = 0.035[1 + (200/d_p)^{1/2}]^{3/2} f_{ck}^{1/2} \tag{3.21}$$

with d_p in mm, taken as not less than 200, and v_{min} and f_{ck} in N/mm². The expression $[1 + (200/d_p)^{1/2}]^{3/2}$ allows approximately for the reduction in shear strength of concrete that occurs as the effective depth increases.

This method is conservative, because the side walls of the steel troughs resist vertical shear during unpropped construction, and so make a permanent contribution. A method that allows for this has been proposed, and may be included in the next Eurocode 4. Before shear resistance from sheeting and concrete together can be relied upon, it is necessary to show that their shear strains are compatible.

At present V_{Ed} for the concrete slab is taken as the whole of the vertical shear, including that initially resisted by the sheeting. Resistance to vertical shear is most likely to be critical in design where span/depth ratios are low, as is the case for beams.

3.3.4 Punching shear

Where a thin composite slab has to be designed to resist point loads (e.g. from a steel wheel of a loaded trolley), resistance to punching shear should be checked. Failure is assumed to occur on a 'critical perimeter' of length c_p, which is defined as for reinforced concrete slabs. For a loaded area a_p by b_p, remote from a free edge, and 45° spread through a screed of thickness h_f, it is as shown in Figure 3.6a:

$$c_p = 2\pi h_c + 2(b_p + 2h_f) + 2(a_p + 2h_f + 2d_p - 2h_c) \tag{3.22}$$

A reinforcing mesh is likely to be present above the sheeting. Let its areas of steel be $A_{s,x}$ and $A_{s,y}$, per unit width of slab. The effective depth of the slab may be taken as h_c (Figure 3.6b), giving the reinforcement ratios as $\rho_x = A_{s,x}/h_c$ and $\rho_y = A_{s,y}/h_c$. The effective ratio is given in EN 1992-1-1 as

$$\rho = (\rho_x \rho_y)^{1/2} \leq 0.02,$$

and the design shear stress as

$$v_{Rd} = \left(\frac{0.18}{\gamma_C}\right)[1 + (200/d)^{1/2}](100\ \rho f_{ck})^{1/3} \geq v_{min} \tag{3.23}$$

where v_{min} is given by Equation 3.21; d is the mean of the effective depths of the two layers of reinforcement, but not less than 200 mm; γ_C has the recommended value 1.50; and the units are as for Equation 3.21.

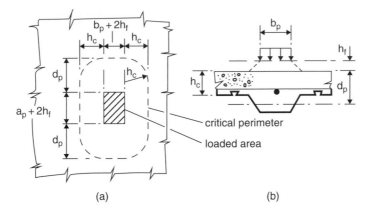

(a)   (b)

Figure 3.6 Critical perimeter for punching shear

The punching shear resistance is

$$V_{Rd} = v_{Rd} c_p d \tag{3.24}$$

It is not clear from EN 1994-1-1 whether account can be taken of contributions from the concrete ribs and the sheeting. None has been assumed here, so Equation 3.24 is likely to give a conservative result.

3.3.5 Bending moments from concentrated point and line loads

Since composite slabs span in one direction only, their ability to carry masonry partition walls or other heavy local loads is limited. Rules are given in EN 1994-1-1 for widths of composite slabs effective for bending and vertical shear resistance, for point and line loads, as functions of the shape and size of the loaded area. These are based on a mixture of simplified analyses, test data and experience.

Where transverse reinforcement is provided with a cross-sectional area of at least 0.2% of the area of concrete above the ribs of the sheeting, no calculations are needed for characteristic concentrated loads not exceeding 7.5 kN, or distributed loading up to 5.0 kN/m². In the next Eurocode 4, it is expected that these rules will take account of the net thickness of the slab above the sheeting, h_c, and the depth of the transverse reinforcement below the top of the slab, d_{sc}. The 7.5 kN exception will be increased to 0.003$d_{sc} h_c$ kN, if greater. For example, if d_{sc} is 30 mm and h_c is 100 mm, 0.2% of reinforcement is sufficient for characteristic point loads up to 9 kN.

There may also be a new rule for line loading, for example, from a partition. It is that transverse reinforcement, as above, is sufficient for line loads not exceeding the greater of 2.5 and 0.01$d_{sc} h_c/L$ kN/m. For example, with span 4 m and d_{sc} and h_c as above, this reinforcement is sufficient for characteristic line loads up to 7.5 kN/m. Presumably, the line of the loading is assumed to be parallel to the span of the slab.

The rules for concentrated loads are now explained, with reference to a rectangular loaded area a_p by b_p, with its centre distance L_p from the nearer support of a slab of span L, as shown in Figure 3.7a. The load may be assumed to be distributed over a width b_m, defined by lines at 45° (Figure 3.7b), where

$$b_m = b_p + 2(h_f + h_c) \tag{3.25}$$

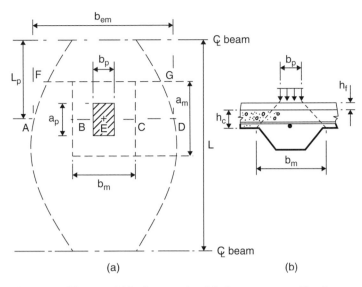

Figure 3.7 Effective width of composite slab, for concentrated load

and h_f is the thickness of finishes, if any. The code does not refer to distribution in the spanwise direction, but it would be reasonable to use the same rule, and take the loaded length as

$$a_m = a_p + 2(h_f + h_c) \tag{3.26}$$

The width of slab assumed to be effective for global analysis and for resistance is given by

$$b_{em} = b_m + kL_p[1 - (L_p/L)] \le \text{width of slab} \tag{3.27}$$

where k is taken as 2 for bending and longitudinal shear (except for interior spans of continuous slabs, where $k = 1.33$) and as 1 for vertical shear. These rules become unreliable where the depth of the ribs is a high proportion of the total thickness, h. Their use is limited in EN 1994-1-1 to slabs with $h_p/h \le 0.6$.

For a simply-supported slab of span L and a point load Q_{Ed}, the sagging moment per unit width of slab on line AD in Figure 3.7a is thus

$$m_{Ed} = Q_{Ed}L_p[1 - (L_p/L)]/b_{em} \tag{3.28}$$

which is a maximum when $L_p = L/2$.

The variation of b_{em} with L_p is shown in Figure 3.7a. If the load is assumed to be uniformly-distributed along line BC, whereas the resistance is distributed along line AD, there is sagging transverse bending under the load. The maximum sagging bending moment is at E, and is given by

$$M_{Ed} = Q_{Ed}(b_{em} - b_m)/8 \tag{3.29}$$

The sheeting has no tensile strength in this direction, because the corrugations can open out, so bottom reinforcement (Figure 3.7b) must be provided. The transverse bending moment will in fact depend on the ratio between the flexural stiffnesses of the slab in the longitudinal and transverse directions. New rules that allow for this have been proposed for the next Eurocode 4.

Vertical shear should be checked along a line such as FG, when L_p is such that FG is above the edge of the flange of the steel beam. It rarely governs design.

3.3.6 Serviceability limit states for composite slabs

3.3.6.1 Cracking of concrete

The lower surface of the slab is protected by the sheeting. Cracking will occur in the top surface where the slab is continuous over a supporting beam, and will be wider if each span of the slab is designed as simply-supported, rather than continuous, and if the spans are propped during construction.

For these reasons, longitudinal reinforcement should be provided above internal supports. The minimum amounts are given by EN 1994-1-1 as 0.2% of the area of concrete above the sheeting for unpropped construction, and 0.4% if propping is used. These amounts may not ensure that crack widths do not exceed 0.3 mm. If the environment is corrosive (e.g. de-icing salt on the floor of a parking area), the slabs should be designed as continuous, with cracking controlled in accordance with EN 1992-1-1.

3.3.6.2 Deflection

Where composite slabs are designed as simply-supported and are not hidden by false ceilings, deflection may govern design. The maximum acceptable deflection is more a matter for the client than the designer, but if predicted deflections are large, the designer may have to allow for the extra weight of concrete in floors that are cast with a horizontal top surface, known as 'ponding', and provide clearance above non-structural partitions.

Limiting deflection/span ratios are given, for guidance, in national annexes. Both the ratios and the load combination for which the deflection is calculated depend on whether the deflection is reversible or not, and whether brittle finishes or partitions are at risk. The deflections of the beams that support the composite slab are also relevant.

As with reinforced concrete slabs, the effects of shrinkage on deflections are rarely considered in current practice. Research has found that in composite slabs, shrinkage is greater at the top surface than at the surface that is 'sealed' by the profiled sheeting. This causes sagging curvature, and the resulting deflection may not be negligible. The next Eurocode 4 may give design guidance for this.

It is known from experience that deflections are not excessive when span-to-depth ratios are kept within certain limits. In EN 1994-1-1, calculations of deflections of composite slabs may be omitted if both:

- the degree of shear connection is such that, based on test results, end slip will not occur under service loading; and
- the ratio of span to effective depth is below a limit given in EN 1992-1-1. The recommended limits for a simply-supported slab range from 14 to 20, and can be modified by a national annex.

The provision of anti-crack reinforcement, as specified above, should reduce deflection by a useful amount. For internal spans of continuous slabs, the Eurocode recommends that the second moment of area of the slab should be taken as the mean of values calculated for the cracked and uncracked sections. Some of these points are illustrated in the worked example in Section 3.4.

3.4   Example: composite slab

The strengths of materials and characteristic actions for this structure are given in Section 3.2, and a typical floor is shown in Figure 3.1. The following calculations illustrate the methods described in Section 3.3. In practice, the calculations may be done by the provider of the sheeting, and presented as 'safe load tables', but here it is assumed that only the manufacturer's test data are available.

For unpropped construction, the sheeting for a span of 4 m would need to be over 100 mm deep. To reduce the floor thickness, it is assumed that the sheets will be propped at mid-span during construction. This requires the supporting beams also to be propped. It will be found in Section 3.11 that propping is needed also to reduce their own deflection.

The profile chosen for trial calculations is shown in Figure 3.8. Its overall depth is 70 mm, but the cross-section is such that the span/depth ratio based on this depth (28.6) is misleading. A more realistic value is 2000/55, which is 36.4. This may be adequate, as there is continuity over the prop between the two 2-m spans.

The next step is to choose a thickness for the composite slab, which will be designed as simply-supported over each 4-m span. The centroid of the sheeting is 30 mm above its lower surface, so the effective depth (d_p) is 120 mm and the span/depth ratio, for a slab 150 mm thick, is 4000/120, or 33.3. This is rather high for simply-supported spans, but the top reinforcement above the supporting beams will provide some continuity. From preliminary calculations, it appears that sheeting of nominal thickness 0.9 mm may be sufficient.

It is instructive to discover why a deeper profile should have been chosen, so these initial choices are not changed.

It is assumed that the supplier of the sheeting has provided the geometric and test data given below, some of which are shown in Figure 3.8.

- Guaranteed minimum yield strength, $f_{yp} = 350 \, \text{N/mm}^2$
- Design thickness, allowing for zinc coating, $t_p = 0.86 \, \text{mm}$
- Effective area of cross-section, $A_p = 1178 \, \text{mm}^2/\text{m}$
- Second moment of area, $I_p = 0.548 \times 10^6 \, \text{mm}^4/\text{m}$
- Characteristic plastic moment of resistance, $M_{pa} = 6.18 \, \text{kN m/m}$

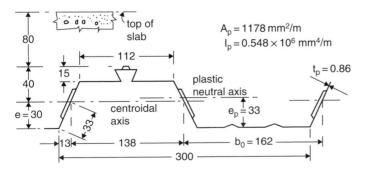

Figure 3.8 Typical cross-section of profiled sheeting and composite slab

- Distance of centroid above base, $e = 30$ mm
- Distance of plastic neutral axis above base, $e_p = 33$ mm
- Characteristic resistance to vertical shear, $V_{pa} = 60$ kN/m (approx.)
- For resistance to longitudinal shear, $m = 184$ N/mm^2
  $k = 0.0530$ N/mm^2
- For partial-interaction design, $\tau_{u,Rd} = 0.144$ N/mm^2
- Volume of concrete 0.125 m^3 per sq. m. of floor
- Weight of sheeting 0.10 kN/m^2

- Weight of composite slab at 19.5 kN/m^3, $g_k = 0.10 + 0.125 \times 19.5 = 2.54$ kN/m^2

These data are illustrative only, and should not be relied upon in engineering practice.

3.4.1 Profiled steel sheeting as formwork

In EN 1991-1-1, the density of 'unhardened concrete' is increased by 1 kN/m^3 to allow for its higher moisture content. This increases the permanent load by 0.125 kN/m^2. The imposed load during construction is 1.0 kN/m^2 (Section 3.3), so the design loads for the sheeting are:

$$\text{Permanent:} \qquad g_d = (2.54 + 0.125) \times 1.35 = 3.60 \text{ kN/m}^2 \qquad (3.30)$$

$$\text{Variable:} \qquad q_d = 1.0 \times 1.5 = 1.5 \text{ kN/m}^2 \qquad (3.31)$$

The top flanges of the supporting steel beams are assumed to be at least 150 mm wide. The bearing length for the sheeting should be at least 50 mm. Assuming that the sheeting is supported 25 mm from the flange tip (Figure 3.9) gives the effective length of each of the two spans as

$$L_e = (4000 - 150 + 50)/2 = 1950 \text{ mm} \qquad (3.32)$$

3.4.1.1 Flexure and vertical shear
The most adverse loading for sagging bending is shown in Figure 3.9, in which the weight of the sheeting alone in span BC is neglected. Elastic analysis gives the maximum design bending moments as:

Sagging, $\qquad M_{Ed} = 0.0959 \times (3.6 + 1.5) \times 1.95^2 = 1.86$ kN m/m

Hogging (both spans loaded), $\qquad \mathbf{M_{Ed} = 0.125 \times 5.1 \times 1.95^2 = 2.42 \text{ kNm/m}}$

From Table 1.2, the partial factor for the sheeting is 1.0, so that the design resistance is $M_{Rd} = M_{pa} = \mathbf{6.18 \text{ kN m/m,}}$ which is ample.

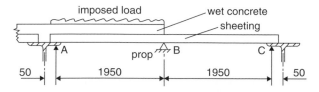

Figure 3.9 Profiled sheeting during construction

Vertical shear rarely governs design of profiled sheeting. Here, the maximum value, to the left of point B in Figure 3.9, is

$$V_{Ed} = 0.625 \times 5.1 \times 1.95 = 6.2 \text{ kN/m}$$

far below the design resistance of about 60 kN/m.

3.4.1.2 Deflection

The characteristic permanent load for the sheeting is $2.66 + 1.0 = 3.66 \text{ kN/m}^2$. It is assumed that the prop does not deflect. The maximum deflection in span AB, if BC is unloaded and the sheeting is held down at C, is

$$\delta_{max} = \frac{wL_e^4}{185E_aI_p} = \frac{3.66 \times 1.95^4}{185 \times 0.21 \times 0.548} = 2.5 \text{ mm} \tag{3.33}$$

This is span/784, which is satisfactory.

3.4.2 Composite slab – flexure and vertical shear

This continuous slab is designed as a series of simply-supported spans. For bending, the reactions from the beams are assumed to be located as in Figure 3.9, so the equivalent span is $L_e = 3.90$ m. For vertical shear, the span is taken as 4.0 m, so that the whole of the slab is included in the design loading for the beams. Plastic section analysis is used, so no account need be taken of the stresses in the sheeting when it acted as formwork.

From Section 3.2, the characteristic loadings are:

- $g_k = 2.54$ (slab) $+ 1.3$ (finishes) $= 3.84 \text{ kN/m}^2$
- $q_k = 5.0$ (imposed) $+ 1.2$ (partitions) $= 6.2 \text{ kN/m}^2$

The design ultimate loadings are:

- permanent: $g_d = 3.84 \times 1.35 = 5.18 \text{ kN/m}^2$
- variable: $q_d = 6.2 \times 1.5 = 9.30 \text{ kN/m}^2$

The mid-span bending moment is:

$$M_{Ed} = 14.48 \times 3.9^2/8 = \textbf{27.6 kN m/m}$$

For the bending resistance, from Equation 3.4,

$$N_{c,f} = 1178 \times 0.35/1.0 = 412 \text{ kN/m} \tag{3.34}$$

The design compressive strength of the concrete is $0.85 \times 25/1.5 = 14.2 \text{ N/mm}^2$, so from Equation 3.5, the depth of the stress block for full shear connection is

$$x = 412/14.2 = 29.0 \text{ mm} \tag{3.35}$$

This is less than h_c (which can be taken as 95 mm for this profile: Figure 3.8), so from Equation 3.6 with $d_p = 120$ mm,

$$M_{Rd} = 412(0.12 - 0.015) = \textbf{43.3 kN m/m} \tag{3.36}$$

The bending resistance is sufficient, subject to a check on longitudinal shear.

The design vertical shear for a span of 4 m is

$$V_{Ed} = 2(5.18 + 9.3) = \textbf{29.0 kN/m}$$

For the shear resistance, from Equation 3.21 with d_p taken as 200 mm,

$$v_{min} = 0.035 \times 2^{3/2} \times 25^{1/2} = 0.49 \ N/mm^2$$

From Equation 3.20 with $b_0 = 162$ mm, $b = 300$ mm (Figure 3.8),

$$V_{Rd} = (162/300) \times 120 \times 0.49 = \mathbf{31.7 \ kN/m} \tag{3.37}$$

which is just sufficient.

3.4.3 Composite slab – longitudinal shear

Longitudinal shear will be checked by both the *m–k* method (Section 2.8.1) and the partial-interaction method, explained in Section 3.3.2. From Equation 2.29, the *m–k* method gives the vertical shear resistance as

$$V_{\ell,Rd} = bd_p \left[\left(\frac{mA_p}{bL_s} \right) + k \right] / \gamma_{Vs} = \mathbf{25.9 \ kN/m} \tag{3.38}$$

The values used are:

$b = 1.0$ m $\qquad\qquad m = 184 \ N/mm^2$

$d_p = 120$ mm $\qquad\quad k = 0.0530 \ N/mm^2$

$A_p = 1178 \ mm^2/m \qquad \gamma_{Vs} = 1.25$

$L_s = L/4 = 1000$ mm

where γ_{Vs} is taken from Table 1.2, and the other values are explained above.

The design vertical shear is **29.0 kN/m** (Section 3.4.2), so the slab is not strong enough, using this method.

3.4.3.1 Partial-interaction method

The mean design resistance to longitudinal shear, $\tau_{u,Rd}$, is taken as $0.144 \ N/mm^2$ for this slab (Section 3.4). Account is taken of the shape of the profile when this value is determined from test data, so the shear resistance per metre width of sheeting is 0.144 kN per mm length. For full shear connection, the plastic neutral axis is in the slab, so the compressive force in the slab, $N_{c,f}$, is 412 kN/m, from Equation 3.34. The required length of shear span to develop this force (in the absence of any end anchorage) is

$$L_{sf} = N_{c,f}/(b\tau_{u,Rd}) = 0.412/0.144 = 2.86 \ m \tag{3.39}$$

The depth of the resulting stress block in the concrete, now denoted x_f, is 29 mm, from Equation 3.35. At a distance L_x ($< L_{sf}$) from an end support, the degree of shear connection is given by

$$\eta = L_x/L_{sf} = N_c/N_{c,f} = x/x_f \tag{3.40}$$

where N_c is the force in the slab and x the depth of the stress block.

Equations 3.16 to 3.18 then become:

$$z = 150 - 0.5x_f - 33 + 3\eta = 102.5 + 3\eta \ \ mm$$

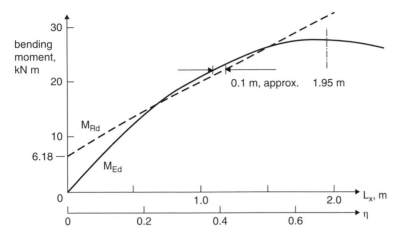

Figure 3.10 Partial-interaction method, for longitudinal shear

$$M_{pr} = 1.25 \times 6.18(1 - \eta) \leq 6.18 \text{ kN m}$$

$$M_{Rd} = 412\eta(z/1000) + M_{pr} \text{ kN m} \tag{3.41}$$

These enable M_{Rd} to be calculated for any value of η between zero and 1.0 (and hence, of L_x, from Equation 3.40). The curve so obtained is given in Figure 3.10. The bending-moment diagram for the loading, also shown, is a parabola with maximum value 27.6 kN m/m, at mid-span (from Section 3.4.2). (A slightly more favourable curve is obtained if account is taken of the increase in lever arm, z, that occurs because the partial-interaction stress block has depth ηx_f, not x_f, but that does not give sufficient resistance here.)

It is evident that the resistance is not quite sufficient where $L_x \approx 1.0$ m ($M_{Rd} = 20.2$ kN m, $M_{Ed} = 21.0$ kN m). It would be sufficient if the curve for M_{Rd} could be moved about 0.1 m to the left. This can be achieved by the use of end anchorage, as shown in Section 3.6.3.2. Since $\tau_{u,Rd}$ is 0.144 N/mm², an extra 'shear span' of 0.1 m would increase N_c by 14.4 kN. Assuming that one 19-mm stud connector will be welded through each trough to the supporting beam, there are 3.3 per metre, so each stud has to provide $14.4/3.3 = 4.4$ kN of anchorage. It will be shown in Section 3.11.2 that the reduction in the resistance of these studs to longitudinal shear (about 40 kN) by a force of 4.4 kN in the perpendicular direction is negligible, so the studs can provide sufficient anchorage.

Another source of increased shear resistance is the friction between the sheeting and the slab above an end support. It can only be used if it was not relied upon in the analysis of test results that led to the value of $\tau_{u,Rd}$ given by the provider of the sheeting. This is explained in Section 3.3.2. The information may not be available. To illustrate the method, it is now assumed that end friction can be used.

The whole of the vertical shear is assumed to be resisted by the concrete. The coefficient of friction recommended in EN 1994-1-1 is 0.5, so the additional resistance is $0.5V_{Ed}$, or 14.5 kN here. Coincidentally, this is precisely the force required, so end anchorage need not be relied upon.

Use of the partial-interaction method therefore enables the slab to be verified for longitudinal shear.

3.4.4 Local effects of point load

The design point load is Q_{Ed} = 4.0 × 1.5 = **6.0 kN** on any area 50 mm square, from Equation 3.3. The slab should be checked for punching shear and local bending. It is assumed that the thickness h_f of floor finish is at least 20 mm, so the data for Figure 3.6 are:

$$b_p = a_p = 50 \text{ mm} \quad h_f = 20 \text{ mm} \quad d_p = 120 \text{ mm}$$

The small ribs at the top of the sheeting (Figure 3.8) are neglected, so the thickness of slab above the sheeting is taken as h_c = 95 mm.

3.4.4.1 Punching shear
From Equation 3.22 in Section 3.3.4:

$$c_p = 2\pi h_c + 2(b_p + 2h_f) + 2(a_p + 2h_f + 2d_p - 2h_c) = 1023 \text{ mm}$$

Assuming that A193 mesh reinforcement (7-mm bars at 200-mm spacing, both ways, with area A_s = 193 mm²/m) is provided, resting on the sheeting, the effective depths in the two directions are 76 mm and 83 mm. The reinforcement ratios are:

$$\rho_x = 0.193/76 = 0.0025; \quad \rho_y = 0.193/83 = 0.0023, \quad \text{so } \rho = (\rho_x \rho_y)1/2 = 0.0024$$

From Equation 3.23,

$$v_{Rd} = 0.12 \times 2^{1/2} \times (0.24 \times 25)^{1/3} = 0.31 \text{ N/mm}^2$$

From Equation 3.24,

$$V_{Rd} = v_{Rd}c_p d = 0.31 \times 1.023 \times 79 = \textbf{25 kN}$$

This is a conservative value, because the ribs and sheeting have been ignored, but it far exceeds Q_{Ed}.

3.4.4.2 Local bending
From Equations 3.25 and 3.26 in Section 3.3.5,

$$a_m = b_m = 50 + 2(20 + 80) = 250 \text{ mm}$$

The most adverse situation for local longitudinal sagging bending is when the load is at mid-span, so L_p (Figure 3.7) is 1.95 m. From Equation 3.27, the effective width of slab is

$$b_{em} = b_m + 2L_p[1 - (L_p/L)] = 0.25 + 3.9 \times 0.5 = 2.20 \text{ m}$$

From Equation 3.28, the sagging moment per unit width with $L_p = L/2$ is

$$m_{Ed} = Q_{Ed}L_p[1 - (L_p/L)]/b_{em} = \textbf{2.66 kN m/m}$$

which is well below the resistance of the slab, **43.3 kN m/m.**
The transverse sagging moment under the load is given by Equation 3.29:

$$M_{Ed} = Q_{Ed}(b_{em} - b_m)/8 = 1.46 \text{ kN m}$$

The breadth of non-composite slab assumed to resist M_{Ed} may be given in the next Eurocode 4. It will clearly exceed a_m. So conservatively, using a_m (0.25 m), the moment per unit width is

$$m_{Ed} = 1.46/0.25 = \mathbf{5.85 \ kN \, m/m}$$

For the A193 mesh defined above, the effective depth is 76 mm and the force at yield is

$$193 \times 0.500/1.15 = 83.9 \text{ kN}$$

The depth of the concrete stress block is

$$x = 83.9/(0.85 \times 25/1.5) = 5.9 \text{ mm}$$

so the lever arm is $76 - 2.95 \approx 73$ mm and

$$m_{Rd} = 83.9 \times 0.073 = \mathbf{6.12 \ kN \, m/m}$$

which exceeds m_{Ed}.

It will be found later that, for this slab, other limit states govern much of the slab reinforcement; but in regions where they do not, mesh of area 193 mm^2/m would probably be used, as it provides more than 0.2%, and so satisfies the empirical rule given in Section 3.3.5.

3.4.5   Composite slab – serviceability

3.4.5.1   Cracking of concrete above supporting beams

Following Section 3.3.6, continuity across the steel beams should be provided by reinforcement of area 0.4% of the 'area of concrete on top of the steel sheet'. For the profile of Figure 3.8, it is not obvious whether h_c should be taken as 80 mm or 95 mm. The choice can be based on the direction of the tensile force. Here, 95 mm is used. Hence,

$$A_s = 0.004 \times 1000 \times 95 = 380 \ \text{mm}^2/\text{m} \tag{3.42}$$

The detailing is best left until fire resistance has been considered (Section 6.2).

3.4.5.2   Deflection

The characteristic load combination is used. The reinforcement of area A_s, calculated above, will provide continuity over the supports, and so reduce deflections. For this situation, EN 1994-1-1 permits the second moment of area of the slab, I, to be taken as the mean of the 'cracked' and 'uncracked' values for the section in sagging bending, and the use of a mean value of the modular ratio. This is $n = 20.2$, from Section 3.2. With the method of transformed sections, these values of I in 'steel' units are 8.27 m^2 mm^2 and 13.5 m^2 mm^2, respectively, so the mean value is

$$I = 10.9 \ \text{m}^2 \text{mm}^2/\text{m} \tag{3.43}$$

The self-weight of the slab is 2.54 kN/m^2, so the load on the prop at B (Figure 3.9), treated as the central support of a two-span beam, is

$$F = 2 \times 0.625 \times 2.54 \times 1.95 = 6.2 \text{ kN/m}$$

This is assumed to act as a load at mid-span of the composite slab, applied when the props are removed. There is in addition a load of 1.3 kN/m^2 from finishes (g), 1.2 kN/m^2 from partitions (q), and an imposed load q of 5.0 kN/m^2.

The mid-span deflection (for a simply-supported slab) is

$$\delta = \frac{L^3}{48EI}\left[F + \frac{5(g+q)L}{8}\right] = 3.4 + 9.9 = 13.3 \text{ mm} \tag{3.44}$$

with $L = 3.90$ m and $E = 210$ kN/mm^2. Hence, $\delta/L = 13.3/3900 = 1/293$.

The previous deflection of the sheeting at this point is zero, if the props do not deflect during concreting, and the total deflection at the quarter-span points is about 12 mm.

The ratio 1/293 is within the range of limits recommended in the British national annex to EN 1990 (BSI, 2002c). Examples are $L/300$ for removable partitions and plastered ceilings, $L/250$ for flexible floor coverings, and $L/200$ for suspended ceilings.

However, the ends of the slab follow the deflections of their supporting beams. It will be found that when these are added, the total deflection may be excessive.

3.4.6 Example: composite slab for a shallow floor using deep decking

The thickness of the floor slab designed above is 150 mm. In Section 3.11 it will be supported on a steel beam 406 mm deep, so the overall span/depth ratio of this 9 m span is low, at about 16. Where minimum depth is required, a different type of construction, known as a 'slim floor', can be used. The decking is much deeper (Section 3.12), and can be supported on the bottom flange of the steel beams.

An example of this type of slab is now given (Figure 3.11), simply supported on the composite beam designed in Section 3.13. The decking is ComFlor 225 (Figure 3.12). Its properties are given in the *ComFlor Manual* (Tata Steel UK, 2016). The thickness of the normal-density concrete above it has to be related to the steel section chosen, and is assumed to be 130 mm. The overall depth of slab, 355 mm, is high for a 4 m span, so the spacing of the steel frames is increased here from 4 m to 5 m.

If each point of support for the decking were 20 mm from its end, the effective span would be $5 - 0.32 - 0.04 = 4.64$ m. It is now assumed to be 4.7 m. Unpropped construction is assumed for the slab, with the beams propped. During construction the span/depth ratio is $4700/225 = 20.9$.

The sheeting has a design thickness of 1.21 mm, its weight is 0.17 kN/m^2 (Tata Steel UK, 2016), and the weight of the wet concrete is 4.66 kN/m^2. The ends of the troughs

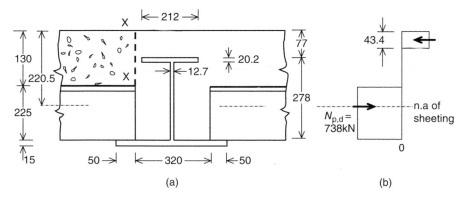

Figure 3.11 (a) Cross-section of composite slab at supporting steel beam. (b) Stress blocks for plastic resistance of slab to sagging bending

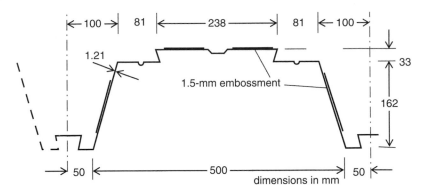

Figure 3.12 Cross-section of ComFlor 225 deep decking (simplified)

are closed. The weight of the concrete around the web of the beam will be included in design of the beam. It is

$$0.32[0.355 \times 25 - 4.66] = 1.35 \text{ kN/m}$$

The characteristic floor loadings are as given in Sections 3.2 and 3.3, and are listed in the columns 'SLS' of Table 3.1. The imposed loading during construction, 1.0 kN/m^2, is replaced after completion by the specified variable load, 5.0 kN/m^2. The partial factors and design loadings for ULS verification are also shown.

The ComFlor brochure assumes C30/37 concrete, and gives maximum spans that depend on the fire resistance period (60 min here), on whether propped or unpropped, and on the unfactored loading additional to self-weight. That is $5 + 1.2 + 1.3 = 7.5 \text{ kN/m}^2$ here, for which the maximum span is 5.5 m, so the flexural and vertical and longitudinal shear resistances will be sufficient for a span of 4.7 m. Manufacturers do these calculations for designers because the profile of the sheeting is so complex. The loading will have been assumed to be uniformly distributed.

As explained in Section 3.2, creep is allowed for approximately by using a reduced elastic modulus for concrete $(E_{cm}/2)$ for all loading. For the C30/37 concrete, $E_{cm} = 33 \text{ GPa}$, so the modular ratio is $210/16.5 = 12.7$. Deflections in service are considered first. For sagging bending of the steel sheeting, the brochure gives $I_p = 10.9 \times 10^{-6} \text{ m}^4/\text{m}$, with the

Table 3.1 Characteristic and design loads per unit area of slab

Limit state and partial factor	Load for steel sheeting			Load for composite slab		
	SLS	γ	ULS	SLS	γ	ULS
Profiled sheeting	0.17	1.35	0.23			
Composite slab	4.66	1.35	6.29			
Construction load	1.0	1.5	1.5	−1.0	1.5	−1.5
Floor and ceiling finishes				1.3	1.35	1.76
Allowance for partitions				1.2	1.5	1.80
Imposed loading				5.0	1.5	7.5
Totals, kN/m²	5.83		8.02	6.5		9.56

neutral axis 90.5 mm below the top of the profile. Excluding the construction loading, the mid-span bending moment during construction is $M_{Ek} = 4.83 \times 4.7^2/8 = 13.3$ kN m/m. With $E_a = 210$ GPa, the deflection of the sheeting is

$$\delta = 5ML^2/(48EI) = 13.4 \text{ mm}$$

It should be noted that if the mean deflection is 10 mm and the finished surface of the slab is horizontal, what is known as 'ponding' increases the weight of the concrete by 0.2 kN/m², which has not been allowed for here.

For the composite section, elastic theory gives the neutral axis depth as 85.3 mm below the top of the slab, and in 'steel' units, $I = 66.0 \times 10^{-6}$ m⁴/m. The unfactored loading is 7.5 kN/m² (above) and the deflection is 3.4 mm. The total deflection, 16.8 mm plus the deflection of the steel beam, is discussed later.

For the ultimate limit state, the whole of the loading, 17.6 kN/m², is assumed to be resisted by the composite section, so

$$M_{Ed} = 17.6 \times 4.7^2/8 = \textbf{48.6 kN m/m}$$

Reinforcement may be provided in the bottom of each trough to increase resistance to bending and to fire, and there will be mesh reinforcement in the slab. If these are ignored, the maximum tensile force for resistance to bending is $A_p f_{yp}$, which is

$$N_{p,d} = 2108 \times 350/1000 = 738 \text{ kN/m}$$

This is balanced by a depth of slab in compression of $x = N_{p,d}/f_{cd} = 738/17 = 43.4$ mm. The plastic stress blocks for M_{Rd} are shown in Figure 3.11b. The lever arm is

$$z = 220.5 - 43.4/2 = 199 \text{ mm}$$

and

$$M_{Rd} = N_{p,d}z = 738 \times 0.199 = \textbf{147 kN m/m}$$

This resistance is more than sufficient, and the design is likely to be governed by deflection.

Vertical shear is unlikely to govern. Assuming that it is resisted by the sheeting and that the webs do not buckle, the formula $\tau = V/bz$ provides a rough check. The total thickness of the webs is $b = 3.33 \times 1.21 = 4.0$ mm/m. The distance between the flanges is approximately $z = 0.2$ m, and $\tau_{Rd} = 350/1.732 = 202$ MPa.

Hence,

$$V_{Rd} \approx 202 \times 4.0 \times 0.2 = \textbf{162 kN/m}$$

This far exceeds the vertical shear:

$$V_{Ed} = 17.6 \times 4.7/2 = \textbf{41.4 kN/m}$$

If an accurate calculation is needed, account should be taken of the embossments in the webs, and web buckling should be considered.

The design manual gives no information on longitudinal shear. The 'safe load' tables will have taken account of information from testing, but usually assume uniformly distributed loading. If there are significant off-centre point loads, further information should be sought.

3.4.7 Comments on the designs of the composite slab

It was remarked in Section 3.4 that, for the normal-depth decking, the chosen depth of slab and size of sheeting could be inadequate for the required span and loading. The consequence was that propping was required during construction. This implies that the beams should also be propped, as discussed in Section 3.11.3.1.

The deep decking used in Section 3.4.6 has a span/depth ratio of 20.9, which seemed appropriate for unpropped construction, but when combined with the deflection of the supporting beams, the total was again found to be excessive (Section 3.13).

These problems arose from the combination of lightweight-aggregate concrete with a heavy imposed floor loading. This shows that the total deflection of composite slabs should be considered early in the design process.

3.5 Composite beams – sagging bending and vertical shear

Composite beams in buildings are usually supported by joints to steel or composite columns. The cheapest joints have little flexural strength, so it is convenient to design the beams as simply-supported. Such beams have the following advantages over beams designed as continuous at supports:

- very little of the steel web is in compression, and the steel top flange is restrained by the slab, so the resistance of the beam is not limited by buckling of steel;
- webs are less highly stressed, so it is easier to provide holes in them for the passage of services;
- bending moments and vertical shear forces are statically determinate, and are not influenced by cracking, creep, or shrinkage of concrete;
- there is no interaction between the behaviour of adjacent spans;
- bending moments in columns are lower, provided that the frame is braced against sidesway;
- no concrete at the top of the slab is in tension, except over supports;
- global analyses are simpler, and design is quicker.

The disadvantages are that deflection at mid-span or crack width at supports may be excessive, and structural depth is greater than for a continuous beam.

The behaviour and design of mid-span regions of continuous beams are similar to those of simply-supported beams, considered in this chapter. The other aspects of continuous beams are treated in Chapter 4.

3.5.1 Effective cross-section

The presence of profiled steel sheeting in a slab is normally ignored when the slab is considered as part of the top flange of a composite beam. Longitudinal shear in the slab (explained in Section 1.6) causes shear strain in its plane, with the result that vertical cross-sections through the composite T-beam do not remain plane when it is loaded.

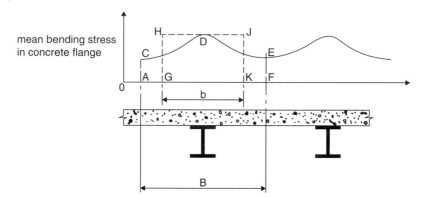

Figure 3.13 Use of effective width to allow for shear lag

At a cross-section, the mean longitudinal bending stress through the thickness of the slab varies across the width of the flange, as sketched in Figure 3.13.

Simple bending theory can still give the correct value of the maximum stress (at point D) if the true flange width B is replaced by an effective width, b (or b_{eff}), such that the area GHJK equals the area ACDEF. Research based on elastic theory has shown that the ratio b/B depends in a complex way on the ratio of B to the span L, the type of loading, the boundary conditions at the supports, and other variables.

At a cross-section of maximum bending moment, the compressive force in the slab is usually independent of the effective width, being governed by the tension in the steel member. The width assumed has little influence on the plastic moment of resistance, so simplified values are used in both Eurocodes 2 and 4.

For the mid-span region of a simply-supported beam in a building, EN 1994-1-1 gives the effective width as $L_e/8$ on each side of the steel web, where L_e is the distance between points of zero bending moment. The width of steel flange occupied by shear connectors, b_0, can be added, so

$$b = L_e/4 + b_0 \tag{3.45}$$

provided that a width of slab $L_e/8$ is present on each side of the shear connectors. For elastic global analysis, this width is assumed for the whole span, to avoid having to take account of non-uniform stiffness (EI) within a span. For analysis of cross-sections, a slightly lower width is specified for the end quarter of each span. This has little effect in practice, as bending resistance of end regions is unlikely to govern the design.

Where profiled sheeting spans at right angles to the span of the beam (as in the worked example here), only the concrete above its ribs can resist longitudinal compression (e.g. its effective thickness in Figure 3.8 is 80 mm). Where ribs run parallel to the span of the beam, the concrete within ribs can be included, although it is rarely necessary to do so.

Longitudinal reinforcement within the slab is usually neglected in regions of sagging bending.

3.5.2 Classification of steel elements in compression

Because of local buckling, the ability of a steel flange or web to resist compression depends on its slenderness, represented by its width/thickness ratio, c/t. In design to EN 1994-1-1, as in EN 1993-1-1, each flange or web in compression is placed in one of four classes. The highest (least slender) class is Class 1 (plastic). The class of a cross-section of a composite beam is the lower of the classes of its web and compression flange (excluding a flange restrained by concrete), and this class determines the design procedures that are available.

This well-established system is summarized in Table 3.2. The Eurocodes allow several methods of plastic global analysis, of which rigid-plastic analysis (plastic hinge analysis) is the simplest. This is considered further in Section 4.3.3.

The Eurocodes give several idealized stress–strain curves for use in plastic section analysis, of which only the simplest (rectangular stress blocks) are used in this book. The class boundaries are defined by limiting slenderness ratios that are proportional to $(f_y)^{-0.5}$, where f_y is the nominal yield strength of the steel. This allows for the influence of yielding on loss of resistance during buckling. The ratios in EN 1993-1-1 for steel with $f_y = 355\,\text{N/mm}^2$ are given in Table 3.2 for uniformly compressed flanges of rolled I-sections with outstands of width c and thickness t. The root radius is not treated as part of the outstand.

Encasement of webs in concrete to improve resistance to fire is now rarely done. It also prevents rotation of a flange towards the web, which occurs during local buckling, and so enables higher c/t ratios to be used at the class 2/3 and 3/4 boundaries, as shown. At the higher compressive strains that are relied upon in plastic hinge analysis, the encasement is weakened by crushing of concrete, so the c/t ratio at the class 1/2 boundary is unchanged.

The class of a steel web is strongly influenced by the proportion of its clear depth, d, that is in compression, as shown in Figure 3.14. For the class 1/2 and 2/3 boundaries, plastic stress blocks are used, and the limiting d/t ratios are given in EN 1993-1-1 as functions of α, defined in Figure 3.14. The curves show, for example, that a web with $d/t = 40$ moves from class 1 to class 3 when α increases from 0.7 to 0.8. This high rate of change is significant in the design of continuous beams (Section 4.2.1).

For the class 3/4 boundary, elastic stress distributions are used, defined by the ratio ψ. Pure bending (no net axial force) corresponds not to α = 0.5, but to ψ = −1. In a composite T-beam in hogging bending, the elastic neutral axis is normally higher than the

Table 3.2 Classification of cross-sections (BSI, 2014b), and methods of analysis

Slenderness class and name	1 Plastic	2 Compact	3 Semi-compact	4 Slender
Method of global analysis	Plastic[d]	Elastic	Elastic	Elastic
Analysis of cross-sections	Plastic[d]	Plastic[d]	Elastic[a]	Elastic[b]
Maximum ratio c/t for flanges of rolled I-sections:[c]				
uncased web	7.32	8.14	11.4	No limit
encased web	7.32	11.4	16.3	No limit

a) hole-in-the-web method enables plastic analysis to be used;
b) with reduced effective width or yield strength;
c) for S 355 steel ($f_y = 355\,\text{N/mm}^2$);
d) elastic analysis may be used, but is more conservative.

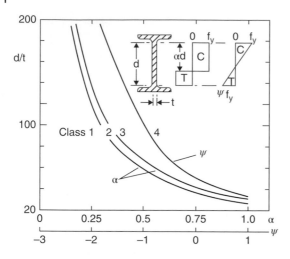

Figure 3.14 Class boundaries for webs, for $f_y = 355$ N/mm²

plastic neutral axis, and its positions for propped and unpropped construction are different, so the curve for the class 3/4 boundary is not comparable with the others in Figure 3.14.

For simply-supported composite beams, the steel compression flange is restrained from local buckling (and also from lateral buckling) by its connection to the concrete slab, and so is in Class 1. The plastic neutral axis for full interaction is usually within the slab or steel top flange, so the web is not in compression, when flexural failure occurs, unless partial shear connection (Section 3.5.3.1) is used. Even then, α is sufficiently small for the web to be in Class 1 or 2. This may not be so for the much deeper plate or box girders used in bridges.

During construction of a composite beam, the steel beam alone may be in a lower slenderness class than the completed composite beam, and may be susceptible to lateral buckling. Design for this situation is governed by EN 1993-1-1 for steel structures.

3.5.3 Resistance to sagging bending

3.5.3.1 Beams with cross-sections in Class 1 or 2

The methods of calculation for sections in Class 1 or 2 are in principle the same as those for composite slabs, explained in Section 3.3.1, to which reference should be made. The main assumptions are as follows:

- the tensile strength of concrete is neglected;
- plane cross-sections of the structural steel and reinforced concrete parts of a composite section each remain plane;

and, for plastic analysis of sections only:

- the effective area of the structural steel member is stressed to its design yield strength f_{yd} ($=f_y/\gamma_A$) in tension or compression;
- the effective area of concrete in compression resists a stress of $0.85f_{cd}$ (where $f_{cd} = f_{ck}/\gamma_C$) constant over the whole depth between the plastic neutral axis and the most compressed fibre of the concrete.

This definition of f_{cd} in Eurocode 4 differs from the one used in Eurocode 2. It is not known whether the current revision of the two codes will remove this anomaly. Its complex origin is explained elsewhere (Johnson, 2012), and is associated with another difference, concerning rectangular stress blocks. In Eurocode 2, they extend part of the way to the neutral axis (usually 80%). In a composite member this can lead to complex calculations (see Section 3.5.3.1(3) below), so in Eurocode 4 they extend to the plastic neutral axis of the cross-section. This optimistic assumption is offset by the factor 0.85, above.

In deriving the formulae below, it is assumed that the steel member is a rolled I-section, of cross-sectional area A_a, with yield strength not exceeding 355 MPa, and the slab is composite, with profiled sheeting that spans between adjacent steel members. The composite section is in Class 1 or 2, so that the whole of the design load can be assumed to be resisted by the composite member, whether the construction is propped or unpropped. This is because the inelastic behaviour that precedes flexural failure allows internal redistribution of stresses to occur.

The effective section is shown in Figure 3.15a. As for composite slabs, there are three common situations, as follows. The first two occur only where full shear connection is provided. They are explained assuming a simply-supported beam. Hogging bending in continuous beams is covered in Section 4.2.

(1) *Neutral axis within the concrete or composite slab*

The stress blocks are shown in Figure 3.15b. The depth x_c, assumed to give the position of the plastic neutral axis, is found by resolving longitudinally:

$$N_{c,f} = A_a f_{yd} = b_{eff} x_c (0.85 f_{cd}) \tag{3.46}$$

This method is valid when $x_c \leq h_c$.

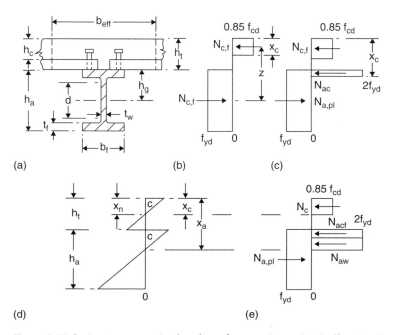

(a)   (b)   (c)

(d)   (e)

Figure 3.15 Resistance to sagging bending of composite section in Class 1 or 2

For the transverse sheeting shown, h_c is the net depth of the concrete slab. Where the slab is supported by secondary beams, and the sheeting runs parallel to the main beam, the gross depth can be used for that beam, because the loss of cross-section caused by the top rib is negligible.

Taking moments about the line of action of the force in the slab,

$$M_{pl,Rd} = A_a f_{yd} = (h_g + h_t - x_c/2) \tag{3.47}$$

where h_g defines the position of the centre of area of the steel section, which need not be symmetrical about its major (y–y) axis.

(2) *Neutral axis within the steel top flange*

If Equation 3.46 gives $x_c > h_c$, then it is replaced by:

$$N_{c,f} = b_{eff} h_c (0.85 f_{cd}) \tag{3.48}$$

This force is now less than the yield force for the steel section, denoted by

$$N_{a,pl} = A_a f_{yd} \tag{3.49}$$

so the plastic neutral axis is at depth $x_c > h_t$, and is at first assumed to lie within the steel top flange (Figure 3.15c). The condition for this is

$$N_{ac} = N_{a,pl} - N_{c,f} \leq 2b t_f f_{yd} \tag{3.50}$$

The distance x_c is most easily calculated by assuming that the strength of the steel in compression is $2f_{yd}$, so that the force $N_{a,pl}$ and its line of action can be left unchanged. Resolving longitudinally to determine x_c:

$$N_{a,pl} = N_{c,f} + N_{ac} = N_{c,f} + 2b_f(x_c - h_t)f_{yd} \tag{3.51}$$

Taking moments about the line of action of the force in the slab,

$$M_{pl,Rd} = N_{a,pl}(h_g + h_t - h_c/2) - N_{ac}(x_c - h_c + h_t)/2 \tag{3.52}$$

If x_c is found to exceed $h_t + t_f$, the plastic neutral axis lies within the steel web, and $M_{pl,Rd}$ can be found by a similar method.

(3) *Partial shear connection*

The symbol $N_{c,f}$ was used in paragraphs (1) and (2) above for consistency with the treatment of composite slabs in Section 3.3.1. In design, its value is always the lesser of the two values given by Equations 3.46 and 3.48. It is the force that the shear connectors between the section of maximum sagging moment and each free end of the beam (a 'shear span') must be designed to resist, if full shear connection is to be provided.

Let us suppose that the shear connection is designed to resist a force N_c, smaller than $N_{c,f}$. If each connector has the same resistance to shear, and the number in each shear span is n, then the degree of shear connection is defined by:

$$\text{degree of shear connection} = \eta = n/n_f = N_c/N_{c,f} \tag{3.53}$$

where n_f is the number of connectors required for full shear connection.

The plastic moment of resistance of a composite slab with partial shear connection had to be derived in Section 3.3.1(3) by an empirical method, because the flexural properties of profiled sheeting are so complex. For composite beams, simple plastic theory can be used (Slutter and Driscoll, 1965).

The depth of the compressive stress block in the slab, x_c, is given by

$$x_c = N_c/(0.85f_{cd}b_{eff}) \tag{3.54}$$

and is always less than h_c. The distribution of longitudinal strain in the cross-section is intermediate between the two distributions shown (for stress) in Figure 2.2c, and is shown in Figure 3.15d, in which C indicates compressive strain. The neutral axis in the slab is at a depth x_n slightly greater than x_c, as shown.

In design of reinforced concrete beams and slabs, it is generally assumed that x_c/x_n is between 0.8 and 0.9. The less accurate assumption $x_c = x_n$ is made for composite beams and slabs to avoid the complexity that otherwise occurs in design when $x_c \approx h_c$ or, for beams with non-composite slabs, $x_c \approx h_t$. This introduces an error in M_{pl} that is on the unsafe side, but is negligible for composite beams. It is not negligible for composite columns, where it is allowed for (Section 5.6.5.1).

There is a second neutral axis within the steel I-section. If it lies within the steel top flange, the stress blocks are as shown in Figure 3.15c, except that the concrete block for the force $N_{c,f}$ is replaced by a shallower one, for force N_c.

Resolving longitudinally,

$$N_{ac} = N_{a,pl} - N_c$$

The depth of the neutral axis in the steel is found from

$$x_a = h_t + N_{ac}/(2b_f f_{yd}) \tag{3.55}$$

Taking moments about the line of action of N_c,

$$M_{Rd} = N_{a,pl}(h_g + h_t - x_c/2) - N_{ac}(x_a + h_t - x_c)/2 \tag{3.56}$$

If the second neutral axis lies within the steel web, the stress blocks are as shown in Figure 3.15e, and M_{Rd} can be found by a method similar to that for Equation 3.56.

Use of partial shear connection in design   The curve ABC in Figure 3.16 shows a typical relationship between $M_{Rd}/M_{pl,Rd}$ and degree of shear connection η, found by using the preceding equations for assumed values of η. When N_c is taken as zero, then

$$M_{Rd} = M_{pl,a,Rd}$$

where $M_{pl,a,Rd}$ is the resistance of the steel section alone.

The curve is not valid for very low degrees of shear connection, for reasons explained in Section 3.6.2. Where it is valid, it is evident that a substantial saving in the cost of shear

Figure 3.16 Design methods for partial shear connection

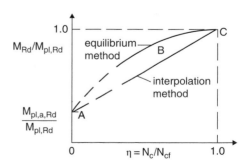

connectors can be obtained (e.g. by using $\eta = 0.7$) when the required bending resistance M_{Ed} is only slightly below $M_{pl,Rd}$.

Where profiled sheeting is used, there is sometimes too little space in the troughs for n_f connectors to be provided within a shear span, and then partial-connection design with $\eta < 1$ becomes essential.

Unfortunately, curve ABC in Figure 3.16 cannot be represented by a simple algebraic expression. In practice, it is therefore sometimes replaced (conservatively) by the line AC, given by

$$N_c = N_{c,f}(M_{Rd} - M_{pl,a,Rd})/(M_{pl,Rd} - M_{pl,a,Rd}) \tag{3.57}$$

In design, M_{Rd} is replaced by the known value M_{Ed}, and $M_{pl,Rd}$, $M_{pl,a,Rd}$ and $N_{c,f}$ are easily calculated, so this equation gives directly the design compressive force in the slab, N_c, and hence the number of connectors required in each shear span:

$$n = n_f N_c/N_{c,f} = N_c/P_{Rd} \tag{3.58}$$

where P_{Rd} is the design resistance of one connector. There is a worked example in Section 3.5.3.3.

The design of shear connection is further discussed in Section 3.6.

Variation in bending resistance along a span In design, the bending resistance of a simply-supported beam is checked first at the section of maximum sagging moment, which is usually at mid-span. For a steel beam of uniform section, the bending resistance elsewhere within the span is then obviously sufficient; but this may not be so for a composite beam. Its bending resistance depends on the number of shear connectors between the nearer end support and the cross-section considered. This is shown by curve ABC in Figure 3.16, because the x-coordinate is proportional to the number of connectors.

Suppose, for example, that a beam of span L is designed with partial shear connection and $n/n_f = 0.5$ at mid-span. Curve ABC is redrawn in Figure 3.17a, with the bending resistance at mid-span, $M_{Rd,max}$, denoted by B. Length BC of this curve is not now valid, because shear failure would occur in the right-hand half span. If the connectors are

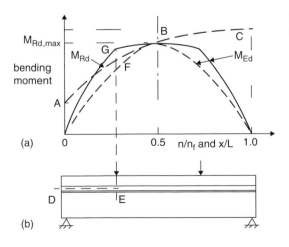

Figure 3.17 Variation of bending resistance along a span

uniformly spaced along the span, as is usual in buildings, then the axis n/n_f is also an axis x/L, where x is the distance from the nearer support, and n is the number of connectors effective in transferring the compression to the concrete slab over a length x from a free end. Only these connectors can contribute to the bending resistance $M_{Rd,x}$ at that section, denoted E in Figure 3.17b. In other words, bending failure at section E would be caused (in the design model) by longitudinal shear failure along length DE of the interface between the steel flange and the concrete slab.

Which section would in fact fail first depends on the shape of the bending-moment diagram for the loading. For uniformly distributed loading, the curve for $M_{Ed,x}$ is parabolic, and curve OFB in Figure 3.17a shows that failure would occur at or near mid-span. The addition of significant point loads (e.g. from small columns) at the quarter-span points changes the curve from OFB to OGB. Failure would occur near section E. A design with uniformly spaced connectors in which M_{Ed} at mid-span was equated to $M_{Rd,max}$ would be unsafe.

This is why design codes would not allow the $0.5n_f$ connectors to be spaced uniformly along the half span, for this loading. The number required for section E would be calculated first, and spaced uniformly along DE. The remainder would be located between E and mid-span, at wider spacing. Spacing of connectors is further discussed in Section 3.6.1.

3.5.3.2 Non-linear and elastic resistances to bending of beams

The elastic resistance $M_{el,Rd}$ is rarely required in design of structures for buildings. It is found by elastic analysis of the cross-section taking account of unpropped construction, if used, and of creep. Shrinkage can normally be neglected. Eurocode 4 gives the stress limits for ultimate limit states as the usual design yield and compressive strengths, f_{yd}, f_{sd} and f_{cd}. The elastic resistance can be as low as $0.7M_{pl,Rd}$. Rules for the shear connection are given in Eurocode 4 Part 2, *Bridges*. They are based on the distributions of longitudinal shear found by elastic theory.

For design where $M_{el,Rd} < M_{Ed} < M_{pl,Rd}$, Eurocode 4 gives an interpolation method for finding the degree of shear connection for beams with Class 1 and 2 sections, in a clause 'non-linear resistance'. To allow for unpropped construction, the design bending moment is subdivided:

$$M_{Ed} = M_{a,Ed} + M_{c,Ed} \tag{3.59}$$

where subscripts a and c refer to the moments resisted by the steel and composite cross-sections, respectively. The stresses caused by $M_{a,Ed}$ are deducted from the three stress limits given above. Elastic analysis then finds the bending moment $kM_{c,Ed}$ that causes the first of the three reduced stress limits to be reached. Then,

$$M_{el,Rd} = M_{a,Ed} + kM_{c,Ed} \tag{3.60}$$

Thus, the elastic resistance is assumed to be reached while the moment M_c is increasing towards $M_{c,Ed}$, after the whole of $M_{a,Ed}$ has been applied. The inelastic behaviour that then follows is not analysed.

The elastic analysis also gives the compressive force in the concrete flange, $N_{c,el}$, when the bending moment is $M_{el,Rd}$. The higher compressive force in the slab needed for the design of the shear connection, N_c, is found by the interpolation method shown by the

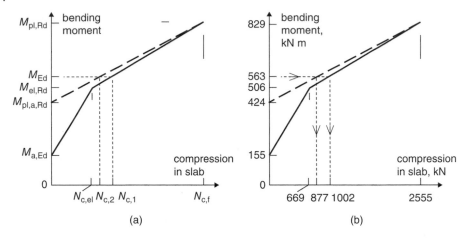

Figure 3.18 Compression force in a concrete flange for $M_{el,Rd} < M_{Ed} < M_{pl,Rd}$ by the non-linear and interpolation methods. (a) Notation. (b) Results from the worked example

dashed line in Figure 3.18. For propped construction, $M_{a,Ed}$ is taken as zero. There is a worked example in Section 3.5.3.3.

The method leads to a lower requirement for shear connection than plastic analysis does, but is not worth using where the shear connection is already governed by the 'minimum' rules.

3.5.3.3 Example: non-linear resistance to sagging bending

The main example for a simply-supported composite beam is in Section 3.11. Its shear connection is designed by the partial-interaction method that uses plastic analysis of cross-sections, and is governed by the rules for minimum degree of shear connection. The non-linear resistance method is less attractive, because elastic analysis is also required. In this example, it is used for the beam of Section 3.11 (Figure 3.31), where calculations for the following results, shown in Figure 3.18b, will be found:

- plastic resistance moment with full shear connection, $M_{pl,Rd} = \mathbf{829\,kN\,m}$
- design bending moment at mid-span (ULS), $M_{Ed} = \mathbf{563\,kN\,m}$
- compressive force in the slab at $M_{pl,Rd}$, $N_{c,f} = \mathbf{2555\,kN}$

From Section 3.11.3, the elastic properties for the assumed modular ratio (20.2) are as follows:

- for the steel beam, $10^{-6}I_a = 215\,mm^4$, with neutral axis at the centre of the 406 mm depth
- for the composite beam, $10^{-6}I_c = 636\,mm^4$, with neutral-axis depth, $x_c = 176\,mm$

The factored distributed loading per unit length is given in Section 3.11.1 as 60.9 kN/m, of which $1.35(10.2 + 2.2) = 16.74$ kN/m acts on the steel beam alone. Hence, from Equation 3.59,

$$M_{a,Ed} = (16.74/60.9) \times 563 = \mathbf{155\,kN\,m}$$

$$M_{c,Ed} = 563 - 155 = 408\ kN\,m$$

The first of the three stress limits to be reached is yield in the steel bottom flange. These stresses are:

For $M_{a,Ed}$: $\sigma_{a,bot} = M_{a,Ed}(h_a/2)/I_a = 155 \times 203/215 = 146$ MPa
For $M_{c,Ed}$: $\sigma_{c,bot} = M_{c,Ed}(h_t + h_a - x_c)/I_c = 408(150 + 406 - 176)/636 = 244$ MPa

The stress available for M_c is $355 - 146 = 209$ MPa, so $k = 209/244 = 0.86$

From Equation 3.60, $M_{el,Rd} = 155 + 0.86 \times 408 = \mathbf{506\ kN\,m}$

For the composite cross-section, the elastic neutral axis is 176 mm below the top of the slab, so the distance from it to the centre of the compressed region of the slab is $176 - 80/2 = 136$ mm. The elastic stress at this level from $kM_{c,Ed}$ is

$$0.86 \times 408 \times 136/(636 \times 20.2) = 3.71\ \text{MPa}$$

which is far below the design stress of 17 MPa. The effective width is 2.25 m (Figure 3.29) so the compressive force in the slab is

$$N_{c,el} = 3.71 \times 80 \times 2.25 = \mathbf{669\ kN}$$

Entering M_{Ed} on the y-axis of Figure 3.18b and following the route shown gives the design compressive force in the slab:

$$N_{c,1} = \mathbf{1002\ kN}$$

Hence, the degree of shear connection is:

$$\eta = 1002/2555 = \mathbf{0.39}$$

This method is now compared with a calculation by the interpolation method for partial shear connection, using the same data.

From Figure 3.16, $M_{pl,a,Rd}$ is required. The plastic section modulus for the steel cross-section is given in Figure 3.31 as $10^{-6}W_{pl,a} = 1.194\ \text{mm}^3$, so

$$M_{pl,a,Rd} = 355 \times 1.194 = 424\ \text{kN\,m}$$

Using the dashed line in Figure 3.18b,

$$N_{c,2} = 2555(563 - 424)/(829 - 424) = \mathbf{877\ kN}$$

so the degree of shear connection is:

$$\eta = 877/2555 = \mathbf{0.343}.$$

These two results agree quite well, considering that both methods are approximations. In design to the current Eurocode 4, the rules for minimum shear connection give $\eta_{min} = 0.51$ (Section 3.11.2), which governs. This is for unpropped construction, 6 mm slip capacity, $M_{Ed} = M_{Rd}$ and 8.6 m span. For this situation, the proposed new and more liberal rules are likely to reduce η_{min} by about 15%. That value, 0.43, would still govern.

3.5.3.4 Beams with cross-sections in Class 3 or 4

The elastic resistance to bending of a beam of semi-compact or slender section is usually governed by the maximum stress in the steel section, and is calculated as $M_{el,Rd}$ by the elastic method given above. This is the upper limit to M_{Ed}. It is fortunate that in design for buildings, it is almost always possible to ensure that sections in sagging bending are in Class 1 or 2. This is more difficult for hogging bending, as explained in Section 4.2.1.

3.5.4 Resistance to vertical shear

In a simply-supported steel beam, bending stresses near a support are within the elastic range even when the design ultimate load is applied; but in a composite beam, maximum slip occurs at end supports, so bending stresses cannot be found accurately by simple elastic theory based on plane sections remaining plane.

Vertical shear stresses are calculated from rates of change ($d\sigma/dx$) of bending stresses σ, and so cannot easily be found near an end of a composite beam. It is at present conservatively assumed in Eurocode 4 that vertical shear is resisted by the steel beam alone, exactly as if it were not composite. The web thickness of most rolled steel I-sections is sufficient to avoid buckling in shear, and then design is simple. The shear area A_v for such a section is given in EN 1993-1-1 (BSI, 2014b) as

$$A_v = A_a - 2b_f t_f + (t_w + 2r)t_f \tag{3.61}$$

with root radius r and other notation as in Figure 3.15a. This shows that some of the vertical shear is resisted by the steel flanges.

The shear resistance is calculated by assuming that the yield strength in shear is $f_{yd}/\sqrt{3}$ (von Mises yield criterion), and that the whole of area A_v can reach this stress:

$$V_{pl,a,Rd} = A_v(f_{yd}/\sqrt{3}) \tag{3.62}$$

This is a 'rectangular stress block' plastic model, based essentially on test data.

The maximum slenderness of an unstiffened web for which shear buckling can be neglected is given in Eurocode 4 as

$$h_w/t_w \leq 72\,\varepsilon$$

where h_w is the clear distance between the flanges.

Where the steel web is encased in concrete in accordance with rules given in EN 1994-1-1, shear buckling can be neglected if

$$d/t_w \leq 124\,\varepsilon \tag{3.63}$$

The dimensions d and t_w are shown in Figure 3.15a, and

$$\varepsilon = (235/f_y)^{1/2} \tag{3.64}$$

with f_y in N/mm^2 units. This allows for the influence of yielding on shear buckling.

It has been shown by tests on simply-supported beams with solid slabs that some of the shear ($V_{c,Rd}$) is resisted by the slab (Vasdravellis and Uy, 2014). Following parametric studies, a design model has been proposed in the form

$$V_{Rd} = V_{pl,a,Rd} + V_{c,Rd}$$

With full shear connection, the increase above $V_{pl,a,Rd}$ is proportional to the ratio of slab thickness to depth of the steel beam, h_c/h_s. It reaches about 20% for $h_c/h_s = 0.4$, almost independent of the width of the concrete flange. The increase is less where there is partial shear connection.

In continuous beams, the contribution from the slab would also be influenced by whether it is continuous across the end support, by how much it is cracked, and by local details of the shear connection.

Interaction between bending and vertical shear can influence the design of continuous beams, and is treated in Section 4.2.2. The large openings in webs often required for services can reduce their resistance to vertical shear. Section 3.14 gives a design method that may be included in the next Eurocode 4.

3.5.5 Resistance of beams to bending combined with axial force

Bending combined with axial compression occurs most often in columns, and is treated in Chapter 5, where the theory assumes that cross-sections have biaxial symmetry. It therefore cannot be used for most composite beams. The axial forces that occur in beams for buildings (e.g. from wind loading) are usually small, except in an accidental load combination that includes removal of a column. Axial forces can be quite large in bridges with inclined columns or tied arches.

There is guidance from research, but not yet from Eurocode 4. Tests and parametric studies on beams with cross-sections in Class 1 or 2 have been reported for composite T-beams with solid slabs, covering the four possible interactions between axial force and bending (tensile and compressive force, sagging and hogging bending). They all led to convex interaction curves (commonly found in structural engineering), to which the interaction lines in Figure 3.19 are conservative approximations. Detailed results are available (Vasdravellis et al., 2013).

The first quadrant in the graph is similar in shape to the simplified rules for composite columns. The symbols N_{pl} and M_{pl} are for values calculated from plastic theory as used for composite sections, ignoring concrete in tension, and with steel, including reinforcement, at yield. Measured mean values of strengths of materials were used, without partial factors.

Symbols N and M are for test values, with bending moments calculated assuming that the axial force acts at the neutral axis of the cracked reinforced composite section when subjected to bending only. Partial shear connection with $n/n_f \approx 0.6$ was provided for specimens in sagging bending, and full shear connection for hogging bending: values typical of practice for solid slabs.

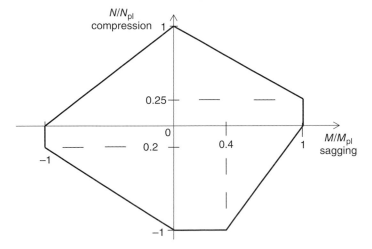

Figure 3.19 Resistance of a beam cross-section to bending combined with axial force

If relying on these results in practice, design of the shear connection should ensure that in regions of maximum bending moment, the axial force is appropriately shared between the steel and concrete. There are provisions for this subject, 'load introduction', in the Eurocode 4 clauses on columns (Section 5.6.6 of this book). In these tests, axial compression was applied to the steel beam and part of the width of the concrete flange. Axial tension was applied to the steel beam only.

The compressive strength of the concrete in the tests was quite low, about 24 MPa. The yield strength of the steel flanges was reported as 365 MPa. In designs with very different strengths, or with a composite slab flange, the applicability of these results should be considered.

Design methods for beams with cross-sections in slenderness Class 3 or 4 that also resist axial force are given in Eurocode 4: Part 2 (BSI, 2005c). There are clauses on tension members that are not limited to biaxial symmetry. For beams with axial compression, reference is made to Eurocode 3-1-5 (BSI, 2006b). Further details and a worked example are available (Hendy and Johnson, 2006).

3.6   Composite beams – longitudinal shear

3.6.1   Critical lengths and cross-sections

As noted in Section 3.5.3.2, the bending moment at which yielding of steel first occurs in a simply-supported composite beam can be below 70% of the ultimate moment. If the bending-moment diagram is parabolic, then at ultimate load, partial yielding of the steel beam can extend over half of the span.

At the interface between steel and concrete, the distribution of longitudinal shear is influenced by yielding, and also by the spacing of the connectors, their load/slip properties, and shrinkage and creep of the concrete slab. For these reasons, no attempt is made in design to calculate this distribution. Wherever possible, connectors are uniformly spaced along the span.

It was shown in Section 3.5.3 that this cannot always be done. For beams with all critical sections in Class 1 or 2, uniform spacing is allowed by EN 1994-1-1 along each *critical length*, which is a length of the interface between two adjacent *critical cross-sections*. These are:

- sections of maximum bending moment;
- supports;
- sections subjected to concentrated loads or reactions;
- places where there is a sudden change of cross-section of the beam; and
- free ends of cantilevers.

There is also a definition for tapering members.

Where the design ultimate loading is uniformly distributed, a typical design procedure, for half the span of a beam, whether simply-supported or continuous, would be as follows.

(1) Determine the compressive force required in the concrete slab at the section of maximum sagging moment, as explained in Section 3.5.3. Let this be N_c.

(2) Determine the tensile force in the concrete slab at the support that is assumed to contribute to the bending resistance at that section (i.e., zero for a simple support, even if crack-control reinforcement is present; and the yield force in the longitudinal reinforcement, if the span is designed as continuous). Let this force be N_t.

(3) If there is a critical cross-section between these two sections, determine the force in the slab at that section. The bending moment will usually be below the yield moment, so elastic analysis of the section can be used.

(4) Choose the type of connector to be used, and determine its design resistance to shear, P_{Rd}, as explained in Section 2.5.

(5) The number of connectors required for the half span is:

$$n = (N_c + N_t)/P_{Rd} \tag{3.65}$$

The number required within a critical length where the change in longitudinal force is ΔN_c, is $\Delta N_c/P_{Rd}$.

An alternative to the method of step (3) would be to use the shear force diagram for the half span considered. Such a diagram is shown in Figure 3.20 for the length ABC of a span AD, which is continuous at A and has a heavy point load at B. The sagging moment will be a maximum where the vertical shear is zero, so the critical sections are A, B and C. The total number of connectors is shared between lengths AB and BC in proportion to the areas of the shear force diagram, OEFH and GJH.

In practice it might be necessary to provide extra connectors along BC, because codes limit the maximum spacing of connectors, to prevent uplift of the slab relative to the steel beam, and to ensure that the steel top flange is sufficiently restrained from local and lateral buckling.

3.6.2 Non-ductile, ductile and super-ductile stud shear connectors

The use of uniform spacing of shear connectors is possible because all connectors have some ductility, or *slip capacity*. Its determination from push tests is explained in Section 2.5, where the definition of characteristic slip capacity is given.

The slip capacity of headed stud connectors increases with the diameter of the shank and has been found to be about 6 mm for 19-mm studs in solid concrete slabs (Johnson and Molenstra, 1991). Studs with at least this slip capacity are named 'ductile' in Eurocode 4. Higher values have been found in tests with single studs centrally placed in the trapezoidal troughs of profiled steel sheeting. Those with characteristic slip exceeding 10 mm are known as 'super-ductile'.

Figure 3.20 Vertical shear in a beam with an off-centre point load

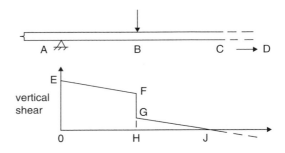

Slip enables longitudinal shear to be redistributed between the connectors in a critical length, before any of them fail. More slip is required where the degree of shear connection is low, and also as the critical length increases (a scale effect). A connector that is 'ductile' (has sufficient slip capacity) for a short span becomes 'non-ductile' in a long span, and then a higher degree of shear connection is required.

One of the conditions in Eurocode 4 associated with its rules for the shear resistance of welded stud connectors is that mesh reinforcement be provided at least 30 mm below the head of a stud. This has been found to be inconvenient in practice and is widely ignored. This and other changes in practice and in decking profiles has led to some reductions in the specified shear resistances. Profiled sheeting severely limits the density of studs that can be provided, so there has been pressure from the industry for Eurocode 4 to provide less restrictive rules for super-ductile studs.

The degree of shear connection, η, is defined in Section 3.5.3.1(3). To allow for the effect of span on the slip capacity required, the current Eurocode 4 gives rules for the minimum permitted degree of shear connection. This increases with span, and the rules take some account of the improved performance in a composite slab. Eurocode 4 does not recognize other parameters that recent research has found to be relevant. These include:

- whether construction is propped or unpropped;
- the ratio M_{Ed}/M_{Rd} for the beam concerned, which in practice is often well below 1.0;
- the level in a composite slab at which reinforcement (usually welded mesh) is provided;
- imposed loadings not uniformly distributed over a span;
- unusually high ratios of variable to permanent loading on the beam;
- the beneficial effect of regular circular holes in the beam web.

A report from the Steel Construction Institute in the UK (Couchman, 2015) proposes revised and additional rules for the next Eurocode 4, mainly in the form of graphs relating minimum degree of shear connection (n/n_f) to the span of the beam. Some of these are shown in Figure 3.21. They are all more liberal than the general rule now in Eurocode 4 (also shown), but have many restrictions to their scope. The account of them here is limited as follows:

- simply-supported beams of span up to 22 m, with uniformly distributed loading;
- beams of grade S355 steel, with solid webs, equal flanges and in slenderness Class 1 or 2;
- 19-mm welded studs, with length after welding at least 95 mm;
- solid concrete slabs, or composite slabs with profiled sheeting transverse to the supporting beams, with trapezoidal troughs and depth not exceeding 80 mm, excluding any small top rib. This excludes deep decking and re-entrant troughs;
- beams with a concrete flange on both sides of the web, which excludes edge beams.

Some of the subjects excluded here are covered by additional rules given in the Report, but most of them arise from the lack of sufficient evidence from tests to enable current rules in Eurocode 4 to be relaxed.

The lines in Figure 3.21 are for beams propped (P) or unpropped (U) during construction, for slip capacities of 6 mm and 10 mm, and for ratios M_{Ed}/M_{Rd} of 0.8 and 1.0. The ratio 0.8 is an example, as the ratio M_{Ed}/M_{Rd} is included in the formula given for n/n_f.

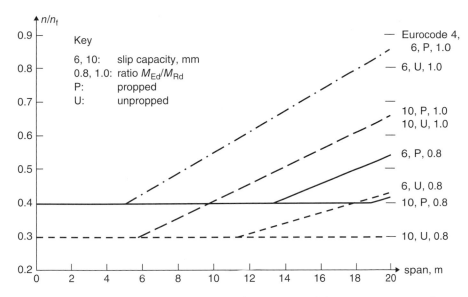

Figure 3.21 Current and proposed rules for minimum degree of shear connection, n/n_f, for $f_y = 355\,\text{MPa}$

For unpropped construction, the minimum degree of shear connection also depends on the ratio of the bending moment in the beam from the weight of the concrete slab to $M_{pl,Rd}$. It is assumed that the slabs supported by the beams are cast unpropped, as they would have to be if the beams are unpropped. With propping, the whole of the load applies shear to the connection, but unpropped beams carry the wet concrete without help from the shear connection, so fewer connectors need be provided.

Where the cross-section of a beam is governed by, for example, resistance to vertical shear, it will have surplus resistance to bending, so that $M_{Ed} < M_{Rd}$. If the shear connection provided is sufficient for resistance M_{Rd}, there will be less inelastic behaviour where M_{Ed} is lower, and less slip required of the connectors. This enables the minimum degree of shear connection to be reduced. Figure 3.21 shows that for some situations, less than half the number of connectors now required could be provided. It may then be necessary to allow for increased deflections caused by slip (Section 3.7.1).

Use of the proposed limits for short spans could lead to a spacing of connectors that exceeds other rules on maximum spacing. They are provided to ensure that sudden longitudinal shear failures do not occur by 'unzipping' of the shear connection, starting from one end of the beam, and also because most other design rules assume that shear transfer across the steel–concrete interface is continuous, rather than intermittent.

Where sheeting runs parallel to an adjacent beam (Figure 2.15), the proposed rule is shown by the line labelled 6, U, 1.0 in Figure 3.21.

Most recommendations in Couchman (2015) are likely to be included in the next Eurocode 4. The report covers the subjects omitted from this brief outline, and further information on partial shear connection is available (Johnson, 2012).

Other types of shear connector No design data are given in EN 1994-1-1 for shear connectors other than headed studs, such as those described in Section 2.4.2. Some of them, such as bars with hoops, have little slip capacity, and some others are at present treated as non-ductile, through lack of sufficient evidence of ductility from tests.

Where partial shear connection is used and the connectors are 'ductile', the bending resistance of cross-sections in Class 1 or 2 may be found by plastic theory (Equation 3.56). Otherwise, elastic theory is required, which is more complex and requires more connectors. Also, ductile connectors may be spaced uniformly along a critical length, whereas for non-ductile connectors, the spacing must be based on elastic analysis for longitudinal shear. Design of shear connection for bridges is based more on elastic behaviour than it is for buildings, because of the repetitive loading and the risk of fatigue failure, so there is less of a barrier to the use of non-ductile connectors.

3.6.3   Transverse reinforcement

The reinforcing bars shown in Figure 3.22 are longitudinal reinforcement for the concrete slab, to enable it to span between the beam shown and those either side of it, but they also enhance the resistance to longitudinal shear in the beam at vertical cross-sections such as B–B. Bars provided for that purpose are known as 'transverse reinforcement', as their direction is transverse to the axis of the composite beam. Like stirrups in the web of a reinforced concrete T-beam, they supplement the shear strength of the concrete, and their behaviour can be represented by a truss analogy.

The design rules for these bars are extensive, as account has to be taken of many types and arrangements of shear connectors, of haunches, of the use of precast or composite slabs, and of interaction between the longitudinal shear stress on the section considered, v, and the transverse bending moment, shown as M_s in Figure 3.22. The loading on the slab also causes vertical shear stress on surfaces such as B–B; but this is usually so much less than the local longitudinal shear stress, that it can be neglected.

The word 'surface' is used here because EFGH in Figure 3.22, although not a plane, is another potential surface of shear failure. In practice, the rules for minimum height of shear connectors ensure that in slabs of uniform thickness, surfaces of type B–B are more critical; but this may not be so for haunched slabs, considered later.

In a symmetrical T-beam, the design longitudinal shear per unit length ('shear flow', denoted v_L) for surface B–B is approximately half of that for the shear connectors, which must be transmitted across surface EFGH. For an L-beam (Figure 3.23) or where the flange of the steel beam is wide, the more accurate expressions should be used:

$$v_{L,BB} = 2v_L b_1/b \quad \text{and} \quad v_{L,DD} = 2v_L b_2/b \tag{3.66}$$

where $2v_L$ is the design shear flow for the shear connection and b is the effective width of the concrete flange.

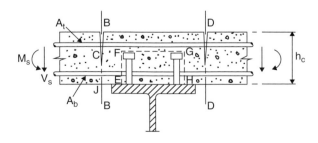

Figure 3.22 Surfaces of potential failure in longitudinal shear

Figure 3.23 T-beam with asymmetrical concrete flange

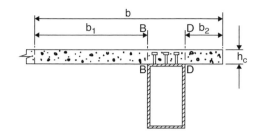

Effective area of reinforcement For planes such as B–B in Figure 3.22, the effective area of transverse reinforcement per unit length of beam, A_e, is the whole of the reinforcement that is fully anchored on both sides of the plane (i.e. able to develop its yield strength in tension). This is so even where the top bars are fully stressed by the bending moment M_s, because this tension is balanced by transverse compression, which enhances the shear resistance in the region CJ by an amount at least equivalent to the contribution the reinforcement would make, in the absence of transverse bending.

3.6.3.1 Design rules for transverse reinforcement in solid slabs

For this subject, EN 1994-1-1 refers to EN 1992-1-1 for concrete structures. Its rules for transverse reinforcement are based on a truss analogy in which the slope of the diagonal members, θ_f, may be chosen, within defined limits, by the designer. Before this choice was introduced, the angle 45° was normally used, and is used here.

Part of a composite beam is shown in plan in Figure 3.24. The truss model for transverse reinforcement is illustrated by triangle ACE, in which CE represents the reinforcement for a unit length of the beam, of area A_e, and v_L is the design shear flow for a cross-section of type B–B (Figure 3.22) above the edge of the steel flange (shown by dashed lines). The shear flow for the connectors, $2v_L$, is applied at point A, and is transferred by concrete struts AC and AE, at 45° to the axis of the beam.

The strut force is balanced at C by compression in the slab and tension in the reinforcement. The model fails when the reinforcement yields. The tensile force in it is equal to the shear on a plane such as B–B caused by the force v_L. The design equation gives the minimum area A_e of reinforcement:

$$v_{L,Ed} < v_{L,Rd} = A_e f_{sd} \tag{3.67}$$

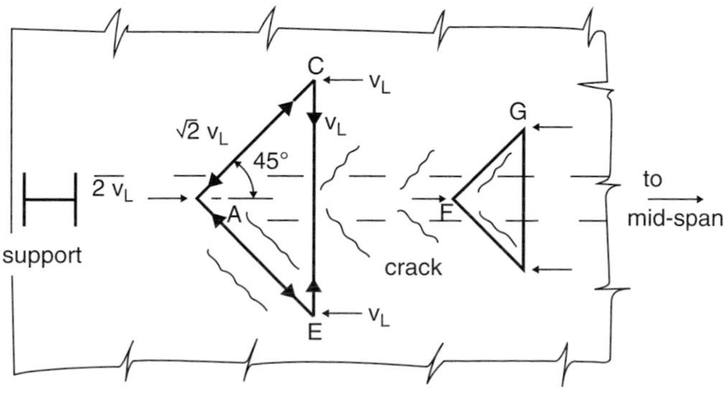

Figure 3.24 Truss model for transverse reinforcement

Another requirement is that concrete struts such as AC do not fail in compression. Their design compressive stress is given in EN 1992-1-1 as

$$0.6(1 - f_{ck}/250)f_{cd} \tag{3.68}$$

with f_{ck} in N/mm². The width of strut per unit length of beam is $1/\sqrt{2}$, and the force in the strut is $\sqrt{2}\,v_{L,Ed}$, so this condition is

$$v_{L,Ed} < 0.6(1 - f_{ck}/250)f_{cd}h_f/2 \tag{3.69}$$

where h_f (the Eurocode 2 symbol) is the slab thickness, for which Eurocode 4 uses h_c. This rule is for normal-density concrete. For lightweight concrete with oven-dry density $\rho\,$kg/m³, the expression is replaced by

$$v_{L,Ed} < 0.5(0.4 + 0.6\rho/2200)(1 - f_{lck}/250)f_{lcd}h_f/2 \tag{3.70}$$

This allows for the reduction in the ratio E_{cm}/f_{lck} as the density of the concrete reduces. The designer's choice of θ_f will depend on whether failure of the reinforcement or the concrete strut is the more critical.

These results are assumed to be valid whatever the length of the notional struts in the slab (e.g. FG), and rely to some extent on the shear flow v_L being fairly uniform within the shear span, because the reinforcement associated with the force $2v_L$ at A is in practice provided at cross-section A, not at some point between A and mid-span. The type of cracking observed in tests where shear failure occurs, shown in Figure 3.24, is consistent with the model.

Haunched slabs Further design rules are required for the transverse reinforcement in haunches of the type shown in Figure 2.1b. These are not discussed here. Haunches encased in thin steel sheeting are considered below.

3.6.3.2 Transverse reinforcement in composite slabs

Where profiled sheeting spans in the direction transverse to the span of the beam, as shown in Figure 3.15a, it can be assumed to be effective as bottom transverse reinforcement for a beam with sheets continuous over the steel top flange. Where they are not, as in the figure, the effective area of sheeting depends on how the ends of the sheets are attached to the top flange.

Where studs are welded to the flange through the sheeting, resistance to transverse tension is governed by local yielding of the thin sheeting around the stud. The design bearing resistance of a stud with a weld collar of diameter d_{do} in sheeting of thickness t is given in Eurocode 4 as

$$P_{p,b,Rd} = k_\varphi d_{do} t f_{yp,d} \tag{3.71}$$

where

$$k_\varphi = 1 + a/d_{do} \le 6.0 \tag{3.72}$$

$f_{yp,d}$ is the yield strength of the sheeting, and dimension a is shown in Figure 3.25. The formula corresponds to the assumption that yielding of the sheet occurs in direct tension along BC and in shear, at stress $f_{yp,d}/2$, along AB and CD. The transverse tension passes

Figure 3.25 Bearing resistance of profiled sheeting, acting as transverse reinforcement

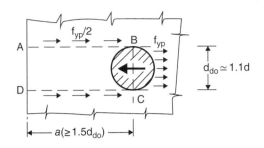

through the flange and into the sheet on its other side through another stud anchor. For pairs of studs at longitudinal spacing s (one near each edge of the steel flange), Equation 3.67 is replaced by

$$v_{L,Ed} < A_e f_{sd} + P_{p,b,Rd}/s \tag{3.73}$$

The resulting reduction in the required area A_e is significant in practice where conventional studs are used; but small-diameter shot-fired pins are less effective, because of the limit $k_\varphi \leq 6$.

The shear $v_{L,Ed}$ in the studs is transferred from the sheeting to the steel flange below. The studs also transfer shear from the sheeting to the concrete slab above, and so improve the resistance of the slab to longitudinal shear. An example is given in Section 3.4.3. These two transfers occur in the cross-sections of a stud on opposite sides of the sheeting, and are assumed in design not to interact.

A stud is provided mainly to transfer longitudinal shear in the beam, so there is a third shear transfer, at right angles to the other two, which occurs in the cross-sections of the stud both below and above the sheeting. The transverse shear $P_{p,b,Rd}$ is usually so much lower than the shear resistance of a stud that interaction between the longitudinal and transverse forces can be neglected.

Where the span of the sheeting is parallel to that of the beam, transverse tension causes the corrugations to open out, so the contribution of the sheeting to transverse reinforcement is ignored.

Level of mesh reinforcement in a composite slab   Section 3.6.4 refers to a detailing rule that 'bottom transverse reinforcement' should be at least 30 mm below the head of a stud. Eurocode 4 is unclear about its application where sheeting is present. Some designers provide steel mesh reinforcement resting on the sheeting. Applying the 30-mm rule would then need the studs to be longer than is otherwise required. To avoid this, the sheeting is deemed to be equivalent to 'bottom' transverse reinforcement. Mesh is often provided to control crack width, and then needs to be 'top' transverse reinforcement, as mesh resting on sheeting is too low to be effective. It is unusual to provide two layers of mesh.

The resistance to longitudinal shear of studs in troughs of sheeting is affected by the position of the mesh. One series of tests (Smith and Couchman, 2010) found that the resistance of studs in the 'favourable' position in 60-mm trapezoidal sheeting was 31% higher with mesh resting on the decking than when it was near the top of the slab. It would be difficult to allow for this benefit within the rules in the current Eurocode 4.

3.6.4 Detailing rules

Where shear connectors are attached to a steel flange, there will be transverse reinforcement, and there may be a haunch (local thickening of the slab, as in Figure 2.1b) or profiled steel sheeting. No reliable models exist for the three-dimensional state of stress in such a region, even in the elastic range, so the details of the design are governed by arbitrary rules of proportion, based essentially on experience.

Several of the rules given in EN 1994-1-1 are shown in Figure 3.26. The left-hand half shows profiled sheeting that spans transversely, and the right-hand half shows a haunch.

The minimum dimensions for the head of a stud, the rule $h \geq 3d$, and the minimum projection above bottom reinforcement, are to ensure sufficient resistance to uplift. The 40 mm dimension shown is reduced to 30 mm where there is no haunch. Its application to composite slabs is discussed in Section 3.6.3.2.

The rule $d \leq 2.5t_f$ is to avoid local failure of the steel flange, caused by load from the connector. For repeated (fatigue) loading, the limit to d/t_f is reduced to 1.5.

The 50-mm side cover to a connector and the $\leq 45°$ rule are to prevent local bursting or crushing of the concrete at the base of the connector; and the 20-mm dimension to the flange tip is to avoid local over-stress of the flange and to protect the connector from corrosion. For haunches of other shapes, the rules for lying studs (Section 2.5.2) may apply.

The minimum centre-to-centre spacing of stud connectors of diameter d is $5d$ in the longitudinal direction, $2.5d$ across the width of a steel flange in solid slabs, and $4d$ in composite slabs. These rules are to enable concrete to be properly compacted, and to avoid local overstress of the slab.

Where precast floor slabs that extend across a steel flange are used (Section 3.10.1), the studs project through holes in the slabs, which are concreted later. To reduce the number of holes, studs can be arranged in groups. The length of the holes is then governed by the $5d$ rule on longitudinal spacing (Figure 3.27a), which shows that two rows of studs need a hole over 160 mm long. The steel flange is shown by dashed lines. Tests have found (Spremic and Markovic, 2016) that if the spacing is reduced to $3d$ (Figure 3.27b), the

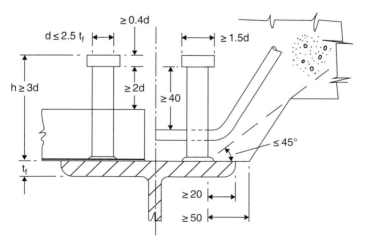

Figure 3.26 Detailing rules for shear connection

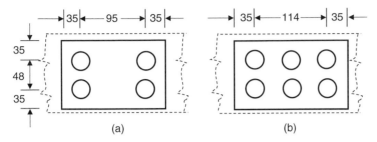

Figure 3.27 Plan of holes in a precast slab and 19-mm stud connectors from the supporting beam with longitudinal spacings of (a) 5*d*, and (b) 3*d*

shear resistance per stud is reduced by about 30%, so three studs at 3*d* spacing have a resistance similar to that of two studs at 5*d* spacing, and need a longer hole. Long holes also affect the detailing of the transverse reinforcement.

The maximum longitudinal spacing of connectors in buildings is limited to the lesser of 800 mm and six times the total slab thickness, because the transfer of shear is assumed in design to be continuous along the span, and also to avoid excessive uplift.

All such rules relevant to stresses should in principle give ratios of dimensions. Where actual dimensions are given, there may be an implied assumption (e.g. that studs are between 16 mm and 22 mm in diameter), or it may be that corrosion or crack widths are relevant.

3.7   Stresses, deflections and cracking in service

A composite beam is usually designed first for ultimate limit states. Its behaviour in service must then be checked. For a simply-supported beam, the most critical service-ability limit state is usually excessive deflection, which can govern the design where unpropped construction is used. Floor structures subjected to dynamic loading (e.g. as in a dance hall or gymnasium) are also susceptible to excessive vibration (Section 3.9).

The width of cracks in concrete needs to be controlled in web-encased beams, and in hogging regions of continuous beams (Section 4.2.5).

Excessive stress in service is not itself a limit state. It may, however, invalidate a method of analysis (e.g. linear-elastic theory) that would otherwise be suitable for checking compliance with a serviceability criterion. Accurate determination of stresses by elastic analysis is so difficult that Eurocode 4 does not set limits to them. Those in the rules for crack-width control represent limits to elastic strains. Where elastic analysis is used, with appropriate allowance for shear lag and creep, the policy is to modify the results, where necessary, to allow for yielding of steel and, where partial shear connection is used, for excessive slip.

If yielding of structural steel occurs in service, in a simply-supported composite beam for a building, it will be in the bottom flange, near mid-span. It is most likely when the ratio of the variable load to the permanent load, q_k/g_k, is low; unpropped construction is used; and the ratio of the design resistance to bending for ultimate limit state to the yield moment, known as the *shape factor*, is high. The shape factor is:

$$Z = M_{pl,Rd}/M_{el,Rk} \tag{3.74}$$

where $M_{el,Rk}$ is the bending moment at which yield of steel first occurs. For sagging bending, and assuming γ_A for steel is 1.0, Z is typically between 1.25 and 1.35 for propped construction, but can rise to 1.45 or above for unpropped construction.

Deflections are usually checked for the characteristic combination of actions, given in Equation 1.8. So, for a beam designed for distributed loads g_k and q_k only, the ratio of design bending moments (ultimate/serviceability) is

$$\mu = \frac{1.35g_k + 1.5q_k}{g_k + q_k} \tag{3.75}$$

The range of this ratio is narrow; for example, from 1.42 at $q_k = 0.8\ g_k$ to 1.45 at $q_k = 2.0\ g_k$.

From these expressions, the stress in steel in service will reach or exceed the yield stress if $Z > \mu$. The values given above show that this is unlikely for propped construction, but could occur for unpropped construction.

Where the bending resistance of a composite section is governed by local buckling, as in a Class 3 section, elastic section analysis is used for ultimate limit states, and then stresses and/or deflections in service are less likely to influence design.

As shown below, elastic analysis of a composite section is more complex than plastic analysis, because account has to be taken of the method of construction and of the effects of creep. In principle, the following three types of loading then have to be considered separately:

- load carried by the steel beam;
- short-term load carried by the composite beam, without creep; and
- long-term load carried by the composite beam, with creep.

However, as an approximation, the composite beam may be analysed for its whole loading using a reduced value for the creep coefficient.

3.7.1 Elastic analysis of composite sections in sagging bending

It is assumed first that full shear connection is provided, so that the effect of slip can be neglected. All other assumptions are as for the elastic analysis of reinforced concrete sections by the method of transformed sections. The algebra is different because the flexural rigidity of the steel section alone is much greater than that of reinforcing bars.

For generality, the steel section is assumed to be asymmetrical (Figure 3.28) with cross-sectional area A_a, second moment of area I_a, and centre of area distance z_g below the top surface of the concrete slab, which is of uniform overall thickness h_t and effective width b_{eff}. The effective slab thickness, h_c, is the net thickness for transverse sheeting and the gross thickness for parallel sheeting. Where there is no small top rib, as in Figure 3.28, these thicknesses are the same.

The modular ratio for short-term loading is $n_0 = E_a/E_{cm}$, where the subscript 'a' refers to structural steel and E_{cm} is the mean value of the elastic modulus for concrete, given in EN 1992-1-1. For long-term loading, a value $3n_0$ is a good approximation. For simplicity, a single value $2n_0$ is permitted for use with both types of loading. From here onwards, the symbol n is used for whatever modular ratio is appropriate, so it is defined by

$$n = E_a/E_{c,eff} \tag{3.76}$$

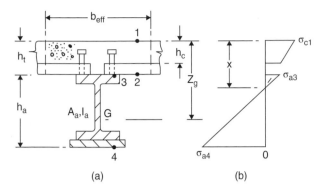

Figure 3.28 Elastic analysis of cross-section of composite beam in sagging bending

where $E_{c,eff}$ is the relevant effective modulus for the concrete. (Note: the symbol n is also used for number of shear connectors.)

It is usual to neglect reinforcement in compression, concrete in tension, and also concrete between the ribs of profiled sheeting, even when the sheeting spans longitudinally. The condition for the neutral-axis depth x to be less than h_c is

$$A_a(z_g - h_c) < b_{eff}h_c^2/(2n) \tag{3.77}$$

The neutral-axis depth is then given by the usual 'first moments of area' equation,

$$A_a(z_g - x) = b_{eff}x^2/(2n) \tag{3.78}$$

and the second moment of area, in 'steel' units, by

$$I = I_a + A_a(z_g - x)^2 + b_{eff}x^3/(3n) \tag{3.79}$$

If Condition 3.77 is not satisfied, then the neutral-axis depth exceeds h_c, as in Figure 3.28, and is given by

$$A_a(z_g - x) = b_{eff}h_c(x - h_c/2)/n \tag{3.80}$$

The second moment of area is

$$I = I_a + A_a(z_g - x)^2 + (b_{eff}h_c/n)\left[h_c^2/12 + (x - h_c/2)^2\right] \tag{3.81}$$

In global analyses, it is sometimes convenient to use values of I based on the uncracked composite section. The values of x and I are then given by Equations 3.80 and 3.81 above, whether x exceeds h_c or not. In sagging bending, the difference between the 'cracked' and 'uncracked' values of I is usually small.

Stresses due to a sagging bending moment M are normally calculated in concrete only at level 1 in Figure 3.28, and in steel at levels 3 and 4. These stresses are, with tensile stress positive:

$$\sigma_{c1} = -Mx/(nI) \tag{3.82}$$

$$\sigma_{a3} = M(h_t - x)/I \tag{3.83}$$

$$\sigma_{a4} = M(h_a + h_t - x)/I \tag{3.84}$$

Deflections Deflections are calculated by the well-known formulae from elastic theory, using Young's modulus for structural steel. For example, the deflection of a simply-supported composite beam of span L due to distributed load q per unit length is

$$\delta_c = 5qL^4/(384E_aI) \tag{3.85}$$

Where the shear connection is partial (i.e. $\eta < 1$, with η from Equation 3.53), the increase in deflection due to longitudinal slip depends on the method of construction. The total deflection δ was given approximately in BS 5950 (BSI, 1990a) as

$$\delta = \delta_c[1 + k(1 - \eta)(\delta_a/\delta_c - 1)] \tag{3.86}$$

with $k = 0.5$ for propped construction and $k = 0.3$ for unpropped construction, where δ_a and δ_c are respectively the deflection for whole load acting on the steel beam alone, and on the composite beam with full shear connection.

This expression is obviously correct for full shear connection ($\eta = 1$), and gives a slightly low result when $\eta = 0$. A modified form of it is likely to be included in the next Eurocode 4 (Lawson et al., 2017).

EN 1994-1-1, unlike BS 5950, allows this increase in deflection to be ignored in unpropped construction where:

- either $\eta \geq 0.5$ or the forces on the connectors found by elastic analysis do not exceed P_{Rd}, where P_{Rd} is their design resistance, and
- for slabs with ribs transverse to the beam, the height of the ribs does not exceed 80 mm.

The arbitrary nature of these rules arises from the difficulty of predicting deflections accurately.

3.7.2 The use of limiting span-to-depth ratios

Calculations using formulae like those derived above are not only long; they are also inaccurate. It is almost as much an art as a science to predict during design the long-term deflection of a beam in a building. It is possible to allow in calculations for some of the factors that influence deflection, such as creep and shrinkage of concrete, but there are others that cannot be quantified. In developing the limiting span–depth ratios for the 1972 British code for concrete structures, nine reasons were identified why deflections of reinforced concrete beams in service were usually less than those calculated by the designers. The theoretical span–depth ratios were increased by 36% to allow for them. Many of the reasons apply equally to composite beams, the most significant of them being the variations in the elasticity, shrinkage and creep properties of the concrete, the stiffening effect of finishes, and restraint and partial fixity at the supports.

The other problem is the difficulty of defining when a deflection becomes 'excessive'. In practice, complaints can arise from the cracking of plaster on partition walls, which can occur when the deflection of the supporting beam is as low as span/800. For partitions and in-fill panels generally, the relevant deflection is that which takes place after their construction, from imposed load and from creep under permanent load, which continues for several years after construction.

Change of slope of a floor, rather than deflection, can be a problem in some laboratory or commercial buildings. For each situation, the design criteria should be agreed between the designer and the future users.

3.8 Effects of shrinkage of concrete and of temperature

In the fairly dry environment of a building, an unrestrained concrete slab can be expected to shrink by 0.03% of its length (3 mm in 10 m) or more. If its ends were fully restrained, this would appear as a tensile strain of about treble the cracking strain, so cracking would occur.

In a composite beam, the slab is restrained by the steel member, which exerts a tensile force on it, through the shear connectors near the free ends of the beam, so its apparent shrinkage is less than the 'free' shrinkage. In beams, these forces on the shear connectors act in the opposite direction to those due to the loads, and so can be neglected in design. This does not apply to cantilevers, where longitudinal shear from shrinkage and from loading act in the same direction.

The stresses due to shrinkage develop slowly, and so are reduced by creep of the concrete, but the increase they cause in the deflection of a composite beam may be significant. An approximate and usually conservative rule of thumb for estimating this deflection in a simply supported beam is to take it as equal to the long-term deflection due to the weight of the concrete slab, excluding finishes, acting on the *composite* member.

In the beam studied in Section 3.11, this rule gives an additional deflection of 5.4 mm, whereas the calculated long-term deflection due to a shrinkage of 0.03% (with a modular ratio $n = 20.2$) is 5.9 mm.

Calculations for the effects of shrinkage have usually assumed that the shortening of a slab is uniform over its depth (e.g. the partial-interaction theory in Appendix A). Where profiled sheeting is present, it seals the under-side of the slab, and the shrinkage strain varies over its depth. This causes an additional sagging curvature, in both slabs and composite beams, which increases deflections.

There are proposals to include guidance or new rules for this in the next Eurocode 4. At present it says, in clause 7.3.1(8), '… the effect of curvature due to shrinkage of normal weight concrete need not be included where the ratio of span to overall depth of the beam is not greater than 20'. This rule may be over-optimistic, particularly for long-span composite beams with composite slabs.

The prediction of total final free shrinkage strain in accordance with EN 1992-1-1 can be unnecessarily complex for composite members in buildings. Annex C of Eurocode 4 therefore gives approximate values of shrinkage strain for two densities of concrete and two environments, which range from 0.02% to 0.05%. It warns against their use 'where shrinkage is expected to take exceptional values', and should probably also refer to composite slabs.

Where the slab is colder than the steel member of a beam, the effect is similar to that of shrinkage, and causes deflection. Such differential temperatures rarely occur in buildings. They are important in beams for bridges and are treated in EN 1994-2.

3.9 Vibration of composite floor structures

In British Standard 6472-1, *Guide to evaluation of human exposure to vibration in buildings* (BSI, 2008) the performance of a floor structure is considered to be satisfactory when the probability of annoyance to users of the floor, or of complaints from them about interference with activities, is low. There can be no simple specification of the dynamic properties that would make a floor structure 'serviceable' in this respect, because the local causes of vibration, the type of work done in the space concerned, and the psychology of its users are all relevant.

A good introduction to this complex subject is available (Wyatt, 1989). It and BS 6472 provided much of the basis for the following introduction to vibration design, which is limited to the situation in the design example – a typical floor of an office building, shown in Figure 3.1.

With the increasing use of floors of longer span, further guidance was needed for sensitive floors, such as those for operating theatres. This led to the report *Design of Floors for Vibration: A New Approach* (Smith et al., 2009). Wyatt's methods are extended, their accuracy is improved, and worked examples are given using both rolled and cold-formed steel members. Comparisons are made with EN and ISO standards, and with the requirements of the National Health Service for hospitals. The use of finite-element analysis is recommended, and guidance is given on its use.

Sources of vibration excitation Vibration from external sources, such as highway or rail traffic, is rarely severe enough to influence design. If it is, the building should be isolated at foundation level.

Vibration from machinery in the building, such as lifts and travelling cranes, should be isolated at or near its source. In the design of a floor structure, it should be necessary to consider only sources of vibration on or near that floor. Near gymnasia or dance floors, the effects of rhythmic movement of groups of people can be troublesome, but in most buildings only two situations need be considered:

- people walking across a floor with a pace frequency between 1.4 Hz and 2.5 Hz; and
- an impulse, such as the effect of the fall of a heavy object.

Typical reactions by floors to pedestrian traffic have been analysed by Fourier series. The fundamental component has an amplitude of about 240 N. The second and third harmonics are smaller, but are relevant to design. Fundamental natural frequencies of floor structures (f_0) often lie within the frequency range of third harmonics (4.2–7.5 Hz). The number of cycles of this harmonic, as a person walks across the span of a floor, can be sufficient for the amplitude of forced vibration to approach its steady-state value.

Pedestrian movement causes little vibration of floor structures with f_0 exceeding about 7 Hz, but these should be checked for the effect of an impulsive load. The consequences that most influence human reactions are then the peak vertical velocity of the floor, which is proportional to the impulse, and the time for the vibration to decay, which increases with reduction in the damping ratio of the floor structure.

Human reaction to vibration The model of BS 6472-1 for the human response to vibration in buildings uses vibration dose values (VDVs), and provides less conservative results than earlier methods based on response to continuous vibration. The method is based on measured or predicted acceleration-time data.

Human sensitivity to vibration depends on whether the acceleration is horizontal or vertical, and also on the frequency. This is allowed for by *weighting factors*, given by curves covering frequencies from 0.1 to 100 Hz. The highest weightings, about 1.0, are for frequencies between 4 and 12 Hz for vertical vibration, and between 1 and 2 Hz for horizontal vibration. Each curve corresponds to an approximately uniform level of human response.

A VDV is essentially a time integral of acceleration, taken over a total time T, either a day or a night. It is given by

$$\text{VDV} = \left[\int_0^T (a_w(t)^4) dt \right]^{1/4} \tag{3.87}$$

where T is in seconds and a_w is the weighted acceleration in m/s^2, so that a VDV has units m/s$^{1.75}$. Its value depends more on peak accelerations than on their duration.

The probability of adverse comment from users of a building on their exposure to vibration is found to depend on the VDV. Guidance can only be given as ranges of values, because some humans are more vibration-sensitive than others. An example of these ranges, from BS 6472-1 and from Smith et al. (2009), is given in Table 3.3 for residential buildings.

As people tend to be more tolerant of vibration and noise in offices and workshops than in residential buildings, multiplying factors are applied to the ranges in Table 3.3. BS 6472-1 gives these as 2 for offices and 4 for workshops.

3.9.1 Prediction of fundamental natural frequency

In composite floors that need checking for vibration, damping is sufficiently low for its influence on natural frequencies to be neglected. For free elastic vibration of a beam or one-way slab of uniform section, the fundamental natural frequency is

$$f_0 = K(EI/mL^4)^{1/2} \tag{3.88}$$

where $K = \pi/2$ for simple supports, $K = 3.56$ for members with both ends fixed against rotation, and m is the mass of the loading resisted by the member.

Values for other end conditions and multispan members are given by Wyatt. The relevant flexural rigidity is EI (per unit width, for slabs), L is the span, and the vibrating mass is m_b per unit length (beams) or m_s per unit area (slabs). Concrete in slabs should normally be assumed to be uncracked, and the dynamic modulus of elasticity should be used for concrete, in both beams and slabs. This modulus, E_{cd}, is typically about 8 kN/mm^2 higher than the static modulus for normal-density concrete, and 3 to 6 kN/mm^2 higher for lightweight-aggregate concretes of dry density not less than 1800 kg/m^3. For composite beams in sagging bending, approximate allowance for dynamic effects can be made by increasing the value of I by 10%. Smith et al. (2009) give values for E_{cd}, but they take no account of the strength of the concrete.

Unless a more accurate estimate can be made, the mass m is usually taken as the mass of the characteristic permanent load plus 10% of the characteristic variable load.

Table 3.3 Vibration dose ranges that might result in various probabilities of adverse comment

Place and time	Low probability of adverse comment	Adverse comment possible	Adverse comment probable
Residential buildings, 16-hour day	0.2 to 0.4	0.4 to 0.8	0.8 to 1.6

A convenient method of calculating f_0 is first to find the mid-span deflection, δ_m say, caused by the weight of the mass m. For simply-supported members of uniform section, this is

$$\delta_m = 5mgL^4/(384EI)$$

Substitution for m in Equation 3.88 gives

$$f_0 = 17.8/\sqrt{\delta_m} \qquad (3.89)$$

with f_0 in Hz and δ_m in millimetres.

Equation 3.89 is useful for a beam or slab considered alone. However, in a typical floor, with composite slabs continuous over a series of parallel composite beams, the total deflection (δ, say) is the sum of deflections δ_s for the slab relative to the beams that support it, and δ_b for the beams. A good estimate of the fundamental natural frequency is then given by

$$f_0 = 17.8/\sqrt{\delta} \qquad (3.90)$$

It follows from Equations 3.89 and 3.90 that

$$\frac{1}{f_0^{\,2}} = \frac{1}{f_{0s}^{\,2}} + \frac{1}{f_{0b}^{\,2}} \qquad (3.91)$$

where f_{0s} and f_{0b} are the natural frequencies for the slab and the beam, respectively, each considered alone. Similar expressions for members with other end conditions and for other vibration modes are given by Smith et al. (2009).

For a single-span layout of the type shown in Figure 3.1, each beam vibrates as if simply-supported, so the length L_{eff} of the vibrating area will be similar to the span, L. The width S of the vibrating area may exceed the beam spacing, s. A cross-section through this area is likely to be as shown in Figure 3.29, with most spans of the composite slab vibrating as if fixed-ended. It follows from Equation 3.88 that:

- for the beam:

$$f_{0b} = (\pi/2)(EI_b/m_bL^4)^{1/2} \qquad (3.92)$$

- for the slab:

$$f_{0s} = 3.56(EI_s/m_ss^4)^{1/2} \qquad (3.93)$$

where m_b and m_s are as defined above, s is the spacing of the beams, and subscripts b and s mean beam and slab, respectively.

deflections not to scale

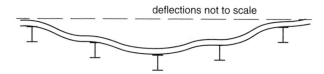

Figure 3.29 Cross-section of vibrating floor structure showing typical fundamental mode

3.9.2 Response of a composite floor to pedestrian traffic

An outline is now given of a method for checking the response to pedestrian traffic. It is assumed that the floor reaches its steady state of damped vibration under harmonic excitation from people walking at frequencies between 1.4 and 2 Hz, and that for the floor, $f_0 > 3$ Hz, to avoid resonance with the first harmonic, of typical amplitude 240 N. There is a worked example in Section 3.11.3.2.

3.9.2.1 Modal mass

The modal mass is the total mass of the vibrating area of floor:

$$M = m_b L_{eff} S \tag{a}$$

For composite floors with shallow decking, in a frame with single-span beams at spacing s, the effective length is given by Smith et al. (2009) as

$$L_{eff} = 1.09[EI_b/(m_b f_0^2)]^{1/4} \quad \text{with } L_{eff} \leq L \tag{b}$$

The effective floor width is slightly affected by f_0, which is found for the floor in Figure 3.1 to be about 6 Hz (Section 3.11.3). For this frequency, Smith et al. (2009) gives the width as

$$S = 0.7(1.15)^{n-1}\left[EI_s/(m_s f_0^2)\right]^{1/4} \quad \text{with } S \leq ns$$

where n is the number of bays of the floor slab, not exceeding four. Here, $n = 4$ is assumed, so that

$$S = 1.065\left[EI_s/(m_s f_0^2)\right]^{1/4} \tag{c}$$

3.9.2.2 Acceleration of the floor

For the general response of the floor to frequent pedestrian traffic, the root-mean-square value of the vertical acceleration is given by Smith et al. (2009). The expression includes a resonance build-up factor, which increases with the walking path length, and can conservatively be taken as 1.0. It also includes two mode shape factors which, for the simply-supported spans in the example here, can be taken as 1.0. Then:

$$a_{w,rms} = 0.1QW/(2.83M\zeta) \tag{d}$$

where Q is the average weight of a person, taken as 746 N, M is the modal mass in kg, ζ is the critical damping ratio, and $a_{w,rms}$ is the acceleration (m/s²) modified by the weighting factor W for vertical vibration. For $f_0 \approx 6$ Hz, $W = 1.0$.

The damping ratio should normally be taken as 0.03 for open-plan offices with composite floors, although Wyatt (1989) reported values as low as 0.015 for unfurnished floors. Hence,

$$a_{w,rms} = 0.1 \times 746 \times 1/(2.83 \times 0.03M) = 879/M \tag{e}$$

The complex response of a floor to pedestrian traffic has been tested and analysed (Ellis, 2001). Account has to be taken of the range of ways in which people walk. This leads to a representative VDV, for use in serviceability verification, given in Smith et al. (2009) as

$$VDV = 0.68a_{w,rms}(n_a T_a)^{1/4} \tag{f}$$

where T_a is the duration of a vibration-producing activity and n_a is the number of repetitions of the activity in the exposure period T (e.g. a 16-hour day).

3.10   Hollow-core and solid precast floor slabs

Section 3.10 gives the changes in the design methods explained earlier, that are needed where the floor slabs are fully or partially precast, rather than composite or cast in situ. The slabs are normally designed as simple spans between the supporting beams, with the width of cracks above the beams controlled by mesh reinforcement. They are usually rectangular in plan, with width that depends on the methods of transportation and erection to be used. Their use speeds up construction and provides earlier access to the floor below, as propping can be avoided. There are three main types of floor construction:

- Solid precast slabs that act as permanent formwork for in situ concrete ('topping'), designed to act compositely with it for all loading applied after setting of the concrete. Their span/depth ratio is obviously higher than that of the finished floor. Their thickness is typically 75–100 mm, so they are best suited to short spans.
- Full-thickness solid slabs without in situ topping. They are relatively heavy and not widely used.
- Full- or partial-thickness slabs with longitudinal voids, known as hollow-core slabs. They usually have longitudinal prestress, and are cast in long lines around pre-tensioned tendons. After hardening they are cut into the lengths required. There may be no other reinforcement. They are typically 150–250 mm deep, and are used for spans up to at least 9 m, as their strength/weight ratio is high.

The scope here is limited to slabs supported on the top flange of the steel beam. They can be supported on an extended bottom flange, as can composite floor slabs, which are treated in Section 3.12, with an example in Section 3.13. Hollow-core slabs with topping are shown in Figure 3.30. Prestressing tendons and other reinforcement are not shown.

The precast concrete is typically Class C50/60. The in situ concrete that surrounds the shear connectors and may provide additional thickness is likely to be of lower grade, such as C30/37. Where it has to fill narrow gaps at the longitudinal edges of the slabs, its aggregate size may be limited to 10 mm.

The current EN 1994-1-1 does not refer to precast slabs. There is a section 'Precast concrete slabs in composite bridges' in EN 1994-2, but it does not include hollow-core slabs. Information on design using precast slabs is widely available, both from their manufacturers and in published form (Rackham et al., 2006; Way et al., 2007; Couchman, 2014). Composite action in beams supporting hollow-core slabs differs in several ways

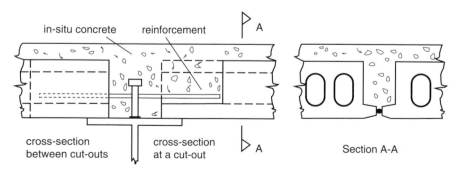

Figure 3.30 Precast hollow-core slab with in situ topping supported on a steel top flange

from that in beams with solid or precast slabs (Leskela, 2017), and the Eurocode rules on limits to partial shear connection are not appropriate. The notes here are based mainly on these publications.

3.10.1 Joints, longitudinal shear and transverse reinforcement

This section is mostly on hollow-core slabs, as its application to solid slabs is straightforward. Allowing for tolerances, the nominal bearing length for a slab on a flange of a supporting steel beam is typically 75 mm. The gap between slabs must be wide enough for stud welding, if done on site, and for concrete to be placed and compacted around shear connectors. For a single row of studs, the required minimum flange width is about 230 mm.

For resistance of the beam to longitudinal shear, reinforcement transverse to the beam is needed. The bars are typically 12 or 16 mm in diameter, with an anchorage length of about 500 mm in in situ concrete. The ends of each hollow-core slab need to be cut back, as shown in Figure 3.30, to enable the bars to be below the heads of the studs, and the in situ concrete to be placed.

The design of the bars, usually placed in alternate cores, follows the usual rules for resistance to longitudinal shear. For solid slabs without chamfered ends, the transverse bars are anchored between or above them. After completion, both the precast and the in situ concrete act as a flange of the composite beam in the usual way, but its effective width and depth can both be less than for an in situ flange.

The effective width recommended is that of the in situ concrete, and so is about 1 m for a T-beam. Where a precast slab is supported on an edge beam, the transverse bars should be looped around the shear connectors.

The slabs are erected with their longitudinal sides butted together. The bottom corners are often chamfered by about 20 mm (Section A–A in Figure 3.30) so that where the soffit (under-side) of the slabs is exposed as a ceiling, small departures from the expected plane surface are less evident. There may be a flexible filler just above the chamfer. To permit transfer of vertical shear between slabs, the gap at the top surface should be wide enough for placing of concrete or grout. For a beam supporting solid slabs, the effective flange thickness is then the depth of the slab less at least 25 mm. It may be possible to use the effective width applicable to in situ construction. Guidance on the shear resistance of grouted joints is given in Way et al. (2007).

If the sides of precast slabs are just butted together over their whole depth, longitudinal compression from composite action of the beam is assumed to be transferred only through the gap between their ends and in the topping above them.

Where solid slabs are used and recommendations on gaps between slabs and minimum spacing of studs are followed, the shear resistance of stud connectors can be taken as that for an in situ slab. For hollow-core slabs it is recommended that their resistance should be reduced by 10%, because the local confinement of the studs may be less than it would be in a solid slab.

3.10.2 Design of composite beams that support precast slabs

3.10.2.1 Stability during construction
In a completed structure, lateral buckling of the top flange of a steel beam is normally prevented by friction with the floor or deck slab, and/or by the shear connection. During

construction of in situ slabs, the formwork provides some lateral restraint. During erection of precast slabs, with no formwork, a steel beam is likely to be loaded on one side only. The resulting torsion and twist increase the risk of lateral buckling, and temporary lateral restraints may be needed. Torsion also occurs where concreting for an unpropped composite slab is done on one side only of a supporting beam. Relevant calculations are difficult, and involve prediction of the imposed (construction) loading on the slabs before the in situ concrete above the beam becomes effective.

There is a rule of thumb: compression flanges of adjacent beams should be tied together at intervals not exceeding 40 times the flange width. Once slabs of similar weight have been placed on both sides of an internal beam, the tendency to twist is much reduced, and there is lateral restraint from friction. It is recommended that full lateral restraint may be assumed for beams of span less than about 160 times the bearing width of the precast units.

For edge beams, the problem can be avoided by making the slabs long enough to extend over the whole width of the steel flange, although their edge details then have to allow for the shear connectors. Another solution is to increase torsional stiffness by using a rectangular hollow steel section, and supporting the slabs on a plate welded to its bottom flange.

Steel I-section beams are torsionally flexible. There are empirical rules that the maximum angle of twist during construction should not exceed 2° and that the lateral movement of the steel top flange should not exceed span/500.

3.10.2.2 Resistance of composite beam to bending

In buildings, beams for floors with precast slabs are usually designed as simply-supported, so only resistance to sagging bending is considered here. The composite cross-section is usually in slenderness Class 1 or 2, so resistance is based on plastic theory. Where full shear connection is used, the plastic neutral axis may be within the steel cross-section. The bending resistance is then found in the usual way, using the effective width and depth of the concrete flange.

Composite action with hollow-core slabs   Where the plastic neutral axis of the composite beam is found to lie within the concrete slab, Couchman (2014) gives an approximate expression for $M_{pl,Rd}$, with the comment that 'this case is not permitted where hollow-core slabs are used'. This arises from the complex stresses in the thin walls between the cores in the slab. They provide the shear connection between the upper and lower flanges of the slab, for the stresses caused by longitudinal composite action in the beam. They are also the webs of the hollow-core slab.

These two horizontal shear forces are at right angles to each other. Interaction between them has been found to reduce the resistance of the slab to vertical shear. Leskela (2017) advises that composite action between beams and hollow-core slabs should be assumed only for serviceability verifications. An alternative (Couchman, 2014) for ultimate limit states is to limit the compressed zone to the region above the cores, and use the linear interaction method of partial shear connection.

Tests have also found that the deflections under loading of a beam with hollow-core slabs can cause additional stresses at the ends of the slabs closest to a beam support. This can reduce the resistance to vertical shear of the walls between the cores. Guidance is available (Couchman, 2014) and a design rule is likely to be given in the next Eurocode 4.

3.11 Example: simply-supported composite beam

In this example, a typical composite T-beam is designed for the floor structure shown in Figure 3.1, using the materials specified in Section 3.2, and the first of the floor designs given in Section 3.4. Ultimate limit states are considered first. An appropriate procedure that minimizes trial and error is as follows. Some of the choices made here lead to problems later, so that solutions to them can be shown.

(1) Choose the types and strengths of the materials to be used.
(2) Ensure that the design brief is complete. For this example it is assumed that:
   - no special provision of openings for services is required (except in Section 3.14);
   - the main source of vibration is pedestrian traffic on the floor, and occupants' sensitivity to vibration is typical of that found in office buildings;
   - the specified fire resistance class is R60.
(3) Make policy decisions. For this example:
   - the steel member to be a rolled universal beam (UB) section;
   - unpropped construction is assumed (it will be found in Section 3.11.3.1 that this causes deflections that in many applications would be excessive);
   - fire resistance could be provided by encasing the web, but not the bottom flange, in concrete, and its weight is to be allowed for – other methods (Chapter 6) would add less weight;
   - nominally-pinned beam-to-column joints are to be used.
(4) With guidance from typical span-to-depth ratios for composite beams, guess the overall depth of the beam. Assuming that the floor slab has already been designed, this gives the depth h_a of the steel section.
(5) Guess the weight of the beam, and hence estimate the design mid-span bending moment, M_{Ed}.
(6) Assume the lever arm to be (in the notation of Figure 3.15)

$$\left(h_a/2 + h_t - h_c/2\right)$$

and find the required area of steel, A_a, if full shear connection is to be used, from

$$A_a f_{yd} \left(h_a/2 + h_t - h_c/2\right) \geq M_{Ed} \tag{3.94}$$

(7) If full shear connection is to be used, check that the yield force in the steel, $A_a f_{yd}$, is less than the compressive resistance of the concrete slab, $b_{eff} h_c (0.85 f_{cd})$. If it is not, the plastic neutral axis will be in the steel – unusual in buildings – and A_a as found above will be too small.
(8) Knowing h_a and A_a, select a rolled steel section. Check that its web can resist the design vertical shear at an end of the beam.
(9) Design the shear connection to provide the required bending resistance at mid-span.
(10) Check deflections and vibration in service.
(11) Design for fire resistance, if not done earlier.

3.11.1 Composite beam – full-interaction flexure and vertical shear

From Section 3.4, the uniform characteristic loads from a 4.0-m width of floor are:

- permanent,

  $g_{k1} = 2.54 \times 4 = 10.2 \, \text{kN/m}$ on steel alone

  $g_{k2} = 1.3 \times 4 = 5.2 \, \text{kN/m}$ on the composite beam

- variable,

  $q_k = 6.2 \times 4 = 24.8 \, \text{kN/m}$ on the composite beam.

The weight of the beam and its fire protection is estimated to be 2.2 kN/m, so the design ultimate loads are:

$$g_d = 1.35 \, (15.4 \, + \, 2.2) = 23.7 \, \text{kN/m}$$

$$q_d = 1.5 \times 24.8 = 37.2 \, \text{kN/m}$$

The depth of the steel H-section for the supporting columns is estimated to be 200 mm, and the effective point of support for the beam is assumed to be 100 mm from the face of the H section, so the simply-supported span is $9.0 - 4 \times 0.10 = 8.6 \, \text{m}$.

The mid-span bending moment for $L = 8.6 \, \text{m}$ is:

$$M_{Ed} = 60.9 \times 8.6^2/8 = \mathbf{563 \, kN\,m} \tag{3.95}$$

Vertical shear forces are required for the design of the columns and are calculated for a span of 9.0 m. It is assumed that external walls and any other load outside a 9-m width of floor are carried by edge beams that span between adjacent columns.

The design vertical shear for columns is

$$V_{Ed} = 60.9 \times 9/2 = \mathbf{274 \, kN} \tag{3.96}$$

It has been assumed that the composite section will be in Class 1 or 2, so that the effects of unpropped construction can be ignored at ultimate limit states.

Deflection of this beam is likely to influence its design, because of the use of steel with $f_y = 355 \, \text{N/mm}^2$, lightweight-aggregate concrete, and unpropped construction. The relatively low span-to-depth ratio of 16 is therefore chosen, giving an overall depth of $8600/16 = 537 \, \text{mm}$. The slab is 150 mm thick (Figure 3.8), so for the steel beam, $h_a \approx 387 \, \text{mm}$. From Expression 3.94 the required area of the steel beam is

$$A_a \approx \frac{563 \times 10^6}{(355/1.0)(194 + 150 - 40)} = 5217 \ \text{mm}^2$$

A suitable rolled I-section appears to be 406×140 UB 46 ($A_a = 5900 \, \text{mm}^2$); but its top flange may be unstable during erection or, at 140 mm, too narrow for the profiled sheeting and shear connection. Also, with profiled sheeting it is usually necessary to use partial shear connection, so a significantly heavier section is chosen, 406×178 UB 60. Its relevant properties are shown in Figure 3.31. For its compression flange, with root radius $r = 10.2 \, \text{mm}$, the flange outstand is given by

$$c = (178 \, - \, 7.8)/2 \, - \, 10.2 = 74.9 \, \text{mm, so} \ c/t = 74.9/12.8 = 5.9$$

which is well below the Class 1 limit of 7.32 (Table 3.2).

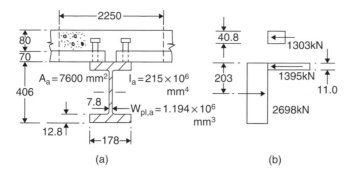

Figure 3.31 Cross-section and stress blocks for composite beam in sagging bending

The weight of this beam, with normal-density web encasement, if provided, at $25 \, \text{kN/m}^3$, is

$$g_k = 0.060 \times 9.81 \; + \; 25 \times (0.406 \times 0.178 \; - \; 0.0076) = 2.20 \, \text{kN/m}$$

so the weight assumed earlier includes sufficient allowance for fire protection.

For the effective flange width, it is assumed that two rows of stud connectors are 0.1 m apart, so from Equation 3.45,

$$b_{\text{eff}} = 8.6/4 + 0.1 = 2.25 \, \text{m} \tag{3.97}$$

3.11.1.1 Full shear connection
Results for full shear connection are useful, even where it is not used. The depth x_c of the plastic neutral axis is first assumed to be less than h_c (80 mm), so x_c is found from Equation 3.46 with $f_{cd} = 25/1.5 = 16.7 \, \text{N/mm}^2$:

$$N_{\text{cf}} = 7600 \times 0.355 = 2.25x(0.85 \times 16.7)$$

whence $x = 84.6$ mm. This exceeds h_c, so the correct N_{cf} is given by Equation 3.48:

$$N_{\text{c,f}} = 2.25 \times 80 \, (0.85 \times 16.7) = \textbf{2555 kN}$$

From Equation 3.49,

$$N_{\text{a,pl}} = A_a f_{\text{yd}} = 7600 \times 0.355 = \textbf{2698 kN} \tag{3.98}$$

Assuming that the neutral axis is in the steel top flange, Equations 3.51 give

$$N_{\text{ac}} = 2698 - 2555 = 143 \, \text{kN}$$

and

$$143 = 2 \times 178(x_c - 150) \times 0.355$$

whence $x_c = 151.1$ mm.

This is less than $h_t + t_f$ (163 mm), so the stress blocks are as shown in Figure 3.15c. The distance of force N_{ac} below force $N_{\text{c,f}}$ is $151.1 - 40 - 1.1/2 = 110.6$ mm, so from Equation 3.52,

$$M_{\text{pl,Rd}} = 2698(0.353 - 0.04) - 143 \times 0.1106 = \textbf{829 kN\,m} \tag{3.99}$$

This exceeds M_{Ed} by 47%, so partial shear connection will be used, giving a concrete stress block much less than 80 mm deep.

3.11.1.2 Vertical shear

From Equation 3.61, the shear area of this rolled section, with $r = 10.2\,\mathrm{mm}$, is

$$A_\mathrm{v} = 7600 - 2 \times 178 \times 12.8 + 12.8(7.8 + 2 \times 10.2) = 3404 \ \mathrm{mm}^2$$

From Equation 3.62 the resistance to vertical shear is

$$V_\mathrm{pl,a,Rd} = 3404(0.355/\sqrt{3}) = \textbf{697 kN} \tag{3.100}$$

which far exceeds V_Ed, as is usual in simply-supported composite beams for buildings when rolled steel I-sections are used.

3.11.1.3 Buckling

The steel web is entirely in tension, so its buckling need not be considered. The steel top flange is restrained by its connection to the composite slab, and so is stable. However, its stability during erection should be checked. It can be helped by the presence of web encasement, which can be cast before the beam is erected. The buckling calculation, to EN 1993-1-1, is not given here. Lateral stability of continuous beams is considered in Section 4.6.3 and in Johnson (2012).

3.11.2 Composite beam – partial shear connection, non-ductile connectors and transverse reinforcement

The minimum degree of shear connection is found next, as it may be sufficient. The condition for stud connectors to be treated as ductile when the span is 8.6 m, given by the line 'Eurocode 4' in Figure 3.21, is

$$n/n_\mathrm{f} \geq 0.51$$

To provide an example of the use of the equilibrium method, the bending resistance is now calculated using $n = 0.51 n_\mathrm{f}$. The notation of Figure 3.15 is used. The force N_c is 0.51 times the full-interaction value. With $N_\mathrm{c,f} = 2555\,\mathrm{kN}$ from Section 3.11.1,

$$N_\mathrm{c} = 0.51 \times 2555 = \textbf{1303 kN} \tag{3.101}$$

For $N_\mathrm{c,f}$, the depth of the compressive-stress block is 80 mm, so now $x_\mathrm{c} = 0.51 \times 80 = 40.8\,\mathrm{mm}$. With reference to the stress blocks in Figure 3.15c, with $N_\mathrm{a,pl} = 2698\,\mathrm{kN}$ (Equation 3.98),

$$N_\mathrm{ac} = 2698 - 1303 = 1395\,\mathrm{kN}$$

Assuming that there is a neutral axis within the steel top flange, the depth of flange in compression is

$$1395/(0.178 \times 2 \times 355) = 11.0\,\mathrm{mm}$$

This is less than t_f (12.8 mm) so the assumption is correct, and the stress blocks are as shown in Figure 3.31b. Taking moments about the top surface of the slab,

$$M_\mathrm{pl,Rd} = 2698 \times 0.353 - 1303 \times 0.020 - 1395 \times 0.156$$

$$= \textbf{709 kN m} \tag{3.102}$$

which exceeds M_Ed (563 kN m).

The interpolation method gives $M_\mathrm{pl,Rd} = 630\,\mathrm{kN\,m}$ with $n/n_\mathrm{f} = 0.51$, so it is significantly more conservative. It will now be shown that rules in Eurocode 4 make it difficult to use $n/n_\mathrm{f} = 0.51$ in this design.

3.11.2.1 Number and spacing of stud shear connectors

It is assumed that 19-mm stud connectors will be used, at 125 mm long. The length after welding is about 5 mm less, so the height of the studs is taken as 120 mm. The design shear resistance P_{Rd} is given by Equation 3.1 as 60.2 kN per stud.

The reduction factor k_t for the resistance of studs in ribs is given by Equation 2.17, in which it is not clear whether the height h_p should include the 15-mm height of the top rib shown in Figure 3.8. The rib can be assumed not to affect k_t, so $h_p = 55$ mm, and other values are: $b_0 = 162$ mm; $h = 120$ mm; $n_r = 2$.

So from Equation 2.17,

$$k_t = (0.7/\sqrt{n_r})(162/55)[(120/55) - 1] = 1.72$$

For 0.9-mm sheeting, and 19-mm studs welded through the sheeting, EN 1994-1-1 gives the following upper limits for k_t, based on tests. They govern here:

for $n_r = 1$, $k_t = 0.85$, so $P_{Rd} = 51.2$ kN
for $n_r = 2$, $k_t = 0.70$, so $P_{Rd} = 42.1$ kN

From Equation 3.101 the number of single studs needed in each half span is

$$n = N_c/P_{Rd} = 1303/51.2 = 25.4, \quad \text{say } 26$$

There is one trough every 300 mm, or 14 in a half span of 4.3 m. The sheeting shown in Figure 3.8 has stiffening ribs in its troughs that require studs to be placed centrally. The number of studs per trough is also limited by the width of the steel flange. It cannot exceed two here, so 28 studs can be provided. This is sufficient only if they can be deemed to be 'single studs'.

For the detailing shown in Figure 3.32, the lateral centre-to-centre spacing of the two studs is 118 mm, or 6.2d. This exceeds the minimum given in EN 1994-1-1, which is 4d. If the separation is large, Equation 2.17 should obviously be used with $n_r = 1$, assuming that there is no interaction between the local stresses around the two studs, and giving $k_t = 0.85$. If it is small (e.g. 4d), then n_r is presumably taken as 2, so $k_t = 0.7$.

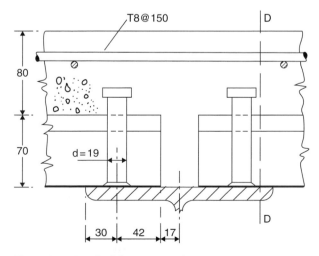

Figure 3.32 Detail of shear connection

The extent to which interaction between two studs 6.2*d* apart would reduce their resistance is unknown. This situation is not clearly covered in the current Eurocode 4. However, details are given in Section 3.6.2 of the proposed new rules for minimum degree of shear connection, based on research since Eurocode 4 was written (Couchman, 2015). Their use here finds that the line '10, U, 0.8' in Figure 3.21 is applicable. For this 8.6-m span it gives $\eta_{min} = 0.30$.

In Section 3.4.3, each stud was used to provide an anchorage force of 4.4 kN for the sheeting, perpendicular to the direction in which resistance is now required. The corrected value for P_{Rd} is, for $n_r = 2$,

$$P_{Rd}^2 = 42.1^2 - 4.4^2, \quad \text{so} \quad \mathbf{P_{Rd} = 42.0 \ kN} \tag{3.103}$$

For 28 studs, the resistance is $N_{Rd} = 28 \times 42 = 1176$ kN and so $n/n_f = 1176/2555 = 0.46$. Recalculation finds that this reduces M_{Rd} from 709 kN m (Equation 3.102) to 693 kN m, so that $M_{Ed}/M_{Rd} = 0.81$.

In conclusion, the shear connection is **two studs in each trough as in Figure** 3.32, **throughout the 8.6 m span,** a total of 56 studs.

3.11.2.2   Design with non-ductile shear connectors
It is unlikely that connectors other than studs would be used where the slab is composite, but it is possible. For example, where sheeting is not continuous across the steel flange, a steel strip of the concrete dowel type (Section 2.4.2) could be welded along its centre. The connection would have to be treated as non-ductile. That is within the scope of Eurocode 4, but the design method is not clearly defined. There are several relevant clauses.

Clause 6.6.1.1(12) says '... the behaviour assumed in design should be based on tests and supported by a conceptual model...'.

Clause 6.6.1.3(5) is more helpful: 'The required number of shear connectors may be distributed ... in accordance with the longitudinal shear calculated by elastic theory for the loading considered. Where this is done, no additional checks on the adequacy of the shear connection are required.'

The shear flow will be proportional to the vertical shear, and a maximum at the ends of the span. There will be inconsistency with the design for bending, which assumed plastic stress blocks for the whole of the loading, 60.9 kN/m, and took no account of the use of unpropped construction.

Elastic analysis of the mid-span cross-section for M_{Ed} (563 kN m) and unpropped construction finds that the lower part of the steel I-section would have yielded. Behaviour there would be elastic-plastic, but not near the ends of the span. Elastic section analysis for the loading resisted by the composite member, 37.2 kN/m, may be a good estimate of the shear flow near the supports. A more conservative method is to find the elastic shear flow corresponding to the design vertical shear, $V_{Ed} = 60.9 \times 4.3 = 262$ kN. This is now done.

Creep reduces the compressive force in the slab, and so is ignored, giving $n = 10.1$. Elastic analysis of the cracked unreinforced composite section, assuming a slab thickness of 95 mm as before, finds the neutral axis to be 128 mm below the top of the slab, with $I = 753 \times 10^{-6}$ m^4. The 'elastic' formula $v_{Ed}b = V_{Ed}A\bar{y}/I$ (where $\bar{y}$ is the distance from the centre of the 'excluded area' A to the neutral axis) finds $v_{Ed}b$ to be 593 kN/m. This is the design shear flow for the non-ductile connectors at the ends of the span. A 'triangular' shear flow distribution over the half-span length, 4.3 m, gives the compressive force in the slab at mid-span as 1274 kN, which is checked by an elastic section analysis for $M_{Ed} = 563$ kN m.

The design with ductile connectors, above, provided a compressive force $28 \times 42 = 1176$ kN at mid-span, and a shear-flow resistance at the ends of $1176/4.3 = 274$ kN/m, less than half of 593 kN/m.

3.11.2.3 Transverse reinforcement

Rules for the use of profiled sheeting as transverse reinforcement are explained in Section 3.6.3.2. The cross-section in Figure 3.32 illustrates compliance with the rule that the sheeting should extend at least $2d_{do}$ beyond the centre of a stud welded through it, where d_{do} is an estimate of the diameter of the stud weld, taken as $1.1d$, or 20.9 mm here. This gives the 42-mm end distance shown. The 30-mm dimension just satisfies the relevant rule shown in Figure 3.26. The clear gap of 34 mm between the ends of the sheeting may be reduced by tolerances, but should not fall below the minimum needed for satisfactory placing of concrete (about 25 mm).

The design longitudinal shear on a plane such as D–D in Figure 3.32 is based on the resistance of the studs, not on the design shear flow. For $P_{Rd} = 42$ kN, the total longitudinal shear resistance is

$$v_L = 2 \times 42/0.3 = 280 \text{ kN/m}$$

There are two planes D–D, and as some longitudinal shear is resisted by the width of flange between them, the design shear for reinforcement of plane D–D is just under half of the total, so assume that $v_{L,Rd}$ must be at least 140 kN/m.

The contribution from the sheeting is calculated next. Using Equations 3.71 and 3.72 with $a = 42$ mm,

$$k_\varphi = 1 + 42/20.9 = 3.0, \quad t = 0.9 \text{ mm}, \quad f_{yp} = 350 \text{ N/mm}^2, \quad \text{and } \gamma_{Ap} = 1.0$$

then

$$P_{p,b,Rd} = 3.0 \times 20.9 \times 0.9 \times 0.350 = 19.7 \text{ kN}$$

For the studs at 0.3 m spacing, $P_{p,b,Rd}/s = 65.8$ kN/m. This must not exceed the tensile strength of the sheeting, which is about 412 kN/m.

Equation 3.73 is now used to find the required area of transverse reinforcement:

$$140 = 0.435A_e + 65.8, \quad \text{whence } A_e = 171 \text{ mm}^2/\text{m} \tag{3.104}$$

For control of cracking of the slab above the beam, it was found (Equation 3.42) that $380 \text{ mm}^2/\text{m}$ is required, and this governs. There is no need to rely on the profiled sheeting.

3.11.3 Composite beam – deflection and vibration

3.11.3.1 Deflection

For unpropped construction, the characteristic load combination for the beam is:

$$\left.\begin{array}{ll} \text{permanent (steel beam)} & g_1 = 10.2 + 2.2 = 12.4 \text{ kN/m} \\ \text{permanent (composite beam)} & g_2 = 5.2 \text{ kN/m} \\ \text{variable (composite beam)} & q = 24.8 \text{ kN/m} \end{array}\right\} \tag{3.105}$$

For a simply-supported span of 8.6 m with distributed load w kN/m and second moment of area I mm^4, the mid-span deflection is

$$\delta = \frac{5wL^4}{384EI} = \frac{5 \times 8.6^4 \times 10^9 w}{384 \times 210I} = 339 \times 10^6 \frac{w}{I} \text{ mm} \tag{3.106}$$

For the steel beam, $I = 215 \times 10^6$ mm^4, so its deflection during construction is

$$\delta_a = 339 \times 12.4/215 = 19.5 \text{ mm} \quad (\text{span}/441) \tag{3.107}$$

From Section 3.2, the short-term elastic modulus for the concrete is 20.7 kN/mm^2, so for variable loading the modular ratio is $n_0 = 210/20.7 = 10.1$.

The modular ratio for permanent load is around $3n_0$, but for simplicity, creep will be allowed for by using $n = 2n_0$ for all loading. This option is given in Eurocode 4 under certain conditions that are satisfied here.

The second moment of area of the composite section is calculated using Equations 3.77 to 3.81. From Figs 3.28 and 3.31, relevant values are:

$$A_a = 7600 \text{ mm}^2 \quad z_g = 203 + 150 = 353 \text{ mm}$$

$$b_{eff} = 2250 \text{ mm} \quad I_a = 215 \times 10^6 \text{ mm}^4$$

The minimum thickness of the slab is 80 mm, but for over 90% of its area it is 95 mm thick (Figure 3.8). For deflection and vibration, mean values of I are appropriate, so h_c is taken here as 95 mm.

Assuming that the neutral-axis depth exceeds h_c, Equation 3.80 gives x, now written as x_c, as

$$x_c = 176 \text{ mm} \tag{3.108}$$

From Equation 3.81,

$$10^{-6}I = 215 + 7600(0.353 - 0.176)^2 + \frac{2250 \times 95}{20.2}\left(\frac{0.095^2}{12} + 0.128^2\right)$$

$$= 215 + 240 + 181 = 636 \text{ mm}^4 \tag{3.109}$$

(Numerical values are of more convenient size in this calculation if $10^{-6}I$ is calculated, rather than I. With other values in N and mm units, bending moments are then conveniently in kN m as required, rather than N mm. Using units m^2 mm^2 for I has the same result.)

The deflection of the composite beam due to permanent load is

$$\delta_g = 339 \times 5.2/636 = 2.8 \text{ mm}$$

and its deflection due to variable load is

$$\delta_q = 339 \times 24.8/636 = 13.2 \text{ mm} \tag{3.110}$$

The total deflection is thus $19.5 + 2.8 + 13.2 = 35.5$ mm.

No account has yet been taken of any increase in deflection due to slip. From Section 3.7.1, EN 1994-1-1 permits it to be neglected where $n/n_f > 0.5$, which is not satisfied by the ratio 0.46 used here. It gives an alternative opt-out: where 'forces resulting from an elastic behaviour and which at the shear connectors ... do not exceed P_{Rd}'. This is for serviceability loadings.

The force on a shear connector close to a support is now calculated, using the characteristic load applied to the composite beam. This is $5.2 + 24.8 = 30\,\text{kN/m}$, so the maximum vertical shear is

$$V_{\text{Ek}} = 4.3 \times 30 = 129\ \text{kN}$$

Using the well-known result:

$$v_{\text{L}} = V_{\text{Ek}}(A_{\text{c}}/n)(x - h_{\text{c}}/2)/I \quad \text{(i.e. } v = VA\bar{y}/I\text{)}$$

with

$$A_{\text{c}} = 2250 \times 95\ \text{mm}^2 \quad n = 20.2 \quad h_{\text{c}} = 95\ \text{mm}$$

and x and I from Equations 3.108 and 3.109 gives the longitudinal shear flow at a support:

$$v_{\text{L}} = 275\ \text{kN/m}$$

For pairs of studs at 0.3 m spacing, the shear per connector is

$$P_{\text{Ek}} = 275 \times 0.3/2 = 41.2\ \text{kN}$$

From Equation 3.103, $P_{\text{Rd}} = 42\,\text{kN}$, so the alternative condition of EN 1994-1-1, clause 7.3.1(4) is just satisfied.

The reader may inquire why the shear per stud for a loading on the composite member of 30 kN/m, 41.2 kN, is almost the same as the resistance provided, 42 kN per stud, for an ultimate loading of 60.9 kN/m. The reason is that these calculations for a serviceability limit state do not allow any redistribution of force per stud along a half span. This doubles the maximum force per stud, in this case. The elastic model with full interaction and the ultimate-strength model with partial interaction happen to give similar compressive forces in the slab, for a given bending moment. The force per stud is then unaltered when the bending moment is halved.

This beam is evidently close to the borderline for deflection due to slip that underlies the rules in EN 1994-1-1. The effect of slip on deflection is now estimated, using Equation 3.86 with $k = 0.3$, $n/n_{\text{f}} = \eta = 0.46$, and $\delta_{\text{c}} = 16.0\,\text{mm}$, as found above.

A load of 12.4 kN/m caused the steel beam to deflect 19.5 mm, so δ_{a}, for the total load on the composite beam, is

$$\delta_{\text{a}} = 19.5(5.2 + 24.8)/12.4 = 47.2\ \text{mm}$$

From Equation 3.86,

$$\delta = 16[1 + 0.3 \times 0.54(47.2/16 - 1)] = 21.0\ \text{mm}$$

Hence, slip could increase a deflection of 16 mm to 21 mm, and the total deflection to 40.5 mm (span/212).

This exceeds the limit recommended in Section 3.7.2 (span/300). Another problem is that the composite slab was assumed in Section 3.4 to be propped at the centre of its 4-m span. Its deflection was calculated ignoring any deflection of the supporting beams during construction. This implies that the beams are also propped, so it is evident that, with this design, they should be. This reduces their deflection during construction to

$\delta_a \times I_a/I$. Therefore, from Equation 3.107,

$$\delta_{a,\text{propped}} = 19.5 \times 215/636 = 6.6 \text{ mm}$$

The total deflection is now $6.6 + 2.8 + 13.2 + 5.0$ (slip) $= 27.6$ mm.

This is span/312, which is less than the recommended limit. In practice, deflections would be slightly reduced by the stiffness of the concrete in the bottom 55 mm of the slab, and by the stiffness of the beam-to-column connections.

It was assumed in Section 3.11.1 that the beams would be built unpropped. The change to propped construction alters the redistributions of elastic stresses caused by yielding of steel before ultimate loading is reached, but does not alter the verifications for flexure and vertical shear given there, because they are based on plastic theory.

Maximum bending stress in the steel section   It is clear from the preceding results that yielding of the steel member under service loading is unlikely, so no allowance is needed for the effect of yielding on deflection. The maximum bending stress in the steel occurs in the bottom fibre at mid-span. For propped construction it is given by

$$\sigma = My/I = wL^2y/8I$$

where w and I are as above, and y is the distance of the bottom fibre below the neutral axis. From Equation 3.108, the neutral-axis depth is 176 mm, and the overall depth is 556 mm, so $y = 380$ mm and

$$\sigma = 42.4 \times 8.6^2 \times 380/(8 \times 636) = 234 \text{ N/mm}^2$$

There will be small additional stresses from the unsupported lengths between props, but the total will be well below the yield strength of the beam, 355 N/mm^2.

3.11.3.2   Vibration

The method given in Section 3.9 is used. It is assumed that the source of vibration is pedestrian traffic. The target value for the vibration dose value (VDV) from Table 3.3 for 'low probability of adverse comment' is assumed to be 0.3, multiplied by 2 for an assumed office location.

Fundamental natural frequency   From Equations 3.105, the permanent load per beam is 17.6 kN/m. The total imposed (variable) load is 24.8 kN/m, and only a tenth of this will be included, because vibration is likely to be worse where there are few partitions and little imposed load. The design load is thus 20.1 kN/m for beams at 4 m centres, giving a vibrating mass:

$$m_b = 20\ 100/9.81 = 2050 \text{ kg/m}$$

For floor slabs, the mass of the steel beam is omitted. It is $2200/9.81 = 224$ kg/m, so that

$$m_s = (2050 - 224)/4 = 456 \text{ kg/m}^2$$

For vibration, the short-term modular ratio, $n_0 = 10.1$, is used, and I is increased by 10% (Section 3.9.1). Values corresponding to those from Equations 3.108 and 3.109 are, from Table 4.5:

$$x = 129 \text{ mm}, \quad 10^{-6}I_b = 1.1 \times 751 = 826 \text{ mm}^4$$

For the slab, the 'uncracked' value found in Section 3.4.5 is too low, because the modular ratio $n = 20.2$ was used. Similar calculations for $n_0 = 10.1$, with a 10% increase, give

$$10^{-6} I_s = 20.5 \ \text{mm}^4/\text{m}$$

From Equation 3.92 for the beam, with $L = 8.6$ m,

$$f_{0b} = \frac{\pi}{2} \left(\frac{210\,000 \times 826}{512 \times 4 \times 8.6^4} \right)^{1/2} = 6.2 \ \text{Hz}$$

From Equation 3.93 for the slab,

$$f_{0s} = 3.56 \left(\frac{210\,000 \times 20.5}{456 \times 4^4} \right)^{1/2} = 21.6 \ \text{Hz}$$

From Equation 3.91,

$$f_0 = 5.96 \ \text{Hz}$$

This is below 7 Hz, so no check need be made for impulsive loads.

Response of the composite floor From Equation b in Section 3.9.2, the effective length of the area of floor assumed to be vibrating is

$$L_{\text{eff}} = 1.09[210\,000 \times 826/(512 \times 4 \times 5.96^2)]^{1/4} = 7.62 \ \text{m}$$

From Equation c, the effective width is

$$S = 1.065[210\,000 \times 20.5/(456 \times 5.96^2)]^{1/4} = 4.30 \ \text{m}$$

From Equation a, the modal mass is

$$M = 2050 \times 7.62 \times 4.30 = 67\,200 \ \text{kg}$$

From Equation e, the weighted r.m.s. vertical acceleration is

$$a_{\text{w,rms}} = 703/67\,200 = 0.0131 \ \text{m/s}^2$$

It is simplest now to find the VDV for an assumed total duration of the walking activity. For example, if it occurs for a total of 8 h in a 16-h day, $n_a T_a = 8$ h. From Equation f in Section 3.9.2,

$$\text{VDV} = 0.68 \times 0.0131(8 \times 3600)^{0.25} = 0.12$$

This is far below the target value, 0.6, probably because the weight of the source of vibration (people walking) is so much less than the design loading for this floor.

3.12 Shallow floor construction

The sheeting for composite slabs is often continuous across a supporting steel beam. Eurocode 4 gives rules for the shear resistance of studs placed within the troughs of sheeting up to 85 mm deep. Allowing 15 mm for a small top rib (Figure 2.17), this limits sheeting depth to 100 mm, and the maximum span of a composite slab to about 4 m.

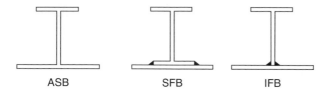

ASB   SFB   IFB

Figure 3.33 Cross-sections of asymmetric steel beams, for use with precast slabs or deep decking

For longer spans, new sheeting profiles have been developed. An example is ComFlor 225 (Tata Steel UK, 2016), one wavelength of which is shown in Figure 3.12. It is used in the worked examples in Sections 3.4.6 and 3.13. To reduce the overall depth of floor structures, this sheeting is usually supported on an extended bottom flange of the steel beam (Figure 3.33). Where the beam is designed as composite, the sheeting does not affect the shear resistance of studs.

In the UK there is a trade name, *Slimdek*, for deep decking used for these slabs. Spans of up to 6 m (unpropped) or 9 m (propped) are possible. Another name, *Slimflor*, is used for the steel beams that support them (Rackham et al., 2006). There is a range of asymmetric steel beams (ASBs) that are rolled I-sections with a bottom flange 110 mm wider than the top flange. This simplifies the erection of deep decking or precast slabs, which need a bearing length of at least 50 mm at each end. The beams have an overall depth of 280 or 300 mm, and raised ribs on their top surface to increase shear-bond strength. Sheet steel diaphragms at each end of the sheet close off the spaces between the ribs, so that the whole of the steel web can be encased in concrete (Figure 3.34).

To widen the range of products (Mullett and Lawson, 1999), asymmetric steel beams are also made by welding a wide bottom-flange plate to a rolled I-section ('SFB') or to a rectangular hollow section ('RHSFB'). The high torsional stiffness of the RHSFB beams make them suitable for use as edge beams.

Other fabricated sections ('IFB') are made (ArcelorMittal, 2016) by welding a plate to part of a rolled IPE or HE steel section (the terminology used in continental Europe). These range in depth from about 130 mm to 370 mm, and have bottom flanges about 200 mm wider than the top flange. The overall depths of the sheeting and of the steel beam, and the thicknesses of their top cover of concrete are interrelated, and need to be

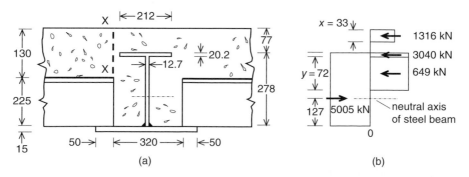

(a)   (b)

Figure 3.34 (a) Cross-section of an asymmetric beam with 8.6 m span, supporting deep decking. (b) Stress blocks for the partial-interaction resistance to sagging bending, 843 kN m

considered early in the design process. Fabricated sections are more versatile than the ASB range, and are likely to replace it.

The beams can be used to support deep decking or various types of precast slab (Section 3.10). The references cited here give detailed guidance, as the subject is not included in the current Eurocode 4. Rackham et al. (2006), for example, refers to the preceding British code, BS 5950, and the beams there do not use composite action, except for serviceability verifications under certain conditions. The worked example in Section 3.13 may need revision when the proposals for the next Eurocode 4 become known.

3.13 Example: composite beam for a shallow floor using deep decking

Section 3.4.6 gives an alternative design for the composite slab, using ComFlor 225 sheeting and the beam cross-section shown in Figure 3.34a. The spacing of the beams was increased from 4 m to 5 m. The asymmetric steel section is needed to enable the sheeting to be placed on the bottom flange.

The steel cross-section chosen is one half of an IPE 0 550, welded to a 420×15 mm plate to provide the bottom flange. Details of the section are taken from the manufacturer's brochure (ArcelorMittal, 2016). The dimensions are shown in the figure, and other properties of the steel beam are: $A_a = 14\,100$ mm^2; $I_y = 218.3 \times 10^{-6}$ m^4; and section modulus $W_y = 1317 \times 10^{-6}$ m^3 for the top flange.

Its weight is 1.09 kN/m, to which must be added the weight of the extra concrete that surrounds its web, found as 1.35 kN/m in Section 3.4.6, so the unfactored weight is 2.44 kN/m.

The loads per unit length for the beam are taken from Table 3.1 in Section 3.4.6 and assembled in Table 3.4. The composite slab was cast unpropped, while the beams had sufficient propping for stresses in them from this process to be negligible.

Casting of the slab encases the compressed part of the steel section, so the beam can be treated as Class 1 (plastic) and designed for the total loading, 91.2 kN/m.

Depending on the construction sequence, and also at edge beams, unbalanced loading can occur. Propping may not prevent twisting of the steel beam. This is discussed in Section 3.10, and is not considered here.

Table 3.4 Characteristic and design loads per unit length of beam

Limit states, and partial factor	Load for composite beam		
	SLS	γ	ULS
Composite slab 4.83×5	24.15	1.35	32.6
Steel beam and adjacent concrete	2.44	1.35	3.3
Floor and ceiling finishes 1.3×5	6.5	1.35	8.8
Partitions 1.2×5	6.0	1.5	9.0
Imposed loading 5×5	25.0	1.5	37.5
Totals, kN/m	64.1		91.2

The effective span of the beam is taken as 8.6 m (Section 3.11.1), so for the ultimate loading the sagging moment at mid-span is

$$M_{Ed} = 91.2 \times 8.6^2/8 = \textbf{843 kN m}$$

The effective width of the concrete flange is taken as span/4 plus 0.21 m, the width of the steel top flange, giving $b_{eff} = 2.36$ m.

Resistance to bending The full-interaction moment of resistance is found first. For the whole steel section at yield in tension, the tensile force is $A_a f_y = 14\,100 \times 0.355 = 5005$ kN. For the design compressive stress of 17 MPa, the thickness of slab needed for 5005 kN is $5005/(2.36 \times 17) = 125$ mm.

This is less than the 130 mm available, but exceeds 77 mm, so part of the steel section is in compression and the longitudinal tensile force is less than 5005 kN.

Trial calculations with a single plastic neutral axis find it to be 10 mm below the top of the steel section. The compressive force in the slab is 3463 kN. This is 'full shear connection', and gives $M_{pl,Rd} = 940$ kN m. Slip is unlikely here, so the flexural behaviour at 843 kN m will be elastic-plastic.

During unpropped construction of the composite slab, its weight is applied by the webs of the sheeting at intervals of 0.6 m along the edges of the bottom flange of the steel beam, causing stresses in it transverse to the span of the beam. If the beam is also unpropped, the combination of these stresses with the longitudinal bending stress in the flange should be considered. Once the concrete has gained strength, it will transfer vertical shear from the slab to the beam, so if the beam is propped, this interaction is not significant.

Longitudinal shear Longitudinal shear force can be estimated using partial-interaction plastic theory for the resistance to bending. This assumes slip between steel and concrete, and two neutral axes, as shown in Figure 3.34b: one at depth x in the concrete slab, and the other with depth y of the steel web in compression. A trial value for x is assumed for the axis in the concrete, and y is found from the condition of longitudinal equilibrium. The assumed x is varied until the required value for M_{Rd} is obtained. The algebra is simplified by assuming that a tension of 5005 kN acts at the centroid of the steel section, as shown, and that there is a compressive stress of $2f_y$ in the steel top flange and over the depth y of the web. The results are shown in Figure 3.34b.

The compressive force in the slab is 1316 kN. From the equilibrium method for partial shear connection (Figure 3.16), the required degree of shear connection is

$$\eta = 1316/3463 = 0.380$$

For this beam, the simpler interpolation method is more conservative. For the steel section alone, $M_{a,pl,Rd} = 546$ kN m. From Equation 3.57,

$$N_c/N_{c,f} = \eta = (843 - 546)/(940 - 546) = 0.754$$

so this method requires almost twice as much shear connection.

The design of shear connection for slim-floor beams is not in the current Eurocode 4. Seeking guidance, one finds clause 6.3.3(2) on web-encased beams, which implies the use of shear connectors, but the encasement of a slim-floor beam is more like that in a fully-encased composite column. For these, clause 6.7.4.3 permits the use of bond or friction between steel and concrete outside regions of load introduction. Its Table 6.6 gives the design shear strength as 0.30 MPa where the concrete cover is 40 mm, with an increase for greater cover. Here, it is 77 mm (Figure 3.34), and the increased value is

$$\tau_{Rd} = 0.52 \text{ MPa}$$

It was found from tests on two slim-floor beams (Lawson et al., 1997) that shear-bond strengths ranged from 0.85 to 1.14 MPa. Assuming a cube strength of 30 MPa and a partial factor of 1.5 led to the proposed design value:

$$\tau_{Rd} = 0.60 \text{ MPa}$$

In a column cross-section, the shear stress is calculated using the whole of the steel-concrete perimeter. In a slim-floor beam, the bottom flange is not in the compression zone. In Lawson et al. (1997) the bottom flange was not included in the steel perimeter. With that method, the contact length here (Figure 3.34a) is $212 + 199 + 2 \times 258 = 927$ mm.

Over half of the 8.6-m span, the mean shear stress is

$$\tau_{111} = 1316/(927 \times 4.3) = 0.33 \text{ MPa}$$

Clause 6.7.4.3 recommends elastic analysis, which is consistent with the negligible slip associated with bond between surfaces. Shear then varies linearly along a half span, so the maximum near a support is

$$\tau_{Ed} = 2\tau_{av} = 0.66 \text{ MPa}$$

The close agreement between these three values may be coincidence. Information on this subject is sparse. There are unpublished tests using asymmetric beams that led to a design shear bond strength of about 0.4 MPa for concrete with $f_{ck} = 25$ MPa. This is probably conservative because experience finds that the penalty that the method of EN 1990 applies to small sets of test results is over-strict.

Guidance on shear connection in slim-floor beams is expected to be given in the next Eurocode 4. Stud connectors could be provided, but their load/slip property is so different from that for bond that it is doubtful whether the two resistances could be combined.

The shear resistance of a 19-mm stud is found in Section 3.11.2 to be 60.2 kN. The maximum shear flow from 1316 kN is $2 \times 1316/4.3 = 612$ kN/m, so near the ends of the span, shear connection from studs alone would need about ten studs per metre, probably attached to the steel web.

Transverse reinforcement, not shown in Figure 3.34, will always be provided, and designed as usual, both for longitudinal shear in the concrete flange and to limit the width of cracks caused by the superimposed dead and imposed loadings on the slab. It can be above the top of the steel beam, if the concrete cover is thick enough, or pass through holes in its web. It would then also help to resist longitudinal shear.

Deflections In Lawson et al. (1997), reference is made to tests that have shown that deflections can be predicted using the effective modulus method and the uncracked composite section. The design method given in that report limits the effective flange breadth to span/8, rather than the usual span/4, which was used here. The span/8 assumption reduces the longitudinal shear stress on planes such as X–X in Figure 3.34a, and so may reduce the amount of transverse reinforcement required.

Here, the concrete of the composite section has been assumed to be cracked in tension. Elastic analysis for sagging bending finds the depth of concrete in compression to be 129 mm, and the second moment of area to be $I = 541 \, m^2 \, mm^2$, in 'steel' units. The area of concrete in compression has been assumed to be rectangular, so the top flange and part of the web of the steel beam have been assumed to be both steel and concrete, for simplicity. As this region is close to the neutral axis, the resulting over-estimate of the second moment of area is small, and will be offset by the stiffness of uncracked concrete in tension.

For this simply-supported beam of span 8.6 m, and propped construction, the central deflection from the loading of 64.1 kN/m is 40.2 mm, of which 19.4 mm is due to the imposed load of 31 kN/m. From Section 3.4.6, the deflections of the slab midway between the beams are 13.4 mm during its unpropped construction and 3.4 mm from imposed load. Thus, the total deflection is 53.6 mm (span/160) and the imposed load deflection is 22.8 mm (span/377).

These deflections will usually be judged to be too high. For example, in Lawson et al. (1997) it is recommended that total deflections should not exceed 50 mm, and imposed-load deflection should not exceed span/360. Much depends on the intended use of this part of the building. The dead-load deflection may be hidden by a false ceiling, and partitions can be designed for the live-load deflection. In reality, the flexural stiffness of so-called 'simple' beam-to-column joints will reduce deflections, but this will slightly increase the bending moments in external columns. This example is not typical of the many multistorey buildings where floor loadings are lower than is assumed here.

The purpose of this alternative design was to reduce floor-to-ceiling heights. The depth of the floor structure at the beams has been reduced from 556 mm (Section 3.11.1) to 370 mm here, but between the beams it is now 355 mm and was 150 mm. The increase in beam spacing from 4 m to 5 m reduces the number of columns needed. On the other hand, deflections are larger, and the weight of steel in the floor slab and beams, excluding reinforcement, has gone up from $0.25 \, kN/m^2$ to $0.66 \, kN/m^2$. As always, design is a matter of balancing competing priorities.

3.14 Composite beams with large web openings

Floor structures for multistorey buildings are made as shallow as possible, but have to allow space for service pipes and ducts. Where slim-floor construction is used, the ducts pass below the beams. Where steel beams extend below the slab, large rectangular or circular openings in their webs are often required for services. Cellular beams are manufactured by cutting and re-welding rolled sections and use regular circular openings. Fabricated beams have discrete openings that are sized according to the bending moment and shear force.

The current Eurocode 4 does not include design rules for webs adjacent to an opening. During drafting there were many sets of rules in use for beams with openings of different shapes: rectangular, circular, hexagonal, and so on, and practice was changing fast. Highly asymmetric steel sections were coming into use, with relatively slender webs and large openings.

Research over the last 20 years has led to an agreed basis for design. Rules applicable to steel beams will be included in the next Eurocode 3, with additional rules for composite beams in Eurocode 4. A full worked example would be too complex to include here. The scope of the following outline of the design method is limited as follows:

- simply-supported composite beams with symmetrical steel I-sections and a composite slab;
- uniformly spaced shear connectors providing partial shear connection;
- closely spaced, unstiffened rectangular openings along most of the span, at mid-depth of the web;
- uniformly distributed loading.

It is an ultimate-limit-state verification. The action effects and resistances here are factored values. The subscript 'd' has been omitted from their symbols. Those that are resistances include the subscript R.

The main problem is to transfer vertical shear across an opening by Vierendeel action. In regions of high shear the edges of openings may need stiffeners to increase the Vierendeel bending resistance.

A comprehensive 80-page report, *Design of Composite Beams with Large Web Openings*, is available (Lawson and Hicks, 2011). It makes use of an earlier report (Lawson et al., 2006). These publications are expected to be the basis of the rules in the next editions of Eurocodes 3 and 4. They include design methods for many types of opening, such as those that are isolated, stiffened, or of shapes other than rectangular, and for both full and partial shear connection.

Analysis of Vierendeel behaviour in a short length of beam   At a web opening, a steel I-beam is reduced to two T-sections, each of which resists vertical shear by double-curvature bending. In this simplified explanation, for the scope given above, it is assumed that the T-sections are in slenderness Class 2 to Eurocode 3.

The method is similar to plastic hinge analysis of a rigid-jointed steel frame. It identifies locations of potential plastic failure and finds equations between their action effects for given bending moment and vertical shear at the cross-section considered. Action effect is then replaced by resistance at a sufficient number of locations for a failure mechanism to form. This gives the design shear resistance V_R. It is checked that the action effects at the other locations do not exceed their resistances.

Figure 3.35a shows the composite slab (of depth h_s), a web post of width s_0 between two rectangular openings, and the centre-lines of the compressed zone of the slab and of the two steel T-sections. Figure 3.35b shows the length s between the centres of the two openings (of height h_0 and width b_0), the three centre-lines, and internal action effects at mid-depth of the web post.

The cross-section X–X, at the centre-line of one of the openings, is referred to as 'section X'. It is at distance x from the left-hand support for the beam. There are n_x shear connectors within the length x, each with design shear resistance P_R, and n_s connectors within the length s. With partial shear connection, it is assumed that at section X, the shear connectors in length x are at failure, so the compressive force in the slab is

$$N_{c,R} = n_x P_R \tag{3.111}$$

It is assumed to be resisted by uniform compressive stress over the effective width of the slab and a depth z_c. The change in this force in length s is

$$\Delta N_{c,R} = n_s P_R \tag{3.112}$$

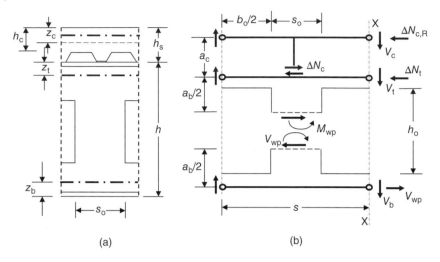

Figure 3.35 (a) Centre-lines of members at a web post in a composite beam with rectangular web openings. (b) Action effects at cross-section X–X and within a web post

The external sagging bending moment and clockwise vertical shear at section X are M_E and V_E (not shown in Figure 3.35).

The axial tension and vertical shear in the bottom tee are constant over length b_0, so its plastic resistance to bending, $M_{bT,N,R}$, allowing for the tension and shear, is the same at both ends of an opening. When it is at flexural failure in double-curvature bending, there is no bending moment at section X. This assumption is made also for the top tee and the slab. These locations of potential plastic hinges are shown as circles in the figure.

The design ultimate load on the beam is w per unit length, so the change in vertical shear in length s is ws. The shear forces V_c, V_t and V_b shown in the figure are assumed to occur at both ends of the length s, and

$$V_c + V_t + V_b = V_E \tag{3.113}$$

One set should be slightly higher, by $ws/2$, and the other slightly lower. This cancels out in the analysis.

The web post is split at its mid-depth, to show longitudinal shear V_{wp} and bending moment M_{wp} at this cross-section. The shear is equal to the change in tension in the bottom tee, so this is also shown as V_{wp}. For longitudinal equilibrium at section X,

$$\Delta N_{c,R} + \Delta N_t = V_{wp} \tag{3.114}$$

The total longitudinal forces in the three members at section X are not shown. They are N_c and N_t (compression positive), and N_b (tension positive), so that

$$N_c + N_t = N_b \tag{3.115}$$

The purpose of the analysis is to find V_E, the maximum vertical shear that section X can resist, for given geometry and properties of materials. Its resistance to bending is explained later. There are too few equilibrium equations for the nine unknowns at section X to be found for given external action effects V_E and M_E. Its resistances V_R and M_R are not reached until several cross-sections are at failure. Current design finds that these are often failure of the shear connection, as assumed above, with plastic failure of the bottom tee in combined tension, bending and shear, and buckling of the web post. This model is used here.

For openings placed centrally within the web, a top tee with the same cross-section as the bottom tee is less highly stressed. The full version of this analysis allows for non-central openings, and then the top tee can govern the resistance. Rules are also given for longitudinal and vertical stiffeners, which may be needed where shear is high, to strengthen the T-sections.

Equilibrium equations The shear connection is usually partial. There will then be a plastic neutral axis within the top tee, so the contribution of the net axial force in the tee to the resistance to bending moment M_E is negligible. The effective width of the slab is known (Eurocode 4), and the compressive stress is $0.85f_{cd}$, so the depth of slab in compression, z_c, depends on N_c. If this reaches the thickness h_c that is available (Figure 3.35a), full shear connection should be assumed, which would alter some of the equations below.

For a slab of overall thickness h_s, and the centre of the top tee at z_t below the top of the steel beam, the lever arm a_c is

$$a_c = h_s + z_t - z_c/2 \tag{3.116}$$

The depth z_c depends on the bending moment, and can be conservatively be replaced here by h_c.

For the steel beam of depth h, the lever arm a_b is

$$a_b = h - z_t - z_b \tag{3.117}$$

Taking moments about the top tee at section X, the tension in the bottom tee is given by

$$N_{b,E}a_b = M_E - N_{c,R}a_c \tag{3.118}$$

For rotational equilibrium of the upper half of length s of the beam,

$$(V_c + V_t)s = \Delta N_{c,R}a_c + V_{wp}a_b/2 + M_{wp} \tag{3.119}$$

Similarly for the bottom half,

$$V_b s + M_{wp} = V_{wp}a_b/2 \tag{3.120}$$

The web-post moment is found by eliminating $(V_c + V_t)$ and V_{wp} from Equations 3.113, 3.119 and 3.120:

$$2M_{wp} = (V_E - 2V_b)s - \Delta N_{c,R}a_c \tag{3.121}$$

Where $M_{wp} \geq 0$, as is usual, the maximum bending moment in the web post, at height $h_0/2$ above its centre, is

$$M_{wp,max} = M_{wp} + V_{wp}h_0/2 \tag{3.122}$$

As web-post failure can determine the shear resistance, the next step is to find if the shear resistance is sufficient with $M_{wp} = 0$ and the bottom tee at flexural failure.

Resistance with plastic hinges in the bottom tee At a given cross-section, V_E and M_E are known. Equations 3.114 and 3.121 with $M_{wp} = 0$ give $V_{b,E}$:

$$V_{b,E} = (V_E - \Delta N_{c,R} a_c / s)/2 \tag{3.123}$$

Equations 3.111 and 3.118 give $N_{b,E}$. The plastic resistances of the T-sections are given by Eurocode 3: $N_{T,pl,R}$, $V_{T,pl,R}$ and $M_{T,pl,R}$, say. The effect of axial tension $N_{b,E}$ on the bending resistance of the bottom tee is given by Eurocode 3:

$$M_{bT,N,pl,R} = M_{T,pl,R} \left[1 - \left(N_{b,E}/N_{T,pl,R} \right)^2 \right] \tag{3.124}$$

Obviously, this effect increases towards mid-span. If, as is usual, $V_b \leq V_{T,pl,R}/2$, then $M_{bT,N,pl}$ needs no further reduction to allow for shear.

The bending moment at the critical cross-section of the bottom tee is $V_{b,E} b_0/2$ to the left of section X, combined with shear $V_{b,E}$ and tension $N_{b,E}$, which depends on the bending moment M_E (Equation 3.118). For a hinge at this cross-section, $V_{b,E} = (2/b_0)M_{bT,N,pl,R}$. Then, from Equation 3.123,

$$V_R = (4 / b_0)M_{bT,N,pl,R} + \Delta N_{c,R} a_c / s$$

If a higher shear resistance is needed, it can be found as follows.

Resistance assuming $M_{wp} > 0$, limited by web-post buckling The web post has thickness t and width s_0. It is assumed that the bending moment $M_{wp,max}$, from Equation 3.122, does not exceed its elastic resistance to bending ($M_{wp,pl,R} = f_y t s_0^2/6$). The following simplified method can be used to check buckling.

An effective value is used for the horizontal shear in the web post. For simplicity, it is assumed that $M_{wp,max}$ acts at both of its ends, so that $V_{wp,eff} = 2M_{wp,max}/h_0$. From Equation 3.122

$$V_{wp,eff} = 2M_{wp}/h_0 + V_{wp} \tag{3.125}$$

The web-post shear is found by eliminating M_{wp} from Equations 3.119 and 3.121 and using Equation 3.113:

$$V_{wp} = (V_E s - \Delta N_{c,R} a_c)/a_b \tag{3.126}$$

Eliminating M_{wp} from Equations 3.121 and 3.122

$$2M_{wp,max} = (V_E - 2V_b)s - \Delta N_{c,R} a_c + V_{wp} h_0 \tag{3.127}$$

For buckling, the compressive force in the web post is assumed to be $V_{wp,eff}$, acting in a strut of cross-section $s_0 t$ and length h_0, which can be assumed to be pin-ended. Its slenderness is given by the usual expression, $(N_{pl}/N_{cr})^{1/2}$. The buckling stress σ_c can be found from the relevant curve in Eurocode 3, giving a buckling force $\sigma_c s_0 t$. Hence, for web post buckling,

$$V_{wp,eff,R} = \sigma_c s_0 t \tag{3.128}$$

When the bottom tee is also assumed to be at failure,

$$V_b = M_{bT,N,pl,R}/(b_0/2) \tag{3.129}$$

The vertical shear resistance is now found by eliminating $M_{wp,max}$, V_{wp}, V_b and $V_{wp,eff}$ (which now equals $V_{wp,eff,R}$) from Equations 3.125 to 3.129 and replacing V_E by V_R:

$$V_R = \frac{\sigma_c s_0 t(h_0/s) + 2M_{bT,N,pl,R}/(b_0/2)}{1 + (h_0/a_b)} + \Delta N_{c,R}(a_c/s) \tag{3.130}$$

where $\Delta N_{c,R}$ is the design resistance of the shear connectors within length s.

This is the key expression for the shear resistance, determined by web buckling, failure of the bottom tee and failure of the shear connection, as has been assumed above. The shear V_E must not exceed V_R.

Sharing of vertical shear between the three members When $V_E = V_R$, Equation 3.129 gives the shear in the bottom tee. The shear in the slab depends on $\Delta N_{c,R}$, which is given by Equation 3.112, and assumes the shear connection to be at failure. At high degrees of shear connection, other failure modes are possible.

The vertical shear in the slab may reach the shear resistance to Eurocode 2 of an effective breadth of the slab that depends on its thickness. Where the length of a rectangular opening is several times the depth of the top tee, tests show that the shear connection may not be fully effective. Where the top tee forms plastic hinges at the corners of an opening, the shear connectors above one corner have to resist vertical tension that may cause local pullout failure. For these reasons, there are rules that limit $\Delta N_{c,R}$, the change in slab tension over length s.

These vertical forces apply bending moment from the slab to the steel beam. It has been found that when estimating the vertical shear in the slab, this bending moment can be ignored, as it has been in Figure 3.35b. Then, for rotational equilibrium of the length s of the slab,

$$V_{c,E}s = \Delta N_{c,R}a_c$$

This gives the shear in the slab, and then, from Equation 3.113, the shear in the top tee is

$$V_{t,E} = V_R - V_{c,E} - V_{b,R}$$

If the opening is closer to the top flange of the beam than to the bottom flange, or if the degree of shear connection is low, causing net axial compression in the top tee, design may be governed by its failure, rather than that of the bottom tee. Rules are given for this alternative.

Resistance to bending moment M_E The preceding analysis assumes failure of the shear connection in length x and that the shear connection is just sufficient for the external bending moment M_E. It takes account of the tension in the bottom tee, $N_{b,E}$, given by Equation 3.118 and caused by M_E. That tension is allowed for in the process of assuming $V_{b,R}$ for Equation 3.130, explained above, so sufficient resistance to bending at section X is provided.

Design procedure for this example A typical design process would be to determine the cross-section of the beam and find from its loading the distributions $M_E(x)$ and $V_E(x)$ along the span. The required openings are dimensioned. At a cross-section of type X, the vertical shear must not exceed V_R from Equation 3.130. Several checks of this type would be needed.

4

Continuous Beams and Slabs, and Beams in Frames

4.1 Types of global analysis and of beam-to-column joint

A continuous beam in a building may be interrupted by its supporting internal columns, and both are usually part of a framed structure. The type of beam-to-column joint used influences the global analysis of all the members in the vicinity. The terminology and definitions for joints used in EN 1994-1-1 follow those given in EN 1993-1-8, *Design of joints* (BSI, 2005a), where this distinction is made:

- *Connection*: location at which two members are interconnected, and assembly of connection elements and – in case of a major axis joint – the load introduction into the column web panel.
- *Joint*: assembly of basic components that enables members to be connected together in such a way that the relevant internal forces and moments can be transferred between them.

Thus, the joint in Figure 5.2b, between a beam and an interior column, consists of two connections. Each connection consists of a beam end-plate, a column flange and six bolts, plus a share of the column web panel. This panel is part of both connections.

The column is part of a frame in the plane of the diagram. It could also be part of another frame, the beams of which would be attached to its web (e.g. as in Figure 5.2c). This 'joint' then has two more 'connections'; but in practice this four-connection assembly is designed and detailed as two joints, one in each frame, with little interaction between the two designs.

The term 'joint' is sometimes used less precisely, both here and in the Eurocodes, where the context is obvious.

Beam-to-column joints are classified both by stiffness and by strength. The categories are:

- for stiffness: rigid, semi-rigid, and nominally pinned;
- for strength: full-strength, partial-strength, and nominally pinned.

Calculations determine the two classes of a connection. Where all connections at a joint have the same classes, these are the classes of the joint. Where they have different stiffnesses, the joint is semi-rigid. Where they have different strengths, the joint is partial-strength. To identify whether the behaviour of the joints need be allowed for in the global analysis, Eurocode 3-1-8 relates methods of analysis to classes of joint as shown in Table 4.1.

Composite Structures of Steel and Concrete: Beams, Slabs, Columns and Frames for Buildings,
Fourth Edition. Roger P. Johnson.

Table 4.1 Global analysis, and types of joint and joint model

Method of global analysis	Classification of joint		
Elastic	Nominally pinned	Rigid	Semi-rigid
Rigid-plastic	Nominally pinned	Full-strength	Partial-strength
Elastic-plastic	Nominally pinned	Rigid and full-strength	Semi-rigid and partial-strength
			Semi-rigid and full-strength
			Rigid and partial-strength
Type of joint model	Simple	Continuous	Semi-continuous

The most widely used joint models in current practice are 'continuous' for bridges, and 'simple', for buildings. The scope of this book is limited to the types of joint used in the examples, which are either 'nominally pinned', 'rigid and partial-strength' or 'rigid and full strength', and to global analyses that are either elastic (with redistribution of moments) or rigid-plastic.

A summary of the relevant requirements for these joints, based on EN 1993-1-8, is given in Table 4.2. For classification, Eurocode 4 permits cracking and creep in connected members to be neglected.

Table 4.2 Properties of beam-to-column joints in composite frames

Type of joint	Stiffness	Strength
Nominally pinned	Capable of transmitting the internal forces, without developing significant moments that might adversely affect the members or the structure as a whole	As given for stiffness
	Capable of accepting the resulting rotations under the design loads	
Rigid and partial strength	As for full-strength joints, below	Has a design resistance less than the resistance of at least one of the members connected
Rigid and full strength	Has sufficient rotational stiffness to justify global analysis based on full continuity and ignoring joint rotation	Has design resistances not less than the resistances of the members connected
	Has rigidity such that, under the design loads, the rotations of any necessary plastic hinges do not exceed their rotation capacities	Capable of transmitting the forces and moments calculated in design

The use of nominally-pinned joints simplifies analysis of the structure, because at the assumed location of the pin, there is no bending moment. Action effects in beams are then independent of the properties of the columns that support them. Rigid joints transmit bending moments as well as shear forces, and the bending moments depend on the relative stiffnesses of the members joined.

If the columns of the framed structure shown in Figure 3.1 were not encased in concrete, and the beam-to-column joints were nominal pins, the columns would be designed to EN 1993. However, that frame would still satisfy the definition of 'composite frame' in EN 1994-1-1:

> *'composite frame*: a framed structure in which some or all of the elements are composite members and most of the remainder are structural steel members.'

The columns shown in Figure 3.1 are composite, and some of the top reinforcement in each floor slab may be continued into the columns, to control cracking above the ends of the beams.

In EN 1994-1-1, the definition of 'composite joint' is:

> *'composite joint*: a joint between a composite member and another composite, steel, or reinforced concrete member, in which reinforcement is taken into account in design for the resistance and the stiffness of the joint.'

It follows that if the top reinforcement is 'taken into account in design', the joint is composite. It is permitted, however, to ignore light crack-control reinforcement in design for ultimate limit states, and then to design the structural steel components of the joint to EN 1993.

The example used in Chapter 5 is the two-bay nine-storey frame shown in Figure 5.1. If nominally-pinned joints are used, the member ABC is designed as a beam with two simply-supported spans. If rigid joints are used at B, normally it is the column, not the beam, which is continuous through the joints at B, and the member ABC consists of two *beams in a frame*. The bending moments are found from analysis of the frame, or of a local region of it.

If, however, the line of columns (B, E, etc.) is replaced by a wall, then member ABC can be continuous at B, with no transfer of bending moment to the supporting wall. If nominally-pinned joints are used at A and C, the model for analysis is a two-span *continuous beam*, as in the example in Section 4.6, rather than beams in a frame. Such a beam does not contribute to the resistance of the frame to horizontal loading, for example, from wind. In the example in Chapter 5, this resistance is provided by reinforced concrete walls at each end of the building, to which horizontal loads are transferred by the floor slabs.

The rest of Chapter 4 refers to 'continuous beams'. In general, it applies also to 'beams in frames'. For those, the difference lies in the location and design of the joints, and in the discontinuity in the bending-moment diagram at a supporting internal column, caused by the flexural stiffness of the column.

For a given floor slab and design load per unit length of beam, the advantages of continuous beams over simple spans are:

- shallower beams can be used, for given limits to deflections;
- cracking of the top surface of a floor slab near internal columns can be controlled, so that the use of brittle finishes (e.g. terrazzo) is feasible;
- the floor structure has a higher fundamental frequency of vibration, and so is less susceptible to vibration caused by movements of people;
- the structure is more robust (e.g. in resisting the effects of fire or explosion).

The principal disadvantage is that design is more complex. Actions on one span cause action effects in adjacent spans. Even where the steel section is uniform, the stiffness and bending resistance of a composite beam vary along its length, because of cracking of concrete, changes in effective width, and variation in longitudinal reinforcement in the concrete flange.

It is not possible to predict accurately the stresses or deflections in a continuous beam, for a given set of actions. Apart from the variation over time caused by the shrinkage and creep of concrete, there are the effects of cracking of concrete. In reinforced concrete beams, these occur at cross-sections of both sagging and hogging bending, and so have little influence on distributions of bending moment. In composite beams as used in buildings, significant tension in concrete occurs only in hogging regions. It is influenced by the sequence of construction of the slab, the method of propping used (if any), and by effects of temperature, shrinkage and longitudinal slip.

The flexural stiffness (EI) of a fully cracked composite section can be as low as a quarter of the 'uncracked' value, so a wide variation in flexural stiffness can occur along a continuous beam of uniform section. This leads to uncertainty in the distribution of longitudinal moments, and hence in the amount of cracking to be expected. The response to a particular set of actions also depends on whether it follows another set of actions that caused cracking in a different part of the beam.

For these reasons, and also for economy, design is based as far as possible on predictions of ultimate strength (which can be checked by testing) rather than on analyses based on elastic theory. Methods have been developed from simplified models of behaviour. The limits set to the scope of some models may seem arbitrary, as they correspond to the range of available research data, rather than to known limitations of the model.

Almost the whole of Chapter 3, on simply-supported beams and slabs, applies equally to the sagging moment regions of continuous members. The properties of hogging moment regions of beams are treated in Section 4.2, which applies also to cantilevers. Then follow the global analysis of continuous beams and the calculations of stresses and deflections.

Both rolled steel I or H sections and small plate or box girders are considered, with or without web encasement and composite slabs. The use of slabs composite with the bottom flange of a steel beam is rare in buildings, although it occurs in bridges. The depth of a beam can be reduced by partial embedment of the steel section within the concrete slab (Lawson et al., 1997, 1999). This is 'slim-floor construction', treated in Sections 3.12 and 3.13. Elsewhere, it is assumed that the concrete slab is above the steel member.

The use of precast or prestressed concrete floor slabs in composite frames provides an alternative to composite slabs (Hicks and Lawson, 2003; Couchman, 2014). Precast floors are treated in Section 3.10.

4.2 Hogging moment regions of continuous composite beams

4.2.1 Resistance to bending

4.2.1.1 Effective flange width, and classification of cross-sections

Section 3.5.1, on effective cross-sections of beams, is applicable, except that the effective width of the concrete flange is usually less at an internal support than at mid-span. Longitudinal reinforcement will be provided over the whole width of the slab, but only that within the effective width is assumed to contribute to the hogging moment of resistance of the beam. The plastic neutral axis always lies below the slab, so the only contribution from concrete in compression is from the web encasement, if any.

In EN 1994-1-1, the effective width for analysis of cross-sections is as explained in Section 3.5.1, except that for hogging bending the effective span L_e is the approximate length of the hogging moment region, which can be taken as one-quarter of each span. So at a support between spans of length L_1 and L_2, Equation 3.45 for effective width of a T-beam with pairs of stud connectors at lateral spacing b_0 becomes

$$b_{eff} - [(L_1 + L_2)/4]/4 + b_0 = (L_1 + L_2)/16 + b_0 \tag{4.1}$$

provided that at least $b_{eff}/2$ is present on each side of the web. There is a different rule for cantilevers.

For elastic global analysis, the effective width is taken as the mid-span value throughout each span, except that a lower value is used at a support for a cantilever. That does not affect the bending-moment distribution, which for a cantilever is independent of stiffness, but it does increase its predicted deflection.

The rules for the classification of steel elements in compression (Section 3.5.2) strongly influence the design of hogging moment regions. The proportions of rolled steel I-sections are so chosen that when they act in bending, most webs are in Class 1 or 2. But in a composite section, addition of longitudinal reinforcement in the slab rapidly increases the depth of steel web in compression, αd in Figure 3.14. This figure shows that when $d/t > 60$, an increase in α of only 0.05 can move a web from Class 1 to Class 3, which can reduce the design moment of resistance of the section by up to 30%. This anomaly has led to a rule (BSI, 2004) that allows a web in Class 3 to be replaced (in design) by an 'effective' web in Class 2. This method, known as 'hole-in-the-web', is referred to later. It does not apply to compression flanges, which can usually be designed to be in Class 1 or 2, even where plate girders are used.

Design of hogging moment regions is based on the use of full shear connection (Section 4.2.3).

4.2.1.2 Plastic moment of resistance in hogging bending

A cross-section of a composite beam in hogging bending is shown in Figure 4.1a. The numerical values are for the cross-section that is used in the following worked example and the diagram is to scale for these values (except for b_{eff}). The steel bottom flange is in compression, and its class is easily found, as explained in Section 3.5.2. To classify the web, the distance x_a of the plastic neutral axis above G, the centre of area of the steel section, must first be found.

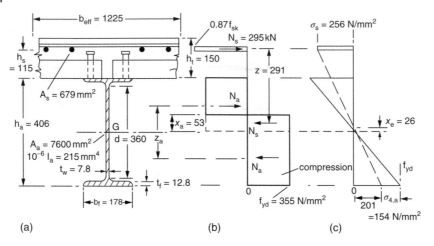

Figure 4.1 Cross-section and stress distributions for composite beam in hogging bending

Let A_s be the effective area of longitudinal reinforcement within the effective width b_{eff} of the slab. Welded mesh is normally excluded, because it may not be sufficiently ductile to ensure that it will not fracture before the design ultimate load for the beam is reached. The design tensile force in this reinforcement is

$$N_s = A_s f_{sk}/\gamma_S = A_s f_{sd} \tag{4.2}$$

where f_{sk} is its characteristic yield strength.

If there were no tensile reinforcement, the bending resistance would be that of the steel section,

$$M_{pl,a,Rd} = W_{a,pl} f_{yd} = N_a z_a \tag{4.3}$$

where $W_{a,pl}$ is the plastic section modulus and f_{yd} is the design yield strength. For rolled sections it is not necessary to calculate the forces N_a in the stress blocks of depth $h_a/2$, nor the lever arm z_a, because values of $W_{a,pl}$ are tabulated; but for plate girders N_a and z_a have to be calculated.

The simplest way of allowing for the force in the reinforcement is to assume that the stress in a depth x_a of web changes from tension to compression, where x_a is given by

$$x_a t_w (2 f_{yd}) = N_s \tag{4.4}$$

provided that (as is usual)

$$x_a \leq h_a/2 - t_f$$

The depth of web in compression is given by

$$\alpha d = d/2 + x_a \tag{4.5}$$

Knowledge of α, d/t_w, and f_y enables the web to be classified, as shown in Figure 3.14 for $f_y = 355 \text{ N/mm}^2$. If, by this method, a web is found to be in Class 4, the calculation should be repeated using the elastic neutral axis, as the curve that separates Class 3 from Class 4 is based on the elastic behaviour of sections. This is why, in Figure 3.14, the ratio ψ is used, rather than α.

Concrete-encased webs in Class 3 are treated as if in Class 2 (Table 3.2), because the encasement helps to stabilize the web.

The lever arm z for the two forces N_s in Figure 4.1b is given by

$$z = h_a/2 + h_s - x_a/2$$

where h_s is the height of the reinforcement above the interface. If both the compression flange and the web are in Class 1 or 2, this is the appropriate model, and the moment of resistance is

$$M_{Rd} = M_{pl,a,Rd} + N_s z \qquad (4.6)$$

If the flange is in Class 1 or 2 and the (uncased) web is in Class 3, it is still possible to use plastic section analysis, by neglecting a region in the centre of the compressed part of the web, which is assumed to be ineffective because of buckling. The calculations are more complex, as explained elsewhere (Johnson, 2012), because this assumption changes the position of the plastic neutral axis, and in plate girders may even move it into the steel top flange. This 'hole-in-the-web' method is not available where the compression flange is in Class 3 or 4.

Example: plastic resistance to hogging bending Figure 4.1a shows a cross-section in a region of hogging moment where the steel section is 406×178 UB 60 with $f_{yd} = 355 \text{ N/mm}^2$ and dimensions as shown. Its plastic section modulus, from tables, is $W_{a,pl} = 1.194 \times 10^6 \text{ mm}^3$. At an internal support between two spans each of 9.0 m, the longitudinal reinforcement is T12 bars, with $f_{sk} = 500 \text{ N/mm}^2$, at 200 mm spacing. The thickness of slab above the profiled sheeting is 80 mm, so the reinforcement ratio is $36\pi/(200 \times 80) = 0.71\%$.

The top cover to 8-mm transverse bars is 20 mm so the T12 bars are $20 + 8 + 12/2 = 34$ mm below the top of the slab, increased to 35 mm to allow for the ribs on both bars (Figure 4.1), giving

$$h_s = 150 - 35 = 115 \text{ mm}$$

The class of the section and its design resistance to hogging moments are now found. From Equation 4.1, assuming that $b_0 = 0.1$ m,

$$b_{eff} = (L_1 + L_2)/16 + b_0 = 18/16 + 0.1 = 1.225 \text{ m}$$

so that six T12 bars are effective, and $A_s = 679 \text{ mm}^2$. It is assumed initially that the web is in Class 1 or 2, so that the rectangular stress blocks shown in Figure 4.1b are relevant. The bottom (compression) flange has $c/t = 5.9$ (Section 3.11.1) and so is in Class 1. From Equation 4.2 with $\gamma_S = 1.15$,

$$N_s = A_s f_{sk}/\gamma_S = 679 \times 0.500/1.15 = 295 \text{ kN}$$

From Equation 4.4,

$$x_a = N_s/(2t_w f_{yd}) = 295/(15.6 \times 0.355) = 53 \text{ mm}$$

From Equation 4.5, the ratio α is

$$\alpha = 0.5 + x_a/d = 0.5 + 53/360 = 0.647$$

The ratio d/t is $360/7.8 = 46.1$. The maximum ratio in EN 1993-1-1 for a Class 2 web is $456\varepsilon/(13\alpha - 1)$, where $\varepsilon = (235/355)^{1/2} = 0.814$, so the limit is

$$d/t \leq \frac{456 \times 0.814}{(13 \times 0.647 - 1)} = 50.1$$

and the web is within Class 2. This can also be seen from Figure 3.14.

From Figure 4.1b, the lever arm for the forces N_s is

$$z = h_a/2 + h_s - x_a/2 = 203 + 115 - 27 = 291 \text{ mm}$$

For the steel section,

$$M_{pl,a,Rd} = W_a f_{yd} = 1.194 \times 355 = 424 \text{ kN m}$$

so from Equation 4.6,

$$M_{pl,Rd} = 424 + 295 \times 0.291 = \mathbf{510 \ kN\,m}$$

4.2.1.3 Elastic moment of resistance in hogging bending

In the preceding calculation, it was possible to neglect the influence of the method of construction of the beam, and the effects of creep, shrinkage and temperature, because these are reduced by inelastic behaviour of the steel, and become negligible before the plastic moment of resistance is reached.

Where elastic section analysis is used, creep is allowed for in the choice of the modular ratio n ($= E_a/E_{c,eff}$), and so has no influence on the flexural properties of cross-sections with no concrete in compression. In buildings the effects of shrinkage and temperature on moments of resistance can usually be neglected, but the method of construction should be allowed for. Here, we assume that at the section considered, the loading causes hogging bending moments $M_{a,Ed}$ in the steel member alone and $M_{c,Ed}$ in the composite member. The small difference ($\approx$3%) between the elastic moduli for reinforcement and structural steel is usually neglected.

The distance x_e of the elastic neutral axis of the composite section (Figure 4.1c) above that of the steel section is found by taking first moments of area about the latter axis:

$$x_e(A_a + A_s) = A_s(h_a/2 + h_s) \tag{4.7}$$

The second moment of area of the composite section is

$$I = I_a + A_a x_e^2 + A_s(h_a/2 + h_s - x_e)^2 \tag{4.8}$$

With unpropped construction, the loading at which the steel bottom flange yields in compression (at level 4 in Figure 3.28a) is usually lower than that at which the slab reinforcement yields in tension. The compressive stress due to the moment $M_{a,Ed}$ is:

$$\sigma_{4,a} = M_{a,Ed}(h_a/2)/I_a \tag{4.9}$$

The remaining stress available is $f_{yd} - \sigma_{4,a}$, so the yield moment is:

$$M_{a,Ed} + M_{c,Rd} = M_{a,Ed} + \frac{(f_{yd} - \sigma_{4,a})I}{(h_a/2 + x_e)} \tag{4.10}$$

The design condition is

$$M_{c,Ed} \leq M_{c,Rd} \tag{4.11}$$

The bending moment $M_{a,Ed}$ causes no stress in the slab reinforcement. In propped construction, for which $M_{a,Ed} \approx 0$, the tensile stress σ_s in these bars may govern design. It is

$$\sigma_s = M_{c,Ed}(h_a/2 + h_s - x_e)/I \tag{4.12}$$

and must not exceed f_{sd}.

4.2.1.4 Example: elastic resistance to hogging bending

Let us now treat the composite section shown in Figure 4.1a as if it were in Class 3, and assume that at the ultimate limit state, a hogging moment of 163 kN m acts on the steel section alone, due to the use of unpropped construction.

What is the design resistance of the section to hogging moments?

From Equation 4.7 the position of the elastic neutral axis of the composite section, neglecting concrete in tension, is given by

$$x_e = \frac{679(203 + 115)}{7600 + 679} = 26 \text{ mm}$$

From Equation 4.8, the second moment of area is

$$10^{-6}I = 215 + 7600 \times 0.026^2 + 679(0.203 + 0.115 - 0.026)^2$$
$$= 278 \text{ mm}^4$$

From tables, the elastic section modulus for the steel section is 1.058×10^6 mm³, so the moment $M_{a,Ed}$ causes a compressive stress at level 4 (the bottom flange):

$$\sigma_{4,a} = 163/1.058 = 154 \text{ N/mm}^2$$

The design yield strength is 355 N/mm², so this leaves 201 N/mm² for resistance to the load applied to the composite member. The elastic neutral axis of the composite section is above that of the steel section, so bending of the composite section increases the stress in the steel top flange less than in the bottom flange, which yields first.

The distance of the bottom fibre from the elastic neutral axis is $h_a/2 + x_e = 203 + 26 = 229$ mm, so the remaining resistance is

$$M_{c,Rd} = \sigma I/y = 201 \times 278/229 = 244 \text{ kN m}$$

The design resistance thus depends on the loading on the steel member alone, and is

$$M_{el,Rd} = M_{a,Ed} + M_{c,Rd} = 163 + 244 = \textbf{407 kN m}$$

From the preceding worked example, the shape factor in this case is

$$Z = M_{pl,Rd}/M_{el,Rd} = 510/407 = 1.25$$

As is usual for composite sections, this exceeds the typical value for steel I-sections, 1.15.

4.2.2 Vertical shear, and moment-shear interaction

As explained in Section 3.5.4, vertical shear is assumed to be resisted by the web of the steel section (Equations 3.61 and 3.62). The action effect V_{Ed} must not exceed the plastic shear resistance $V_{pl,Rd}$ (or some lower value if shear buckling, not considered here, can occur).

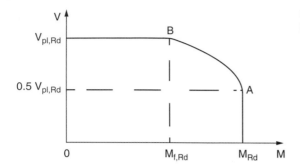

Figure 4.2 Resistance to combined bending and vertical shear

The design rule of EN 1994-1-1 for resistance in combined bending (whether hogging or sagging) and shear is shown in Figure 4.2. It is based on evidence from tests that there is no reduction in bending resistance until $V_{Ed} > 0.5V_{pl,Rd}$ (point A in the figure), and the assumption that the reduction at higher shears follows the parabolic curve AB. At point B the remaining bending resistance $M_{f,Rd}$ is that contributed by the flanges of the composite section, including the reinforcement in the slab. Along curve AB, the reduced bending resistance is given by

$$M_{v,Rd} = M_{f,Rd} + (M_{Rd} - M_{f,Rd})\left[1 - \left(\frac{2V_{Ed}}{V_{pl,Rd}} - 1\right)^2\right] \leq M_{b,Rd} \tag{4.13}$$

where M_{Rd} is the resistance when $V_{Ed} = 0$, and $M_{b,Rd}$ is the resistance to lateral bucking (Section 4.2.4).

When calculating $M_{f,Rd}$, it is usually accurate enough to ignore the reinforcement in the slab. When it is included, or where the steel flanges are of unequal size, only the weaker of the two flanges will be at its design yield stress.

Reference is made in Section 3.5.4 to the additional shear resistance provided by the concrete slab. This would enable $V_{pl,Rd}$ in Figure 4.2 to be replaced by a higher value. A new proposed V-M interaction curve (Vasdravellis and Uy, 2014) would be less convenient than Figure 4.2 because a low value of M_{Ed} would reduce V_{Rd}.

4.2.3 Longitudinal shear

Section 3.6 on longitudinal shear is applicable to continuous beams and cantilevers, as well as to simply-supported spans. Some additional comments, relevant to continuous beams, are now given.

For a typical span with uniformly distributed loading, there are only three critical cross-sections: at the two supports and at the section of maximum sagging moment. Points of contraflexure are not treated as critical sections because their location is different for each load case: a complication best avoided. From Equation 3.65, the number of shear connectors required for a typical critical length is

$$n = (N_c + N_t)/P_{Rd} \tag{4.14}$$

where N_t is the design tensile force in the reinforcement that is assumed to contribute to the hogging moment of resistance, and N_c is the compressive force required in the slab at mid-span, which may be less than the full-interaction value.

When n is calculated, full shear connection is assumed in regions of hogging moment, but as the connectors may be spaced uniformly between a support and the critical section at mid-span, the number provided in the hogging region may not correspond to the force N_t. There is, however, a rule for reinforcement in a hogging region: '... tension reinforcement should be curtailed to suit the spacing of the shear connectors'.

There are several reasons for the apparently conservative rule of EN 1994-1-1 that full shear connection be provided in hogging regions:

(1) To compensate for additional shear that may arise from these simplifications:
 - neglect of the tensile strength of concrete,
 - neglect of strain-hardening of reinforcement,
 - neglect in ultimate-limit-state calculations of reinforcement (e.g. welded mesh) provided for crack-width control.
(2) Because the design resistance of connectors, P_{Rd}, is assumed not to depend on whether the surrounding concrete is in compression or tension. There is evidence that this is slightly unconservative for hogging regions (Johnson et al., 1969), but slip capacity is probably greater, which is beneficial.
(3) For simplicity in design, including design for lateral buckling (Section 4.2.4) and for vertical shear with tension-field action (Allison et al., 1982).

Transverse reinforcement   As explained for regions of sagging bending, this reinforcement should be related to the shear resistance of the connectors provided, even where, for detailing reasons, their resistance exceeds the design longitudinal shear.

4.2.4   Lateral buckling

Conventional 'non-distortional' lateral torsional buckling occurs where the top flange of a simply-supported steel beam of I section has insufficient lateral restraint in the mid-span region. Both flanges are assumed to be restrained laterally at the supports, where the member may be free to rotate about a vertical axis. The top flange, in compression, is prevented by the web from buckling vertically, but if the ratio of its breadth b_f to the span L is low, it may buckle laterally as shown in Figure 4.3a. The cross-section rotates about a longitudinal axis, but maintains its shape.

It has to be checked that lateral-torsional buckling does not occur during casting of the concrete for an unpropped composite beam; but once the concrete has hardened,

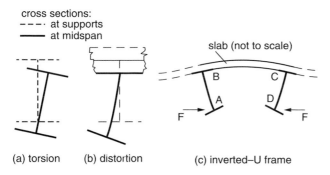

cross sections:
--- · at supports
—— at midspan

(a) torsion   (b) distortion   (c) inverted–U frame

Figure 4.3 Torsional and distortional lateral buckling

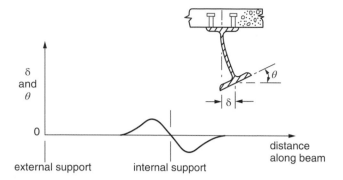

Figure 4.4 Typical deformation of steel bottom flange in distortional lateral buckling

the shear connection prevents buckling of this type. The relevant design methods, being for non-composite beams, are outside the scope of this book.

In hogging regions of continuous beams, the compressed bottom flange of the steel section can only buckle if the web bends, as shown in Figure 4.3b. This is known as 'distortional' lateral buckling, and is the subject of this Section.

The buckle consists of a single half-wave each side of an internal support, where lateral restraint is invariably provided. The half-wave extends over most of the length of the hogging moment region. It is not sinusoidal, as the point of maximum lateral displacement is within two or three beam depths of the support, as shown in Figure 4.4.

It is unlike local flange buckling, where the movement is essentially vertical, not lateral, and where the cross-section of maximum displacement is within one flange width of the support. There is some evidence from tests that local buckling can initiate lateral buckling, but in design they are considered separately, and in different ways. Local buckling is allowed for by the classification system for steel elements in compression (Section 3.5.2). Lateral buckling is avoided by reducing the design moment of resistance at the internal support, M_{Rd}, to a lower value, $M_{b,Rd}$. Local buckling occurs where the breadth-to-thickness ratio of the flange (b_f/t_f) is high; lateral buckling occurs where it is low.

Where, as is usual in buildings, the beam is one of several parallel members, all attached to the same concrete or composite slab, design is usually based on the 'continuous inverted-U-frame' model. The tendency for the bottom flange to displace laterally causes bending of the steel web, and twisting at top-flange level, which is resisted by bending of the slab, as shown in Figure 4.3c.

The design method of Eurocode 4, explained below, is quite complex, so it is followed by a 'simplified verification'. This gives upper limits to the depths of uncased steel I-sections for which no check on lateral buckling is needed. Details are given in Section 4.2.4.4. That approach is widely used, so readers may wish to omit Sections 4.2.4.1 and 4.2.4.2. The worked example in Section 4.6.3 considers the two-span continuous beam both with and without web encasement.

4.2.4.1 Elastic critical moment

Design to EN 1994-1-1 is based on the elastic critical moment M_{cr} at the internal support. The theory for M_{cr} considers the response of a single U-frame (ABCD in Figure 4.3c) to equal and opposite horizontal forces F at bottom-flange level. It leads to the following rather complex expression for M_{cr}:

$$M_{cr} = (k_c C_4/L)\left[\left(GI_{at} + k_s L^2/\pi^2\right) E_a I_{afz}\right]^{1/2} \tag{4.15}$$

where E_a and G are the elastic modulus and shear modulus of steel; I_{at} is the St Venant torsion constant for the steel section; I_{afz} is $b_f^3 t_f/12$ for the steel bottom flange; and L is the span.

Where the steel section is symmetric about both axes, k_c is a property of the composite section (with properties A and I_y) given by

$$k_c = \frac{h_s I_y/I_{ay}}{\frac{h_s^2/4 + (I_{ay} + I_{az})/A_a}{e} + h_s} \tag{4.16}$$

where

$$e = \frac{A I_{ay}}{A_a z_c (A - A_a)} \tag{4.17}$$

and A_a, I_{ay}, and I_{az} are properties of the structural steel section. (It should be noted that in Eurocodes, and here, subscripts y and z refer to the major and minor axes of a steel section, respectively. Former British practice used x and y.) The dimensions h_s and z_c are shown in Figure 4.5.

The term k_s is the stiffness of the U frame, per unit length along the span, which opposes lateral displacement of the bottom flange. It relates a disturbing force F per unit length of beam (Figure 4.3c) to the lateral displacement of a flange, δ, caused by force F, as follows. The rotation at B that would cause displacement δ is δ/h_s; and the bending moment at B is Fh_s. The stiffness k_s is moment/rotation, so

$$k_s = Fh_s/(\delta/h_s) \quad \text{hence,} \quad \delta = Fh_s^2/k_s$$

The flexibility $1/k_s$ is the sum of the flexibilities of the slab, denoted $1/k_1$, and of the steel web, denoted $1/k_2$, so that

$$k_s = k_1 k_2/(k_1 + k_2) \tag{4.18}$$

The stiffness of the slab is represented by k_1. Where the slab is in fact continuous over the beams, even when it is designed as simply-supported, the stiffness may be taken as

$$k_1 = 4E_a I_2/a \tag{4.19}$$

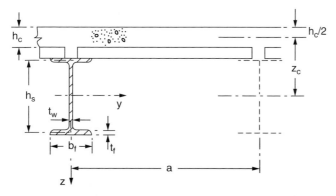

Figure 4.5 Inverted-U frame model for distortional lateral buckling

where a is the spacing of the beams and $E_a I_2$ is the 'cracked' flexural stiffness of the slab above the beams.

The stiffness of the web is represented by k_2. For an uncased web,

$$k_2 = \frac{E_a t_w^{\ 3}}{4(1 - v_a^2)h_s} \tag{4.20}$$

where v_a is Poisson's ratio for steel. For an I-section with the web (only) encased in concrete,

$$k_2 = \frac{E_a t_w b_f^{\ 2}}{16 h_s (1 + 4n t_w / b_f)} \tag{4.21}$$

where n is the modular ratio for long-term effects, and b_f is the width of the steel flange. Equation 4.21 was derived by elastic theory, treating the concrete on one side of the web as a strut that restrains upwards movement of the steel bottom flange below it.

The buckling moment M_{cr} is strongly influenced by the shape of the bending-moment distribution for the span considered. This is allowed for by the factor C_4 in Equation 4.15, values for which were obtained by finite-element analyses. They range from 6.2 for uniform hogging moment, to above 40 where the region of hogging moment is less than a tenth of the span. Values relevant to the design example are given in Figure 4.6.

In Equation 4.15 the term GI_{at} gives the contribution from St Venant torsion of the section. It is usually small compared with $k_s L^2/\pi^2$ and can then be neglected with little loss of economy. The expression then becomes

$$M_{cr} \approx (k_c C_4 / \pi)(k_s E_a I_{afz})^{1/2} \tag{4.22}$$

which is independent of the span L. This enables the values of C_4 to be used for all span lengths.

Equation 4.15 for M_{cr} is valid only where rules for minimum spacing of connectors, bending stiffness of the composite slab, and proportions of the steel I-section, are satisfied. A more detailed explanation of this method, simplified versions of some of its rules and values for C_4 are available (Johnson, 2012).

4.2.4.2 Buckling moment

The value M_{cr} is relevant only for an initially perfect member that remains elastic. Evidence is limited on the influence of initial imperfections, residual stresses, and yielding of steel on this type of buckling, but the Perry-Robertson formulation and the strut curves developed for overall buckling of steel columns provide a suitable basis. The method of EN 1994-1-1 is therefore as follows.

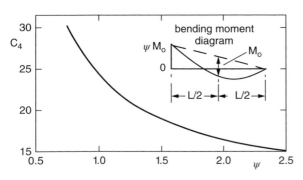

Figure 4.6 Factor C_4 for an end span of a continuous beam

The slenderness $\bar{\lambda}_{LT}$ is given for a Class 1 or 2 section by

$$\bar{\lambda}_{LT} = (M_{Rk}/M_{cr})^{1/2} \tag{4.23}$$

where M_{Rk} is the value that would be obtained for M_{Rd} in hogging bending if the partial factors γ_A and γ_S were taken as 1.0. This is because these factors do not occur in the calculation of M_{cr}. For a Class 3 section, M_{Rk} is the characteristic yield moment.

The buckling moment is given by

$$M_{b,Rd} = \chi_{LT} M_{Rd} \tag{4.24}$$

where χ_{LT} is a function of $\bar{\lambda}_{LT}$ that in practice is taken from the relevant strut curve in Eurocode 3: Part 1.1. For rolled steel sections this curve is given by

$$\chi_{LT} = \left[\Phi_{LT} + \left(\Phi_{LT}^2 - \bar{\lambda}_{LT}^2 \right)^{1/2} \right]^{-1} \quad \text{but} \quad \chi_{LT} \leq 1 \tag{4.25}$$

where

$$\Phi_{LT} = 0.5 \left[1 + \alpha_{LT} \left(\bar{\lambda}_{LT} - \bar{\lambda}_{LT,0} \right) + \beta \bar{\lambda}_{LT}^2 \right] \tag{4.26}$$

and α_{LT} is an imperfection factor. Eurocode 4 refers to clause 6.3.2.3 of Eurocode 3-1-1 on 'lateral-torsional buckling', which recommends use of buckling curve b for rolled sections with $h_a/b_f \leq 2$ and curve c otherwise, with imperfection factors 0.34 and 0.49, respectively. If $\bar{\lambda}_{LT} \leq \bar{\lambda}_{LT,0}$, there is no reduction for buckling. EN 1993-1-1 recommends $\bar{\lambda}_{LT,0} = 0.4$ and $\beta = 0.75$, and the UK's National Annex confirms this choice for rolled sections. Their effect is that lateral buckling does not reduce M_{Rd} until $\bar{\lambda}_{LT} > 0.4$. The treatment of this subject could be clearer, because neither code mentions *distortional* lateral buckling.

Simplified expression for $\bar{\lambda}_{LT}$ For cross-sections in Class 1 or 2, and with some loss of economy, Equation 4.23 can be replaced by

$$\bar{\lambda}_{LT} = 5.0 \left[1 + \frac{t_w h_s}{4 b_f t_f} \right] \left[\left(\frac{f_y}{E_a C_4} \right)^2 \left(\frac{h_s}{t_w} \right)^3 \left(\frac{t_f}{b_f} \right) \right]^{0.25} \tag{4.27}$$

provided that the steel section is symmetrical about both axes. Further details and an account of its derivation are available (Johnson, 2012).

4.2.4.3 Use of bracing
Where the buckling resistance of a beam has to be checked, and has been found using Equation 4.24 to be less than the required resistance, the possibilities are as follows.

(1) Use a steel section with a less slender web or an encased web.
(2) Provide lateral bracing to compression flanges in the hogging moment region.

Lateral bracing is commonly used in bridges, but is less convenient in buildings, where the spacing between adjacent beams is usually wider, relative to their depth. Some examples of possible types of bracing are given in a publication on lateral buckling of haunched composite beams (Lawson and Rackham, 1989).

Little else has been published on the use of bottom-flange bracing for beams in buildings. It interferes with the provision of services, and is best avoided.

4.2.4.4 Exemption from check on buckling

Extensive computations based on $\overline{\lambda}_{LT} = 0.4$ have enabled conditions to be given in EN 1994-1-1 under which no detailed check on resistance to lateral buckling need be made. The principal condition relates to the overall depth h_a of the steel I-section. For IPE sections of steel with $f_y = 355\,\text{N/mm}^2$, $h_a \leq 400\,\text{mm}$ or, if the web is encased, $h_a \leq 600\,\text{mm}$.

The European IPE sections generally have thicker webs than British UB sections, many of which do not qualify for this relaxation. A method for determining which sections qualify is referred to in the UK's national annex to EN 1994-1-1 (Johnson, 2012), and is now explained.

It can be seen from Equation 4.27 that the slenderness $\overline{\lambda}_{LT}$ depends on what is known as the 'section property' of the section, given by

$$F = \left[1 + \frac{t_w h_s}{4 b_f t_f}\right]\left[\left(\frac{h_s}{t_w}\right)^3\left(\frac{t_f}{b_f}\right)\right]^{0.25} \tag{4.28}$$

and on the yield strength of the steel and the distribution of bending moment, represented by C_4. The symbols are shown in Figure 4.5.

The limiting values for F below which exemption applies for British cross-sections have been deduced from the Eurocode limits of F for the European IPE and HE sections. For UB sections in S355 steel the limit is 12.3, increased to 15.8 for web-encased sections. The method is slightly conservative. It is used in Section 4.6.3, and is expected to be included in the next Eurocode 4.

4.2.5 Cracking of concrete

> *Cracking is normal in reinforced concrete structures subject to bending, shear, torsion, or tension resulting from either direct loading or restraint of imposed deformations.*

This quotation from EN 1992-1-1 distinguishes between two types of cracking. These are treated separately in EN 1994-1-1, which follows closely the rules for crack-width control given in EN 1992. Even where no reinforcement is required to resist 'direct loading', it is necessary to limit the widths of cracks that result from tensile strains imposed on the element considered. The origin of these strains can be 'extrinsic' (external to the member), such as differential settlement of the supports of a continuous beam, or 'intrinsic' (inherent in the member), such as a temperature gradient or shrinkage of the concrete.

In reinforced concrete, cracking has little influence on tensile forces caused by direct loading, but it reduces stiffness, and so reduces the tensile force caused by an imposed deformation. For example, if shrinkage of concrete causes tension in a restrained concrete member, the tension is reduced when the member cracks.

Calculations for load-induced cracking are therefore based on the tensile force in the reinforcement after cracking (i.e. on the analysis of cracked cross-sections), whereas calculations for restraint cracking are based on the tensile force in the concrete just before it cracks.

These concepts are more difficult to apply to composite members, where there is local restraint from the axial and flexural stiffnesses of the structural steel component, applied through the shear connectors or by bond. In a web-encased beam, for example, where the steel tension flange is stressed by direct loading, the resulting strains and curvature impose a deformation on the concrete that encases the web.

The presence of structural steel members, and the differences between beam-to-column joints in composite and reinforced concrete frames, made it impossible to cover cracking in Eurocode 4 simply by cross-reference to Eurocode 2 for reinforced concrete. This led to a 'stand-alone' treatment of cracking in a slab that is part of the tension flange of a composite beam, and in the concrete encasement of a steel web.

> *Cracking shall be limited to an extent that will not impair the proper functioning or durability of the structure or cause its appearance to be unacceptable.*

This quotation, also from EN 1992-1-1, refers to function and appearance. For the concrete of composite members for most types of building, the most likely cause of impairment to 'proper functioning' is corrosion of reinforcement, following breakdown of its protection by the surrounding concrete. Design is based on the exposure classes given in EN 1992-1-1. They influence the specifications of type of concrete and minimum cover to reinforcement, as well as the limiting crack width.

Within a building, the relevant class is likely to be X0 (very dry environment) or XC1 (dry environment). For these classes, the limiting surface crack width under the quasi-permanent load combination is recommended to be 0.4 mm, with the comment:

> *For X0, XC1 exposure classes, crack width has no influence on durability, and this limit is set to guarantee acceptable appearance. In the absence of appearance conditions, this limit may be relaxed.*

Tighter control of crack width is normal in bridges, and is sometimes needed in buildings: for example, in the humid environment of a laundry, or in an open-air multi-storey car park. For these and most other environments, the recommended limiting crack width is 0.3 mm for reinforced concrete, or 0.2 mm for some types of prestressed concrete. Prestressing of composite members is rare, because it is difficult to avoid prestressing the steel component also, which increases the size of tendons needed.

The appearance of a concrete surface may be important where a web-encased beam is visible from below, but the top surface of a slab is usually concealed by the floor finish or roof covering. Where the finish is flexible (e.g. a fitted carpet) there may be no need to specify a limit to the width of cracks; but for brittle finishes or exposed concrete surfaces, crack-width control is essential.

Eurocode 2 recommends values from 0.2–0.4 mm for a 'limiting calculated crack width, w_{max}', and also gives methods for calculating a characteristic value, w_k for which the same limiting values are used. The usual interpretation of a characteristic value is one with a 5% probability of exceedence. Crack width is a random variable, but the concept of 'probability of exceedence' is difficult to apply in practice, and a 20% probability of exceedence has been used in the UK and in research (Johnson and Allison, 1983).

Design rules are given in EN 1994-1-1 for the following situations:

(1) where 'the control of crack width is of no interest' and beams are designed as simply-supported although the slab is continuous over supports;
(2) for the control of restraint-induced cracking, based on the tensile strength of the concrete;
(3) for load-induced cracking, with control of crack width to 0.2, 0.3 or 0.4 mm;
(4) for the calculation of estimated crack width and maximum final crack spacing.

For cases (1) to (3), simplified rules are given that do not involve the calculation of crack widths. These are outlined below. For case (4), reference is made to provisions in EN 1992-1-1 for reinforced concrete members. This situation rarely arises in buildings and is not considered further.

4.2.5.1   No control of crack width

'No control' is relevant to serviceability limit states. As it is still necessary to ensure that the concrete retains sufficient integrity to resist shear at ultimate limit states, by acting as a continuum, it is required that the minimum longitudinal reinforcement in a concrete flange in tension shall be not less than:

- 0.4% of the area of concrete, for propped construction, or $\qquad$ (4.29)

- 0.2% of the area of concrete, for unpropped construction. $\qquad$ (4.30)

These bars are likely to yield at cracks, which may be about 0.5 mm wide, but they ensure that several cracks form rather than just one, which could be much wider. The presence of profiled steel sheeting is usually ignored, which may be conservative in some situations.

4.2.5.2   Control of restraint-induced cracking

Uncontrolled cracking between widely-spaced bars is avoided, and crack widths are limited, by:

- using small-diameter bars, which have better bond properties and have to be more closely spaced than larger bars;
- using 'high bond' (ribbed) bars or welded mesh;
- ensuring that the reinforcement remains elastic when cracking first occurs.

The last of these requirements is relevant to restraint cracking, and leads to a design rule for minimum reinforcement, irrespective of the loading, as follows.

Let us assume that an area of concrete in uniform tension, A_{ct}, with an effective tensile strength $f_{ct,eff}$, has an area A_s of reinforcement with characteristic yield strength f_{sk}. Just before the concrete cracks, the force in it is approximately $A_{ct} f_{ct,eff}$. Cracking transfers the whole of the force to the reinforcement, which will not yield if

$$A_s f_{sk} \geq A_{ct} f_{ct,eff} \qquad (4.31)$$

This condition is modified, in EN 1994-1-1, by a factor 0.8 that takes account of self-equilibrating stresses within the member (that disappear on cracking), and by a factor

$$k_c = \frac{1}{1 + (h_c/2z_0)} + 0.3 \leq 1.0 \qquad (4.32)$$

that allows for the non-uniform tension in the concrete prior to cracking. In Equation 4.32, h_c is the thickness of the concrete flange, excluding any ribs, and z_0 is the distance of the centroid of the uncracked composite section (for short-term loading) below the centroid of the concrete flange. For composite beams, the ratio z_0/h_c typically increases with the span. Where it exceeds 1.17, $k_c = 1.0$, and the tension is effectively treated as uniform over the thickness of the concrete flange.

Finally, f_{sk} in Equation 4.31 is replaced by σ_s, the maximum stress permitted in the reinforcement immediately after cracking ($\leq f_{sk}$), which influences the crack width. This leads to the design rule

$$A_s \geq 0.8k_c f_{ct,eff} A_{ct}/\sigma_s \tag{4.33}$$

To use this rule, it is necessary to estimate the value of the tensile strength $f_{ct,eff}$ when the concrete first cracks. If the intrinsic deformation due to the heat of hydration or the shrinkage of the concrete is large, cracking could occur within a week of casting, when $f_{ct,eff}$ is still low. Where this is uncertain, EN 1994-1-1 permits the use of the mean value of the tensile strength, f_{ctm}, corresponding to the specified 28-day strength of the concrete. It is approximately $0.1f_{ck}$, or $0.08f_{cu}$, where f_{cu} is the specified cube strength.

The choice of the stress σ_s depends on the design crack width, w_k, the diameter ϕ of the reinforcing bars, and the value of $f_{ct,eff}$. Figure 4.7 gives the diameters ϕ^* for concrete with the reference strength $f_{ct,0} = 2.9 \, \text{N/mm}^2$ (the value of $f_{ct,eff}$ for class C30/37 concrete). For other concrete strengths, the diameter is given in EN 1994-1-1 by

$$\phi = \phi^* f_{ct,eff}/f_{ct,0}$$

For bars with the usual strength f_{sk}, $500 \, \text{N/mm}^2$, stresses from service loading can be up to about $300 \, \text{N/mm}^2$. Figure 4.7 shows that for efficient use of the strength of bars provided as minimum reinforcement, their diameter cannot exceed about 12 mm.

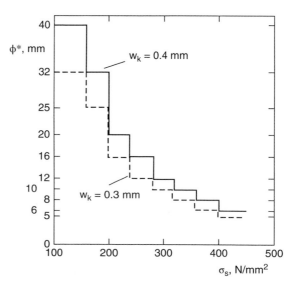

Figure 4.7 Maximum steel stress for minimum reinforcement, high bond bars

4.2.5.3 Control of load-induced cracking

For simply-supported beams in buildings, there may be no cross-sections at risk of load-induced cracking. For continuous beams, cracks are likely to be widest near internal supports, caused by a bending moment to be found from a global analysis.

In EN 1992-1-1, the use of the quasi-permanent load combination (Section 1.3.2.3) is recommended. Imposed loads are then lower than for the characteristic combination, which is used for deciding where minimum reinforcement is required. The resulting bending-moment envelope thus has a lower proportion of each span subjected to hogging bending.

The use of the quasi-permanent combination also implies that there are no adverse effects if the cracks are wider for short periods when heavier variable loads are present. It may sometimes be necessary to check crack widths for a less probable load level, either 'frequent' or 'characteristic'. Where unpropped construction is used, load resisted by the steel member alone is excluded.

Elastic global analysis (Section 4.3.2) is used. The simplest method is to use relative stiffnesses of members based on uncracked unreinforced cross-sections. As cracked regions of hogging moment are less stiff than uncracked mid-span regions, this over-estimates the hogging moments, typically by about 10%. For crack control, this is a 'safe' approximation. Neglect of the effects of shrinkage of concrete has the opposite effect.

The tensile stress in the reinforcement nearest to the relevant concrete surface is calculated by elastic section analysis, neglecting concrete in tension. This stress, $\sigma_{s,o}$, is then increased to a value σ_s by a correction for tension stiffening, given by

$$\sigma_s = \sigma_{s,o} + 0.4 f_{ctm} A_{ct}/(\alpha_{st} A_s) \tag{4.34}$$

where $\alpha_{st} = AI_2/A_a I_a$. The values A and I_2 are for the cracked transformed composite section, and A_a and I_a are for the structural steel section. Elastic properties of the uncracked section are also needed, to find A_{ct}, the area of concrete in tension. This is not divided by the modular ratio. A_s is the area of reinforcement within the area A_{ct}.

The correction to $\sigma_{s,o}$ is relatively small, and with experience can be estimated. It is largest for lightly reinforced slabs (high A_{ct}/A_s) of strong concrete (high tensile strength, f_{ctm}). The basis of Equation 4.34 is that until cracking is fully developed, the curvature of the steel beam equals the mean curvature of the slab. Its actual curvature at cracks is above the mean, which increases the stress in the bars crossing the crack from $\sigma_{s,o}$ to σ_s.

Crack control is achieved by limiting the spacing of the longitudinal reinforcing bars to the values shown in Figure 4.8, which depend on σ_s and w_k, or by limiting the bar diameter in accordance with Figure 4.7. It is not necessary to satisfy both requirements, for they have a common basis. This is that crack-width control relies on the bond-stress/slip property of the surface of the reinforcement, which is almost independent of bar diameter. The higher the stress σ_s, the greater the bar perimeter required, to limit the bond slip at cracks to an acceptable level.

For bars of total area A_s per unit width of slab, of diameter ϕ and at spacing s, the total bar perimeter is $u = \pi\phi/s$, and $A_s = \pi\phi^2/(4s)$. From these equations,

$$u = 4A_s/\phi = 2(\pi A_s/s)^{1/2}$$

Figure 4.8 Maximum bar spacing for high bond bars

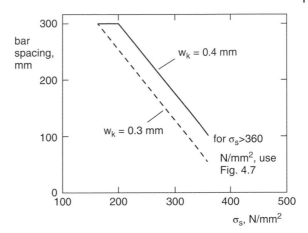

Thus, for a given area A_s, limiting either ϕ or s will give the required value of u. The limits to ϕ and s become more severe as σ_s is increased and as w_k is reduced, as shown in Figures 4.7 and 4.8.

Fuller explanation and discussion of crack-width control for concrete flanges of composite beams is available (Johnson and Allison, 1983; Johnson, 2012).

4.3   Global analysis of continuous beams

4.3.1   General

The subject of Section 4.3 is the determination of design values of bending moment and vertical shear for 'continuous beams' as defined in Section 4.1, caused by the actions specified for both serviceability and ultimate limit states.

Methods based on linear-elastic theory, treated in Section 4.3.2, are applicable for all limit states and all four classes of cross-section. The use of rigid-plastic analysis, also known as plastic hinge analysis, is applicable only for ultimate limit states, and is subject to the restrictions explained in Section 4.3.3. Where it can be used, the resulting members may be lighter and/or shallower, and the analyses are simpler. This is because the design moments for one span are in practice independent of the actions on adjacent spans, of variation along the span of the stiffness of the member, of the sequence and method of construction, and of the effects of temperature and of creep and shrinkage of concrete. Accurate elastic analysis has none of these advantages, and so is used with simplifications.

Section 3.5.1, on effective cross-sections of simply-supported beams, applies also to mid-span regions of continuous beams. For analysis of cross-sections, effective widths of hogging moment regions are generally narrower than those of mid-span regions (Section 4.2.1), but for simplicity, effective widths for global analysis are assumed to be constant over the whole of each span, and are taken as the value at mid-span. This does not apply to cantilevers, where the value at the support is used.

It is assumed in global analysis that the effects of longitudinal slip are negligible. This is appropriate for hogging moment regions, where the use of partial shear connection is not permitted. Slip in a mid-span region slightly reduces the flexural stiffness, but for current levels of minimum shear connection, the uncertainty is probably less than that which results from cracking of concrete in hogging regions.

The sagging curvature of a composite beam or slab caused by shrinkage increases its deflection (Section 3.8). In a continuous beam, shrinkage also causes hogging bending moment at internal supports. There is a worked example in Johnson (2012).

EN 1994-1-1 permits shrinkage effects to be neglected in checks for ultimate limit states under conditions that apply to most composite beams in buildings. For serviceability checks, the stresses that would occur in a simply-supported beam ('primary stresses') can be neglected. In continuous beams with normal-density concrete, shrinkage curvatures can be neglected where the ratio of span to overall depth does not exceed 20. These exemptions are assumed to apply in the worked examples in this book.

4.3.2 Elastic analysis

Elastic global analysis requires knowledge of relative (but not absolute) values of flexural stiffness (EI) over the whole length of the members analysed. Several different values of EI are required at each cross-section, as follows:

(a) for the steel member alone ($E_a I_a$), for actions applied before the member becomes composite, where unpropped construction is used;
(b) for permanent loading on the composite member ($E_a I$), where I is determined, in 'steel' units, by the method of transformed sections using a modular ratio, $n = E_a / E_{c,eff}$, where $E_{c,eff}$ is an effective modulus that allows for creep of concrete;
(c) for variable loading on the composite member, as above, except that the modular ratio is $n_0 = E_a / E_{cm}$, and E_{cm} is the mean secant modulus for short-term loading.

The values (b) and (c) also depend the sign of the bending moment. In principle, separate analyses are needed for the actions in (a), in (b), and for each relevant arrangement of variable loading in (c).

In practice, the following simplifications are made wherever possible.

(1) A value I calculated for the uncracked transformed composite section (denoted I_1 in EN 1994-1-1) is used throughout the span. The transformation is always from concrete to steel, so the appropriate Young's modulus is E_a. There is no requirement to include the reinforcement, as its presence makes little difference. This is referred to as 'uncracked' analysis.
(2) A single value of I, based on a modular ratio that is approximately $\frac{1}{2}[(E_a / E_{c,eff}) + (E_a / E_{cm})]$, is used for analyses of both types (b) and (c).
(3) Where all spans of the beam have cross-sections in Class 1 or 2 only, the influence of method of construction is neglected in analyses for ultimate limit states, and actions applied to the steel member alone are included in analyses of type (b), rather than type (a).

Separate analyses of type (c) are always needed for different arrangements of variable loading. It is often convenient to analyse the member for unit distributed loading on each span in turn, and then obtain the moments and shears for each load arrangement by scaling and combining the results.

The alternative to 'uncracked' analysis is to use, in regions where the slab is cracked, a reduced value of I (denoted I_2 in EN 1994-1-1), calculated neglecting concrete in tension but including its reinforcement. This is known as 'cracked' analysis. Its weakness is that there is no simple or accurate method for deciding which parts of each span are 'cracked'. They are different for each load arrangement, and are modified by the effects of tension stiffening, previous loadings, temperature, creep, shrinkage, and longitudinal slip. A common assumption is that 15% of each span, adjacent to each internal support, is 'cracked'.

In practice, 'uncracked' analysis is usually preferred for ultimate limit states, with allowance for cracking by redistribution of moments. Deflections should be estimated with allowance for cracking, as explained in Section 4.3.2.3.

4.3.2.1 Redistribution of moments in continuous beams

Redistribution is a well-established and simple approximate method for modifying the results of an elastic global analysis for design ultimate loadings, excluding fatigue. It allows for the inelastic behaviour that occurs in all materials in a composite beam before maximum load is reached, and for the effects of cracking of concrete at serviceability limit states. It is also used in the analysis of beams and frames of structural steel and of reinforced concrete, with limitations appropriate to the material and type of member.

It consists of modifying the bending-moment distribution found for a particular loading while maintaining equilibrium between the actions (loads) and the bending moments. Moments are reduced at cross-sections where the ratio of action effect to resistance is highest (usually at the internal supports). The effect is to increase the moments of opposite sign (usually in the mid-span regions).

For continuous composite beams, the ratio of action effect to resistance is higher at internal supports, and lower at mid-span, than for most beams of a single material, and the use of redistribution is essential for economy in design. It is limited by the onset of local buckling of steel elements in compression, as shown in Table 4.3, which is given in EN 1994-1-1, subject to several conditions. For example, each span must be of uniform depth, most of its loading must be uniformly distributed, and it must not be susceptible to lateral-torsional buckling.

The figures show that reliance on cracking to reduce hogging moments increases with the resistance of the cross-section to local buckling.

The hogging moments referred to are the peak values at internal supports, which do not include supports of cantilevers (at which the bending moment is determined by equilibrium and cannot be changed). Where the composite section is in Class 3 or 4, moments due to loads on the steel member alone are excluded. The values in Table 4.3 are based on research and analyses of continuous beams with various proportions and span ratios (e.g. Johnson and Chen, 1991).

Table 4.3 is reproduced here with the title used in the current Eurocode 4, where it is not clear whether its values apply for all limit states. In the revised Eurocode 4, there will be added to the title: '... for ultimate limit state verifications other than fatigue'. There is

Table 4.3 Limits to redistribution of hogging moments, as a percentage of the initial value of the bending moment to be reduced

Class of cross-section in hogging moment region	1	2	3	4
For 'uncracked' elastic analysis	40	30	20	10
For 'cracked' elastic analysis	25	15	10	0

likely to be a new rule: that where uncracked elastic analysis is used for a serviceability limit state, up to 10% redistribution of hogging moments may be used, irrespective of the Class of the cross-sections.

The use of Table 4.3, and the need for redistribution, is illustrated in the following example. The Eurocode also allows limited redistribution from mid-span regions to supports, but this is rare in practice.

4.3.2.2 Example: redistribution of moments

A composite beam of uniform section (apart from reinforcement) is continuous over three equal spans L. The cross-sections are in Class 1. There is no flexural interaction with the supports; for example, they could be of masonry. For the ultimate limit state, the design permanent load is g per unit length, and the variable load is q per unit length, with $q = 2g$. The sagging moment of resistance, M_{Rd}, is twice the hogging moment of resistance, M'_{Rd}. Find the minimum required value for M_{Rd}:

(a) by elastic analysis without redistribution;
(b) by elastic analysis with redistribution to Table 4.3;
(c) by rigid-plastic analysis.

For simplicity, only the middle span is considered, and only symmetrical arrangements of variable load.

Bending moment distributions for the middle span, ABC, given by 'uncracked' elastic analysis, are shown in Figure 4.9 for permanent load plus the following arrangements of variable load:

(1) q on all spans,
(2) q on the centre span only,
(3) q on the end spans only.

The moments are given as multiples of $gL^2/8$. Questions (a) to (c) are now answered.

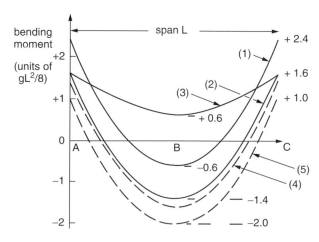

Figure 4.9 Bending moment diagrams with redistribution

(a) Without redistribution, the peak hogging moment, $2.4gL^2/8$, curve (1), governs the design, and since $M_{Rd} = 2 M'_{Rd}$,

$$M_{Rd} \geq 4.8gL^2/8$$

(b) The peak hogging moment at each support is reduced by 40% to $1.44gL^2/8$, curve (4). The corresponding sagging moment is $(0.6 + 0.96)gL^2/8 = 1.56gL^2/8$. Elastic analysis for loadings (2) and (3) gives curves (2) and (3), respectively. Redistribution of 10% is used, so that their peak hogging moments are also $1.44gL^2/8$. This value governs the design, so that

$$M_{Rd} \geq 2.88gL^2/8$$

(c) The method used is explained in Section 4.3.3. Redistribution is unlimited, so that support moments for loading (1) are reduced by 58%, to $1.0gL^2/8$. The corresponding sagging moment is $(0.6 + 1.4)gL^2/8$ (curve 5). Smaller redistributions are required for the other loadings. The available resistances at the supports and at mid-span are fully used, when

$$M_{Rd} = 2.0gL^2/8$$

The preceding three results for M_{Rd} show that the bending resistance required is significantly reduced when the degree of redistribution is increased. A change to a smaller cross-section of course makes it more likely that the design will be governed by a limit state other than flexural failure.

In this example, method (c), widely used for low-rise steel frames, is the only one that can make full use of the sagging moment of resistance. It is rarely available for composite frames. It is thus convenient to use partial shear connection to reduce the resistance to sagging bending.

4.3.2.3 Corrections for cracking and yielding

Cracking of concrete and yielding of steel have some influence on deflections in service and more effect on behaviour at ultimate limit states, because the design loads are higher. In short cantilevers and at some internal supports there may be very little cracking, so deflections may be over-estimated by an analysis where redistribution is used as explained above. Where a low degree of shear connection is used, deflections may be increased by longitudinal slip between the slab and the steel beam.

For these reasons, design codes give simplified methods of elastic analysis for continuous beams. Eurocode 4 does not include a useful method from BS 5950 for the deflection of a span of a uniform beam, allowing for the effects of its end moments, which is now given. Let these hogging moments be M_1 and M_2, for a loading that would give a maximum sagging moment M_0 and a maximum deflection δ_0, if the span were simply-supported. It can be shown by elastic analysis of a uniform member with uniformly-distributed load, that the moments M_1 and M_2 reduce the mid-span deflection from δ_0 to δ_c, where

$$\delta_c = \delta_0[1 - 0.6(M_1 + M_2)/M_0] \tag{4.35}$$

This equation is quite accurate for other realistic loadings. It shows the significance of end moments. For example, if $M_1 = M_2 = 0.42 M_0$, the deflection δ_0 is halved. It is not strictly correct to assume that the maximum deflection always occurs at mid-span, but the error is negligible.

For cracking and yielding, EN 1994-1-1 gives the following modified methods of elastic analysis. Section properties based on the effective flange width are usually used, as they are needed for other calculations.

The general method for allowing for the effects of cracking at internal supports is for beams with all ratios of span lengths, and is applicable for both serviceability and ultimate limit states. It involves two stages of calculation. The 'uncracked' flexural stiffness E_aI_1 is needed for each span, and also the 'cracked' flexural stiffness E_aI_2 at each internal support.

For the characteristic load combinations, the envelope of bending moments due to loading applied to the composite member is first determined using stiffnesses E_aI_1. The regions (if any) are found where the maximum tensile stress in the concrete due to the relevant moment, σ_{ct}, exceeds $2f_{ctm}$, where f_{ctm} is the mean tensile strength of the concrete. In these regions the stiffness E_aI_1 is reduced to E_aI_2. The analyses for bending moments are then repeated using the modified stiffnesses, and the results are used whether the new values σ_{ct} exceed $2f_{ctm}$ or not. For most continuous beams, including those in braced frames, the process can be simplified by assuming that the 'cracked' region extends over 15% of the span on each side of an internal support.

For analyses for deflections in buildings, EN 1994-1-1 gives an alternative to re-analysis of the structure, applicable for beams with critical sections in Classes 1, 2 or 3. It is that at every support where $\sigma_{ct} > 1.5f_{ctm}$, the bending moment is multiplied by a reduction factor f_1, and corresponding increases are made in the sagging moments in adjacent spans. For general use, $f_1 = 0.6$, but a higher value, given by

$$f_1 = (E_aI_1/E_aI_2)^{-0.35} \geq 0.6 \qquad (4.36)$$

is permitted for internal spans with equal loadings and approximately equal length.

Where unpropped construction is used and a high level of redistribution (e.g. 40%) is made in global analyses for ultimate limit states, it is likely that serviceability loads will cause local yielding of the steel beam at internal supports, which increases deflections. In design to EN 1994-1-1, allowance may be made for this by multiplying the moments at relevant supports by a further factor f_2, where:

- $f_2 = 0.5$, if f_y is reached before the concrete slab has hardened;
- $f_2 = 0.7$, if f_y is reached due to extra loading applied after the concrete has hardened.

These methods are used in the example in Section 4.6.5.

The calculation of deflections, with allowance for the effects of slip, is treated in Section 4.4.

4.3.3 Rigid-plastic analysis

For composite beams, use of rigid-plastic analysis can imply even larger redistributions of elastic moments than the 58% found in the example of Section 4.3.2.2, particularly where spans are of unequal length, or support concentrated loads.

Redistribution results from inelastic rotations of short lengths of beam in regions where 'plastic hinges' are assumed in the theory. Rotation may be limited either by crushing of concrete or buckling of steel. Rotation capacity depends on the proportions of the relevant cross-sections, as well as on the shape of the stress–strain curves for the materials.

The formation of a collapse mechanism is a sequential process. The first hinges to form must retain their resistance while undergoing sufficient rotation for plastic resistance moments to be reached at the locations where the last hinges form. Thus, the *rotation capacity* at each hinge location must exceed the *rotation required*. Neither of these quantities is easily calculated, so the requirement is in practice replaced by limitations on the use of the method, based on research. Those given in the Eurocodes include the following.

(1) At each plastic hinge location:
   - lateral restraint to the compression flange should be provided;
   - the effective cross-section should be in Class 1;
   - the cross-section of the steel component should be symmetrical about the plane of its web.
(2) All effective cross-sections in the member should be in Class 1 or Class 2.
(3) Adjacent spans should not differ in length by more than 50% of the shorter span.
(4) End spans should not exceed 115% of the length of the adjacent span.
(5) The member should not be susceptible to lateral-torsional buckling (i.e. $\bar{\lambda}_{LT} \leq 0.4$).
(6) In any span L where more than half of the design load is concentrated within a length of $L/5$, then at any sagging hinge, not more than 15% of the overall depth of the member should be in compression, unless it can be shown that the hinge will be the last to form in that span.

The method of analysis is well known, being widely used for steel-framed structures, so only an outline is given here. The principal assumptions are as follows.

(1) Collapse (failure) of the structure occurs by rotation of plastic hinges at constant bending moment, all other deformations being neglected. No plastic hinges are permitted in composite columns.
(2) A plastic hinge forms at any cross-section where the bending moment due to the actions reaches the bending resistance of the member.
(3) All loads on a span increase in proportion until failure occurs, so the loading can be represented by a single parameter.

The value of this parameter at collapse is normally found by assuming a collapse mechanism, and equating the loss of potential energy of the loads, due to a small movement of the mechanism, with the energy dissipated in the plastic hinges.

For a beam of uniform section with distributed loading w per unit length, the only properties required are the moments of resistance at mid-span, M_p say, and at the internal support or supports, M'_p. Let

$$M'_p/M_p = \mu \tag{4.37}$$

If the beam is continuous at both ends (Figure 4.10a), hinges occur at the ends and at mid-span, and

$$(1 + \mu)M_p = wL^2/8 \tag{4.38}$$

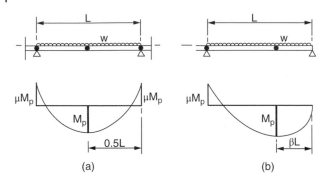

Figure 4.10 Rigid-plastic global analysis

If the beam is continuous at one end only, the bending moment diagram at collapse is as shown in Figure 4.10b. It can easily be shown that

$$\beta = (1/\mu)[\sqrt{(1+\mu)} - 1] \tag{4.39}$$

and

$$M_p = w\beta^2 L^2 / 2 \tag{4.40}$$

These results give the design ultimate load, for an assumed ratio of variable to permanent load, and the corresponding bending-moment distribution, from which shear forces can be found. Checks are then made on longitudinal shear, and that all non-hinge cross-sections have sufficient resistance to bending and vertical shear. Equation 4.40 can also be used to find M_p for a known load w. The method gives no information on ultimate deflections.

4.4 Stresses and deflections in continuous beams

Values of bending stresses at serviceability limit states may be needed in calculations for control of load-induced cracking of concrete (Section 4.2.5.3), and for prediction of deflections where unpropped construction is used. Bending moments are determined by elastic global analysis (Section 4.3.2). Those at internal supports are then modified to allow for cracking and yielding (Section 4.3.2.3). Stresses are found as in Section 3.5.3 for sagging moments, or Section 4.2.1 for hogging moments.

Deflections are much less likely to be excessive in continuous beams than in simply-supported spans or cantilevers, but they should always be checked where design for ultimate limit states is based on rigid-plastic global analysis. For simply-supported beams, the increase in deflection due to the use of partial shear connection can be neglected in certain circumstances (Section 3.7.1) and can be estimated from Equation 3.86. These same rules can be used for continuous beams, where they are a little conservative because partial shear connection is used only in regions of sagging bending moment.

The influence of shrinkage of concrete on deflections is treated in Section 3.8. For continuous beams, the method of calculation is rather complex, because shrinkage causes bending moments as well as sagging curvature; but its influence on deflection is much reduced by continuity, and is often negligible.

4.5   Design strategies for continuous beams

Until experience has been gained, the design of a continuous beam may involve much trial and error. There is no ideal sequence in which decisions should be made, but the following comments on this subject may be useful.

It is assumed that the span and spacing of the beams is known, that the floor or roof slab spanning between them has been designed, and that most or all of the loading on the beams is uniformly distributed, being either permanent (g) or variable (q). The beams add little to the total load, so g and q are known.

One would not be designing a continuous beam if simply-supported spans were satisfactory, so it can be assumed that simple spans of the maximum available depth are too weak, or deflect or vibrate too much, or that continuity is needed for seismic design, or to avoid wide cracks in the slab, or for some other reason.

The provision for services must be considered early. Will the pipes and ducts run under the beams, through openings in the webs, or above the slab? Heavily-serviced buildings needing special solutions (stub girders, haunched beams, etc.) are not considered here. The provision of openings in webs of continuous beams is easiest where the ratio q/g is low (Chung and Lawson, 2001). A low q/g ratio is also the situation where the advantages of continuity over simple spans are greatest. Design rules for openings in webs will be included in the next Eurocodes 3 and 4, but may be limited to simply-supported spans (Section 3.14).

Continuity is more advantageous in beams with three or more spans than where there are only two, and end spans should ideally be shorter than interior spans. The least benefit is probably obtained where there are two equal spans. The example in Section 4.6 illustrates this. Using a steel section that could span 9.0 m simply-supported, it is quite difficult to use the same cross-section for two continuous spans of 9.5 m.

A decision with many consequences is the Class of the composite section at internal supports. Two distinct strategies for beams in buildings are now compared.

(1) Minimal top longitudinal reinforcement is provided in the slab, determined by the need to control crack widths. If the composite cross-section in hogging bending is in Class 1, use of rigid-plastic global analysis may be possible, unless $\bar{\lambda}_{LT} > 0.4$ (Section 4.2.4). If the beam is in Class 2, the hogging bending moment will be redistributed as much as permitted, to enable good use to be made of the available bending resistance at mid-span.

(2) The reinforcement in the slab at internal supports is heavier, with an effective area at least 1% of that of the slab. The composite section at internal supports will certainly be in Class 2, perhaps Class 3. Restrictions on redistribution of moments will probably cause the design hogging moments, M'_{Ed}, to increase (cf. case 1) faster than the increase in resistance, M'_{Rd}, provided by the reinforcement, and further increase in the latter may put the section into Class 4. So the steel section may have to be heavier than for case 1, and there will be more unused bending resistance at mid-span. However, that will allow a lower degree of shear connection to be used. With higher M'_{Ed} the bending-moment diagram for lateral-torsional buckling is more adverse. Deflections are less likely to be troublesome, but the increase in the diameter of the reinforcing bars makes crack-width control more difficult.

The method of fire protection to be used may have consequences for the structural design. For example, web encasement improves the class of a steel web that is otherwise in Class 3, but not if it is in Class 2, and it improves resistance to lateral buckling.

Finally, it has to be decided whether construction will be propped or unpropped. Propped construction allows a shallower steel beam to be used – but it will be less stiff, so the dynamic behaviour may be less satisfactory. Propped construction costs more, and crack-width control is more difficult. The risk of excessive deflection, already quite low, is further reduced.

The design presented next is based on strategy (1) above, using a lightweight-concrete slab. Concrete encasement of the web is considered as an option. The design may not be an optimum solution; its purpose is to illustrate methods.

For buildings, current practice in most European countries is to use simple beam-to-column connections wherever possible, and to treat them as nominal pin joints. Light top reinforcement is provided to limit crack widths, but is ignored in verifications for ultimate limit states. Some designers include its strength when calculating the required number of shear connectors; others do not. The connectors are uniformly spaced along the span.

Nominal pin joints are unlikely to be used in bridge decks, where account must be taken of repeated loading and more adverse exposure.

4.6 Example: continuous composite beam

4.6.1 Data

So that use can be made of previous work, the design problem is identical with that of Chapter 3, except that the building (Figure 3.1) now consists of two bays, each of span 9.5 m. The transverse beams at 4 m centres are assumed to be continuous over a central longitudinal wall. Design would be more complex if they were attached by joints to supporting columns (Section 5.8). One solution is to use paired continuous beams, passing each column on both sides, and supported by it using brackets that transfer little bending moment to the column.

The beams are attached to columns in the outer walls, as before, by nominally-pinned joints located 0.2 m from the centres of the columns. No account is taken of the width of the central wall. Thus, each beam is as shown in Figure 4.11.

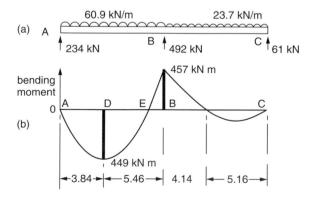

Figure 4.11 Continuous beam with dead load, plus imposed load on span AB only

Table 4.4 Loads and bending moments for a span of 9.3 m

	Characteristic loads		Ultimate loads	
	Load kN/m	$wL^2/8$ kN m	Load kN/m	$wL^2/8$ kN m
Permanent, on steel beam	12.4	134	16.7	181
Permanent, composite	5.2	56	7.0	76
Variable, composite	24.8	268	37.2	402
Total	42.4	458	60.9	659

The use of continuity should offset the increase in span from 9 m to 9.5 m, so it is assumed initially that the design of the slab and the mid-span region of the beam are as before, with the same materials, loads and partial safety factors. The design loads per unit length of beam, represented by the general symbol w, and the corresponding values of the bending moment $wL^2/8$ for a span of 9.3 m, are as given in Table 4.4.

Other design data from Chapter 3 are as follows:

structural steel:	$f_y = 355\,\text{N/mm}^2$,	$f_y/\gamma_A = 355\,\text{N/mm}^2$
concrete:	$f_{ck} = 25\,\text{N/mm}^2$,	$f_{ck}/\gamma_C = 16.7\,\text{N/mm}^2$
bar reinforcement:	$f_{sk} = 500\,\text{N/mm}^2$,	$f_{sk}/\gamma_S = 435\,\text{N/mm}^2$
welded fabric:	$f_{sk} = 500\,\text{N/mm}^2$,	$f_{sk}/\gamma_S = 435\,\text{N/mm}^2$
shear connectors in troughs:	$P_{Rd} = 51\,\text{kN}\ (n_r = 1)$,	$P_{Rd} = 42\,\text{kN}\ (n_r = 2)$
profiled steel sheeting:	details as shown in Figure 3.8; nominal thickness, 0.9 mm	

composite slab: 150 mm thick, with T8 bars at 150 mm (top) above the steel beams, A193 steel mesh generally, and concrete of Grade LC25/28.

composite beam: steel section 406 × 178 UB 60, shown in Figure 3.31, with shear connection as in Figure 3.32. Reference is made earlier to encasement of the steel web for fire resistance. It is not included in these calculations for hogging moment of resistance. The properties of the concrete are given in Table 1.4.

Many properties of the composite cross-section are required, so it is useful to assemble them in a table, for ease of reference. The elastic and plastic properties for major-axis bending, used in Chapter 3 or in this chapter, are given in Table 4.5. The values of A and I for transformed cross-sections are based on $E = 210\,\text{kN/mm}^2$. Values headed 'reinforced' are for a cross-section with 679 mm^2 of top longitudinal reinforcement. As explained later, the thickness of concrete slab above the sheeting is taken as 95 mm for elastic properties and as 80 mm for plastic properties, with the exception shown in Table 4.5. The depths x_c are from the top of the concrete slab.

Other properties of the steel cross-section, not in Table 4.5, are as follows: $10^{-6}W_{a,pl} = 1.194\,\text{mm}^3$, $10^{-6}I_{az} = 12.0\,\text{mm}^4$, $V_{pl,Rd} = 697\,\text{kN}$.

Table 4.5 Properties of cross-sections of a composite beam

Type of section	Cracked un-reinforced (I-section only)	Uncracked unreinforced				Cracked reinforced
Effective width, b_{eff}, mm	–	2250	2250	1225	1225	1225
Modular ratio, n	–	10.1	20.2	10.1	20.2	–
Transformed cross-section, A, mm^2	7600	28 760	18 180	17 300	12 450	8279
Depth x_c of elastic neutral axis, mm	353	129 ($h_c = 95$)	176 ($h_c = 95$)	177 ($h_c = 80$)	231 ($h_c = 80$)	327
Second moment of area, $10^{-6}I$, mm^4	215	751	636	–	508	278
Depth x_c of plastic neutral axis, mm	353	151 ($h_c = 80$)	151 ($h_c = 80$)	–	–	300
$M_{\text{pl,Rd}}$, kN m	424	829 sagging	829 sagging	–	–	510 hogging

4.6.2 Flexure and vertical shear

A rough check on the adequacy of the assumed beam section is provided by rigid-plastic global analysis. The value of $wL^2/8$ that can be resisted by each span is a little less than

$$M_{\text{pl,Rd}} + 0.5M_{\text{pl,a,Rd}} = 829 + 212 = 1041 \text{ kN m}$$

neglecting the reinforcement in the slab at the internal support, B in Figure 4.11. This is well above $wL^2/8$ for the loading (659 kN m, from Table 4.4).

4.6.2.1 Effective width and minimum reinforcement at the internal support
The use of effective widths of a concrete flange is explained in Sections 3.5.1 and 4.2.1.1, and found for hogging regions of this beam to be 1.225 m (Section 4.2.1.2).

For a sagging region of span L the effective span is $0.85\,L$, or 8.1 m here, so $b_{\text{eff}} = 8.1/4 + 0.1 = 2.13$ m, slightly less than the 2.25 m used in Chapter 3. This reduction has little effect on the properties and has been ignored.

It is assumed that the exposure class is X0 or XC1 (Section 4.2.5) and that the limiting crack width is 0.4 mm under quasi-permanent loading. It is assumed initially that the top longitudinal reinforcement at support B is six T12 bars ($A_s = 679$ mm^2) at 200 mm spacing. The cross-section is then as shown in Figure 4.1. It was found in Section 4.2.1.2 that its hogging resistance is $M_{\text{pl,Rd}} = 510$ kN m.

The area A_s may be governed by the rules for minimum reinforcement (Section 4.2.5.2). These require calculation of the distance z_0 in Figure 4.14 for the uncracked unreinforced composite section with $n = 10.1$. Initial cracking is likely to occur above the small top ribs of the sheeting, where the slab thickness is 80 mm, so the assumption $h_c = 95$ mm, made for serviceability checks at mid-span, is not appropriate.

The transformed area of the uncracked section is

$$A = 7600 + 1225 \times 80/10.1 = 17\,300 \text{ mm}^2$$

Taking moments of area about the top of the slab for the neutral-axis depth x_c,

$$17\,300x = 7600 \times 353 + 9700 \times 40; \quad \text{whence} \quad x_c = 177 \text{ mm}$$

Hence,

$$z_0 = 177 - 40 = 137 \text{ mm}$$

From Equation 4.32,

$$k_c = 1/(1 + 80/274) + 0.3 = 1.07, \quad \text{but} \le 1.0$$

The elastic neutral axis is below the slab, so $A_{ct} = 1225 \times 80 = 98\,000 \text{ mm}^2$.

The stress $f_{ct,eff}$ is taken as the mean 28-day tensile strength of the Grade LC 25/28 concrete, given in Table 1.4 as 2.32 N/mm^2.

As a guessed area of longitudinal reinforcement is being checked, it is simplest next to use Expression 4.33 as an equality to calculate σ_s. Hence,

$$\sigma_s = 0.8k_c f_{ct,eff} A_{ct}/A_s = 0.8 \times 1 \times 2.32 \times 98\,000/679 = 268 \text{ N/mm}^2$$

The characteristic crack width is now found. Eurocode 4 permits interpolation between the tops of the 'steps' shown in Figure 4.7. This gives the maximum bar diameter for $w_k = 0.4 \text{ mm}$ and $\sigma_s = 268 \text{ N/mm}^2$ as $\phi^* = 17.2 \text{ mm}$. This is for the reference concrete strength $f_{ct,0} = 2.9 \text{ N/mm}^2$. The correction for concrete strength gives

$$\phi_{max} = 17.2 \times 2.32/2.9 = 13.8 \text{ mm}$$

The 12-mm bars proposed should therefore control crack widths from imposed deformation ('restraint-induced cracking') to better than 0.4 mm.

A similar calculation finds that for $w_k = 0.3 \text{ mm}$, $\phi < 10.6 \text{ mm}$, so that 10-mm bars would be needed.

4.6.2.2 Classification of cross-sections

It is easily shown by the methods of Section 4.2.1 that the steel compression flange is in Class 1 at support B. Calculations in Section 4.2.1.2 that ignored the web encasement found that the web was in Class 2, and that $M_{pl,Rd} = 510 \text{ kN m}$. Web encasement enables a Class 3 web to be upgraded to Class 2; but promotion from Class 2 to Class 1 is not permitted, because crushing of the encasement in compression may reduce the rotation capacity in hogging bending to below that required of a Class 1 section. It is therefore not possible to use plastic global analysis, but $M_{pl,Rd}$ can be used as the bending resistance at support B, subject to checks for vertical shear and lateral buckling.

4.6.2.3 Vertical shear

The maximum vertical shear occurs at support B (Figure 4.11) when both spans are fully loaded. Ignoring the effect of cracking of concrete (which reduces the shear at B) enables results for beams of uniform section to be used. From elastic theory,

$$V_{Ed,B} = 5wL/8 = 5 \times 60.9 \times 9.3/8 = 354 \text{ kN}$$

From Equation 3.100,

$$V_{pl,a,Rd} = 697 \text{ kN}, \quad \text{so} \quad V_{Ed}/V_{Rd} = 0.51$$

This exceeds 0.5, so $M_{pl,Rd}$ should be reduced; but from Equation 4.13, the reduction is obviously negligible.

Redistribution of bending moment from B would reduce $V_{Ed,B}$, but the shear resistance is checked ignoring this, because if the beam is found to be susceptible to lateral buckling, redistribution is not permitted. In global analyses leading to column design, the design vertical shears should be consistent with the design bending moments.

4.6.2.4   Bending moments

Ignoring cracking, the maximum hogging bending moment at B occurs when both spans are fully loaded, and is

$$M_{Ed,B} = wL^2/8 = 60.9 \times 9.3^2/8 = 658 \text{ kNm}$$

From Table 4.3, the maximum permitted redistribution for a Class 2 member is 30%, giving

$$M_{Ed,B} = 658 \times 0.7 = \textbf{461 kNm}$$

which is below $M_{pl,Rd}$ (510 kN m, from Section 4.2.1.2). If lateral buckling governs, the bending resistance must be at least 658 kN m.

The maximum sagging bending moment in span AB occurs with minimum load on span BC. Elastic analysis neglecting cracking gives the results shown in Figure 4.11. The maximum sagging moment, $M_{Ed} = 449$ kN m, is lower than that given by Equation 3.95, 563 kN m, for the simply-supported beam, and the vertical shear at A is lower than that at B.

4.6.3   Lateral buckling

The lateral stability of the steel bottom flange adjacent to support B is checked using the 'continuous U-frame' model explained in Section 4.2.4 and the bending-moment distribution shown for span BC in Figure 4.11b. In practice, a similar check should be made for full loading on both spans. Then, the hogging bending moment at B is higher, but the length of bottom flange in compression is reduced from 4.1 m to 2.3 m. The exemption from these checks given by Eurocode 4 (Section 4.2.4.4) does not apply to the UK's UB cross-sections, so the method of Equation 4.28 is now used.

Simple check using property F   The section property F for the UB section used here is calculated for the following values (all in mm units):

$$t_w = 7.8, \quad h_s = 406 - 12.8 = 393, \quad b_f = 178, \quad t_f = 12.8$$

The result is $F = 13.1$, below the limit of 15.8 for web-encased sections in S355 steel, but above 12.3, so the uncased web is not exempt. The full calculation may be less conservative, so it is now applied to the beam with an uncased web, to illustrate the use of Equations 4.22–4.24.

Elastic critical bending moment for distortional lateral buckling   In Equation 4.22,

$$M_{cr} \approx (k_c C_4/\pi)(k_s E_a I_{afz})^{1/2} \tag{4.22}$$

the term k_s represents the stiffness of the U-frame:

$$k_s = k_1 k_2/(k_1 + k_2) \tag{4.18}$$

Figure 4.12 Cracked section of composite slab, for hogging bending

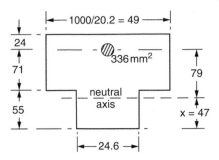

From Equation 4.19, $k_1 = 4E_a I_2/a$, where $E_a I_2$ is the 'cracked' stiffness of the composite slab in hogging bending. To calculate I_2 the trapezoidal rib shown in Figure 3.8 is replaced by a rectangular rib of breadth $162 - 13 = 149$ mm. Using a modular ratio $n = 20.2$, the transformed width of rib is $149/(0.3 \times 20.2) = 24.6$ mm per metre width of slab, since the ribs are at 0.3 m spacing. The transformed section is thus as shown in Figure 4.12.

The position of the neutral axis is given by

$$\tfrac{1}{2} \times 24.6x^2 = 336(126 - x), \quad \text{whence } x = 47 \text{ mm}$$

then

$$10^{-6}I_2 = 336 \times 0.079^2 + 24.6 \times 47 \times 0.47^2/3 = 2.95 \text{ mm}^4/\text{m}$$

From Equation 4.19, with the beam spacing $a = 4.0$ m,

$$k_1 = 4 \times 210\,000 \times 2.95/4 = 0.619 \times 10^6 \text{ N}$$

Equation 4.20 gives k_2 for an uncased web:

$$k_2 = \frac{E_a t_w^{\,3}}{4(1 - v_a^{\,2})h_s} = 210\,000 \times 7.8^3/(3.64 \times 393) = 0.070 \times 10^6 \text{ N}$$

where v_a ($= 0.3$) is Poisson's ratio for steel.

From Equation 4.18,

$$k_s = 0.619 \times 0.07 \times 10^6/(0.619 + 0.07) = 0.063 \times 10^6 \text{ N}$$

This shows that almost the whole of the lateral flexibility of the bottom flange comes from the web.

For the steel bottom flange,

$$I_{afz} = b_f^{\,3}t_f/12 = 178^3 \times 12.8/12 = 6.01 \times 10^6 \text{ mm}^4$$

Factor k_c (Equation 4.16) is concerned with stiffness, so the appropriate depth of slab, h_c, is 95 mm, not 80 mm. Data from Section 4.6.1 and the following further values are now required (see Figure 4.5):

$$z_c = 203 + 150 - 95/2 = 306 \text{ mm}, \quad A = A_a + A_s = 7600 + 679 = 8279 \text{ mm}^2$$

From Equation 4.17,

$$e = \frac{AI_{ay}}{A_a z_c (A - A_a)} = \frac{8279 \times 215.1 \times 10^6}{7600 \times 306 \times 679} = 1128 \text{ mm}$$

The second moment of area of the cracked composite section is found in the usual way (Section 4.2.1.3). The elastic neutral axis is 26 mm above the centroid of the steel section, and

$$10^{-6}I_y = 215 + 7600 \times 0.026^2 + 679 \times 0.292^2 = 278 \text{ mm}^4$$

From Equation 4.16,

$$k_c = \frac{h_s I_y / I_{ay}}{\frac{h_s^2/4 + (I_{ay} + I_{az})/A_a}{e} + h_s} = \frac{393 \times 278/215.1}{\frac{393^2/4 + (215+12) \times 10^6 / 7600}{1128} + 393} = 1.12$$

Where A_s/A_a is small, it is simpler, and conservative, to take k_c as 1.0.

Factor $C4$ is now found, using Figure 4.6. From Figure 4.11, for span BC,

$$M_0 = 23.7 \times 9.3^2/8 = 256 \text{ kN m}, \quad \text{so} \quad \psi = 457/256 = 1.78$$

From Figure 4.6, $C_4 = 17.4$.

The effect of including in Equation 4.15 the St Venant torsion constant for the steel section is negligible, so the simpler Equation 4.22 can be used. It gives:

$$M_{cr} = (1.12 \times 17.4/\pi)(0.063 \times 210\,000 \times 6.01)^{1/2} = \textbf{1749 kN m}$$

Slenderness for distortional lateral buckling, and conclusion The characteristic plastic bending resistance at support B is required for use in Equation 4.23. The design value was found in Section 4.2.1.2 to be 510 kN m. Replacing the γ_S factor for reinforcement (1.15) by 1.0 increases it to $M_{pl,Rk} = 579$ kN m.

The slenderness $\bar{\lambda}_{LT}$ is given by Equation 4.23, which is

$$\bar{\lambda}_{LT} = (M_{pl,Rk}/M_{cr})^{1/2} = (579/1749)^{1/2} = 0.575$$

This exceeds 0.4, so $M_{pl,Rd}$ must be reduced to allow for lateral buckling, which is consistent with the use of section property F, above. A similar calculation for both spans fully loaded also gives $\bar{\lambda}_{LT} > 0.4$.

The simplified expression for $\bar{\lambda}_{LT}$ (Equation 4.27) can be useful as an approximate check. Here, it gives $\bar{\lambda}_{LT} = 0.64$, slightly more conservative than the 0.575 above, as expected.

The reduced resistance, $M_{b,Rd}$ can be found by the method of Eurocode 3 (Equations 4.24–4.26), but even M_{Rd}, 510 kN m, is too low, as M_{Ed} is 658 kN m and cannot be redistributed. Here, it is assumed that web encasement is used, and that redistribution of the bending moments when both spans are fully loaded reduces M_{Ed} at B to 510 kN m. Otherwise, redesign will be needed.

4.6.4 Shear connection and transverse reinforcement

For sagging bending, the resistance required, 449 kN m, is well below the resistance with full shear connection, 829 kN m, so the minimum degree of shear connection may be sufficient. For spans of 9.3 m, and an effective span of $9.3 \times 0.85 = 7.9$ m, this is given by Figure 3.21 as $n/n_f \geq 0.49$.

In Equation 3.57, $N_c/N_{c,f}$ may be replaced by n/n_f. This equation for the interpolation method (Figure 3.16) then gives

$$\boldsymbol{M}_{\mathbf{Rd}} = M_{\mathrm{pl,a,Rd}} + (n/n_f)(M_{\mathrm{pl,Rd}} - M_{\mathrm{pl,a,Rd}}) = 424 + 0.49(847 - 424)$$

$$= \mathbf{631\ kN\,m}$$

which is sufficient.

From Equation 3.103, the resistances of the stud connectors are 51.0 kN and 42.0 kN, for one and two studs per rib ($n_r = 1$ and $n_r = 2$, respectively). From Section 3.11.1, $N_{c,f} = 2555$ kN, so

$$N_c = 0.49 \times 2555 = 1250\ \mathrm{kN}$$

There are 13 troughs at 0.3 m spacing in the 3.84-m length AD in Figure 4.11. The choice of n_r is discussed in Section 3.11.2. Assuming that $n_r = 2$, the number of studs needed is $1250/42 = 30$.

Section 3.11.2 shows that, for the sheeting shown in Figure 3.8, it is not possible to provide more than two studs per trough. Here, only 26 can be provided. This is sufficient if they can be treated as single studs, in view of their quite wide lateral spacing. It is expected that in the next Eurocode 4, the rules for minimum shear connection will be less conservative (Couchman, 2015). As in this shear span, M_{Ed} is only 71% of M_{Rd}, the designer might accept this reduction in the number of studs.

From Figure 4.11, the length DB is 5.46 m. Its sheeting has $5.46/0.3 = 18$ troughs. The force to be resisted is 1250 kN plus 295 kN (Section 4.2.1.2) for the reinforcement at cross-section B, at yield; total, 1545 kN, requiring 37 studs at 42 kN each; but only 36 can be provided. This might be accepted, for the reasons given above.

The transverse reinforcement should be as determined for the sagging region in Chapter 3.

4.6.5 Check on deflections

The limits to deflections discussed in Section 3.7.2 correspond to the characteristic combination of loading (Expression 1.8). Where there is only one type of variable load, as here, this load is simply $g_k + q_k$, but three sets of calculations may be required, because part of the permanent load g acts on the steel member and part on the composite member, and two modular ratios are needed for the composite member.

In practice, it is usually accurate enough to combine the two calculations for the composite member, using a mean value of the modular ratio (e.g. $n = 20.2$ here).

For design purposes, maximum deflection occurs when the variable load is present on the whole of one span, but not on the other span. The three loadings are shown in Figure 4.13, with the bending-moment distributions given by 'uncracked' elastic analyses, in which the beam is assumed to be of uniform section. The data and results are summarized in Table 4.6, where M_B is the hogging moment at support B shown in Figure 4.13.

Following the method of Section 4.3.2.3, the maximum tensile stress in the uncracked composite section at B, σ_{ct}, is now found, using an effective width of 1.225 m (Figure 4.14a). This stress will occur when variable load acts on both spans, and is

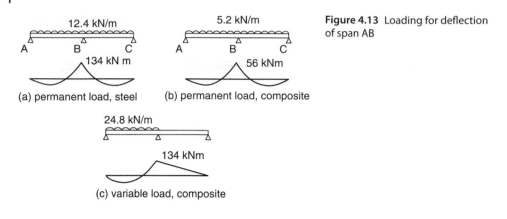

(a) permanent load, steel   (b) permanent load, composite

Figure 4.13 Loading for deflection of span AB

(c) variable load, composite

Table 4.6 Calculation of maximum deflection

	w kN/m	M_0 kN m	M_B kN m	f_1	f_2	M_1 kN m	$10^{-6}I_1$ mm⁴	δ_c mm
g on steel	12.4	134	134	–	–	134	215	10.7
g on composite	5.2	56	56	0.6	0.7	24	636	2.8
q on composite	24.8	268	134	0.6	0.7	56	636	15.9

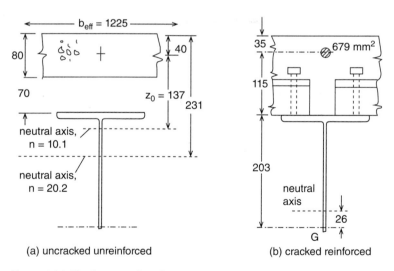

(a) uncracked unreinforced   (b) cracked reinforced

Figure 4.14 Elastic properties of composite section

calculated using $n = 20.2$ for all load on the composite member. Using data from Table 4.5,

$$\sigma_{ct} = \sum \left(\frac{Mx}{nI_1} \right) = \frac{(56 + 268) \times 231}{20.2 \times 508} = 7.3 \text{ N/mm}^2$$

where nI_1 is the 'uncracked' second moment of area in 'concrete' units.

This stress exceeds $1.5f_{lctm}$ ($3.48\,\text{N/mm}^2$). To avoid re-analysis with spans of non-uniform section, the correction factor f_1 given by Equation 4.36 can be used. For these 'external' spans, $f_1 = 0.6$.

The maximum compressive stress in the steel bottom fibre is now calculated, to determine whether the correction factor f_2 for yielding is required. As for σ_{ct}, variable load should be assumed to act on both spans, so the bending moment from load q is doubled to 268 kN m. As the loads on the two spans are now the same, there is no rotation at support B, and the end moments there are $wL^2/8$, given in the column M_0 in Table 4.6.

Using data from Section 4.6.1 and Figure 4.14b,

$$\sigma_{4,a} = \sum \left(\frac{Mx_4}{I_2}\right) = \frac{134 \times 203}{215} + \frac{(56 + 268) \times 229}{278}$$

$$= 126 + 267 = 393\,\text{N/mm}^2$$

where x_4 is the distance of the relevant neutral axis above the bottom fibre. The result shows that yielding occurs ($393 > 355$), but not until after the slab has hardened ($126 < 355$), so from Section 4.3.2.3, $f_2 = 0.7$. The hogging moments M_1 for use in Equation 4.35 are

$$M_1 = f_1\, f_2 M_B$$

and are given in Table 4.6. The other end moment, M_2, is zero.

Equation 4.35 can be used to estimate the mid-span deflections. It uses the deflections δ_0 for each loading acting on a simply-supported span. These are, in general:

$$\delta_0 = \frac{5wL^4}{384EI} = \frac{5 \times 9.3^4 \times 10^9}{384 \times 210}\left(\frac{w}{I_1}\right) = 464 \times 10^6 (w/I_1)\,\text{mm}$$

with w in kN/m and I_1 in mm^4. Using values from Table 4.6, Equation 4.35 gives the total deflection:

$$\delta_c = 464 \times 10^6 \sum [(w/I_1)(1 - 0.6M_1/M_0)]$$

$$= 464[(12.4/215)(1 - 0.6 \times 134/134) + (5.2/636)(1 - 0.6 \times 24/56)]$$

$$+ 464[(24.8/636)(1 - 0.6 \times 56/268)]$$

$$= 10.7 + 2.8 + 15.9 = 29.4\,\text{mm}$$

This result is probably too high, because the factor f_1 may be conservative, and no account has been taken of the stiffness of the web encasement. This total deflection is span/316, less than the guideline of $L/300$ given in the UK's national annex to EN 1990 (BSI, 2002c), for floors with plastered ceilings and/or non-brittle partitions.

The total deflection of the simply-supported span of 8.6 m, for the same loading, was found to be 35.5 mm. This would be $35.5 \times (9.3/8.6)^4 = 48.5\,\text{mm}$ for the present span. The use of continuity at one end has reduced this value by 19.1 mm, or 39%.

Even so, these deflections are fairly large for a continuous beam with a ratio of span to overall depth of only $9300/556 = 16.7$. This results from the use of unpropped construction, high-yield steel, and lightweight concrete. Deflections of a similar propped structure in mild steel and normal-density concrete would be much lower.

4.6.6 Control of cracking

The widest cracks will occur at the top surface of the slab, above an internal support. The reinforcement at this cross-section is 12-mm bars at 200 mm spacing ($A_s = 679$ mm^2). It was shown in Section 4.6.2 that this can control the widths of cracks from imposed deformation to below 0.4 mm. It is now necessary to check the crack width caused by the characteristic loading, using Section 4.2.5.3.

The bending moments M_B given in Table 4.6 are applicable, except that M_B for imposed load must be doubled, as for this purpose it acts on both spans. For deflections, use of the reduction factor $f_1 = 0.6$ probably under-estimated M_B. For checking crack width, any approximation should be an over-estimate. In Table 4.3 for limits to redistribution of moments, it is assumed that cracking causes a 15% reduction in a Class 2 section, so f_1 is taken as 0.85.

The factor f_2 for yielding of steel is not applied, as the yield strength of the steel is likely to be higher than specified, to the small extent needed here. Hence, for cracking,

$$M_B = (56 + 2 \times 134) \times 0.85 = 275 \text{ kN m}$$

For the cracked composite section, $10^{-6}I_2 = 278$ mm^4, from Table 4.5, so the stress in the reinforcement, at distance 292 mm above the neutral axis, is

$$\sigma_{s,o} = 275 \times 292/278 = 289 \text{ N/mm}^2$$

This must be increased to allow for tension stiffening between cracks. From Section 4.2.5.3,

$$\alpha_{st} = AI_2/A_aI_a = 8279 \times 278/(7600 \times 215) = 1.41$$

From Section 4.6.2, the cracked area of concrete is

$$A_{ct} = b_{eff}h_c = 1225 \times 80 = 98\,000 \text{ mm}^2$$

From Equation 4.34, with f_{lctm} from Table 1.4,

$$\sigma_s = \sigma_{s,o} + 0.4 f_{lctm} A_{ct}/(\alpha_{st} A_s)$$

$$= 289 + 0.4 \times 2.32 \times 98\,000/(1.41 \times 679)$$

$$= 384 \text{ N/mm}^2$$

This is below the yield strength, so the existing 12-mm bars are satisfactory if a non-brittle floor finish is to be used, such that cracks will not be visible. If, however, control of crack widths to 0.4 mm were required, reference to Figure 4.7 shows that 8-mm bars would be required. To provide 679 mm^2, their spacing would have to be 90 mm, or bars in pairs at 180 mm. (For this LC25/28 concrete, $f_{lct,eff}$ is taken as 3.0 MPa, so the correction in Section 4.2.5.2 alters the bar diameter to $8 \times 3.0/2.9$, which has no effect here.)

The alternative of increasing the top reinforcement to, say, 10-mm bars at 100 mm (883 mm^2) would require some recalculation. It increases the ultimate hogging bending moment $M_{Ed,B}$ and so makes susceptibility to lateral buckling more likely. It is not obvious whether it would reduce the value of the tensile stress σ_s to below 360 N/mm^2, the limit given in both Figures 4.7 and 4.8 for crack control to 0.4 mm. This illustrates the interactions that can occur in design of a hogging moment region.

4.7   Continuous composite slabs

The concrete of a composite slab floor is almost always continuous over the supporting beams, but the individual spans are often designed as simply-supported (Sections 3.3 and 3.4), for simplicity. Where deflections are found to be excessive, continuous design may be used, as follows.

Elastic theory is used for the global analysis of continuous sheets acting as shuttering. Variations of stiffness due to local buckling of compressed parts can be neglected. Resistance moments of cross-sections are based on tests (Section 3.3).

Completed composite slabs are generally analysed for ultimate bending moments in the same way as continuous beams with Class 2 sections. 'Uncracked' elastic analysis is used, with up to 30% redistribution of hogging moments, assuming that the whole load acts on the composite member. Rigid-plastic global analysis is allowed by EN 1994-1-1 where all cross-sections at plastic hinges have been shown to have sufficient rotation capacity. This has been established for spans not exceeding 3.0 m with reinforcement of 'high ductility' as defined in EN 1992-1-1 for reinforced concrete. No check on rotation capacity is then required.

At internal supports where the sheeting is continuous, resistance to hogging bending is calculated by rectangular-stress-block theory, as for composite beams, except that local buckling is allowed for by using an effective width for each flat panel of sheeting in compression. This width is given in EN 1994-1-1 as twice the value specified for a Class 1 steel web, thus allowing for the partial restraint from the concrete on one side of the sheet. Where the sheeting is not continuous at a support, the section is treated as reinforced concrete.

For the control of cracking at internal supports, EN 1994-1-1 refers to EN 1992-1-1. In practice, the reinforcement to be provided may be governed by design for resistance to fire, or by the transverse reinforcement required for the composite beam that supports the slab.

5

Composite Columns and Frames

5.1   Introduction

A definition of *composite frame* is given in Section 4.1, with discussion of *joints, connections*, and of the relationship between these and the methods of global analysis (Table 4.1).

To illustrate the presentation of continuous beams in the context of buildings, it was necessary to use, in Section 4.6, an atypical example: a two-span beam with a wall as its internal support, with no transfer of bending moment between the wall and the beam.

Where a beam is supported by a column through a joint that is not 'nominally pinned', the bending moments depend on the properties of the joint, the beam, and the column. They form part of a frame, which may have to provide resistance to horizontal loads such as wind (an '*unbraced frame*'), or these loads may be transferred to a bracing structure by the floor slabs.

Where the lateral stiffness of the bracing structure is sufficient, the frame can be designed to resist only the vertical loads (a '*braced frame*'). The lift and staircase regions of multistorey buildings often have concrete walls, for resistance to fire. These can provide stiff lateral restraint, as can the end walls of long narrow buildings. The frames can then be designed as 'braced'. Unbraced frames require stronger members and joints.

Some of the design rules of Eurocodes 3 and 4 for semi-rigid and partial-strength joints are so recent that there is little experience of their use in practice. The scope of this chapter is limited to braced frames with beam-to-column joints that are either '*nominally pinned*' or '*rigid and full strength*'. The structure shown in Figure 5.1 is used as an example. Typical plane frames such as DEF in Figure 5.1a are at 4.0-m spacing, and support composite floor slabs as designed in Section 3.4. For simplicity, each nine-storey frame is assumed to have ten two-span beams, including those at the ground and roof levels, identical with those designed in Section 4.6. The wall that provided their internal support is now replaced by composite columns (B, E, etc.) at 4.0-m spacing, with beam-to-column joints assumed to be rigid and full strength.

The joints to the external columns, near points A, C, D, F, and so on, are nominally pinned. The design bending moments for the columns depend on the assumed location of these pins, which is discussed in Section 5.4.4.2.

The building is assumed to be 60 m long. Lateral support is provided by shear walls at each end (Figure 5.1a), and a lift and staircase tower (the 'core') at mid-length, 4 m wide. Horizontal loads normal to the length of the building are assumed to be transferred to these bracing elements by each floor slab, spanning $(60 − 4)/2 = 28$ m between the core

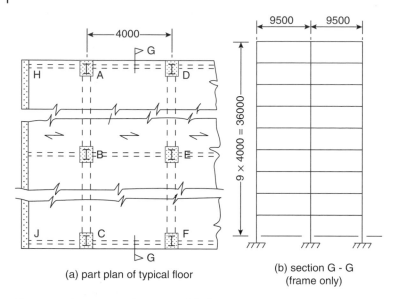

(a) part plan of typical floor

(b) section G - G
(frame only)

Figure 5.1 A composite frame (simplified)

and an end wall. These slabs are quite thin, but as horizontal beams they are about 19 m deep. Their span/depth ratio is so low ($28/19 = 1.47$) that for lateral load they are very stiff, and stresses from wind loading are low.

The effect of horizontal forces in other directions is assumed to be negligible. External walls are assumed to be supported by edge beams such as JC and CF, spanning between the external columns. These beams will not be designed.

This structure is used here as the basis for explanation of behaviour and design methods, and is not fully realistic. For example, it lacks means of escape near its two ends.

The layouts of most multistorey buildings are such that their frames can be designed as two-dimensional. The columns are usually designed with their webs co-planar with those of the main beams, as shown in Figure 5.1a, so that beam–column interaction causes major-axis bending in both members.

For global analysis for gravity loads, each plane frame is assumed to be independent of the others. For each storey-height column length, an axial load N_y and end moments $M_{1,y}$ and $M_{2,y}$ are found for the major-axis frame, and corresponding values N_z, $M_{1,z}$ and $M_{2,z}$ for the minor-axis frame such as HAD in Figure 5.1. The column length is then designed (or an assumed design is checked) for axial load $N_y + N_z$ and for the biaxial bending caused by the four end moments.

In the design of a multistorey composite plane frame, allowance must be made for imperfections. Global imperfections, such as out-of-plumb columns, influence lateral buckling of the frame as a whole ('frame instability'). Member imperfections, such as bow of a column length between floor levels, influence the buckling of these lengths ('member instability'), and may even affect the stability of a frame.

Global analysis is usually linear-elastic, with allowance for creep, cracking, and the moment-rotation properties of the joints. First-order analysis is used wherever possible, but checks must first be made that second-order effects (additional action effects arising from displacement of nodes or bowing of members) can be neglected. If not, second-order analysis is used.

A set of flow charts for the design of such a frame, given elsewhere (Johnson, 2012), is too extensive to reproduce here. However, the sequence of these charts will be followed for the frame shown in Figure 5.1b, which will be found to be free from many of the complications referred to above.

Columns and joints are discussed separately in Sections 5.2 and 5.3. The Eurocode methods for analysis of braced frames are explained in Section 5.4, with a worked example. Details of the design method of EN 1994-1-1 for columns are then given, followed by calculations for two of the columns in the frame, and an alternative design using high-strength materials.

5.2 Composite columns

Steel columns in multistorey buildings need protection from fire. This is often provided by encasement in concrete. Until the 1950s, it was normal practice to use a concrete mix of low strength, and to neglect the contribution of the concrete to the strength and stability of the column. Tests by Faber (1956) and others then showed that savings could be made by using better-quality concrete and designing the column as a composite member. This led to the 'cased strut' method of design. In this method, the presence of the concrete was allowed for in two ways. It was assumed to resist a small axial load, and to reduce the effective slenderness of the steel member, which increased its resistance to axial load. Resistance to bending moment was assumed to be provided entirely by the steel. No account was taken of the resistance of the longitudinal reinforcement in the concrete.

Tests on cased struts under axial and eccentric load showed that this method gave a very uneven and usually excessive margin of safety. It was improved in BS 5950, but was still generally very conservative. Its main advantage was simplicity.

One of the earliest methods to take proper account of the interaction between steel and concrete in a concrete-encased H-section column was due to Basu and Sommerville (1969). It agrees quite well with the results of tests and numerical simulations. It was thought to be too complex for routine use for columns in buildings, but was included in the British code for composite bridges.

The Basu and Sommerville method is based on the use of algebraic approximations to curves obtained by numerical analyses. For Eurocode 4: Part 1.1, preference was given to a method developed by Roik and Bergmann (1992) and others at the University of Bochum. It has wider scope, is based on a clearer conceptual model, and is slightly simpler. It is described in Section 5.6, with a worked example.

5.3 Beam-to-column joints

5.3.1 Properties of joints

Three types of joint between a steel beam and the flange of an H-section steel column are shown in Figure 5.2, and a short end-plate joint is shown in Figure 5.23. They are examples of the only types of joint treated in Eurocode 4. Its rules are extensions of the component method used in EN 1993-1-1 (BSI, 2014b) and EN 1993-1-8 (BSI, 2005a).

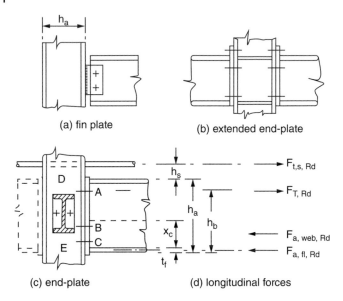

(a) fin plate   (b) extended end-plate

(c) end-plate   (d) longitudinal forces

Figure 5.2 Elevations of beam-to-column joints

Expressions for the stiffnesses of the components are given in Annex A of EN 1994-1-1 (BSI, 2004). The word 'connection', as used in the codes, means a component of a joint. For example, Figure 5.2b shows a joint with two connections, and Figure 5.2c shows a joint with three (or possibly four).

The codes assume that each connection is being designed for some combination of vertical shear and bending moment in the plane of the beam web. Accidental limit states, such as the removal of a column by an explosion, can cause large axial forces in beams. These are outside the scope of Eurocode 4 at present, because the rules are applicable only where an axial force, if present, does not exceed 5% of the design axial resistance ($N_{pl,Rd}$) of the beam, from clause 6.2.3(2) of EN 1993-1-8.

The joints shown in Figures 5.2 and 5.23 are all bolted, because they are made on site, where welding is expensive and difficult to inspect. The column shown in Figure 5.2a is in an external wall. At an internal column, another beam would be connected to the other flange. There may also be minor-axis beams, connected to the column web as shown in Figure 5.2c.

Where the beams are composite and the column is internal, longitudinal reinforcement in the slab will be continuous past the column, as shown in Figure 5.2c. It may be provided only for the control of cracking; but if it consists of individual bars, rather than welded fabric, the tension in the bars may be assumed to contribute to the bending resistance of the joint, as shown in Figure 5.2d. Small-diameter bars may fracture before the rotation of the hogging region of the beam becomes large enough for the resistance of the joint to reach its design value, so these bars should be at least 16 mm in diameter (Anderson et al., 2000).

If there is no beam on the left in Figure 5.2c and hogging bending moment has to be transferred to the column, the tension in the longitudinal bars has to be anchored.

Where the slab has a cantilever extension, the bars can be looped around the column. Eurocode 4 suggests the use of a strut-tie model for the analysis. Tests have now shown that further guidance is needed (ECCS, 2016); for example, the struts may need stirrups to resist shear.

This is not possible where the column is close to an external façade. If there are steel beams that span between adjacent columns, the bars can have hooked ends that pass around studs on these beams. Design rules for these layouts are being developed.

In the fin-plate joint of Figure 5.2a, the bolts are designed mainly for vertical shear, and the flexural stiffness is low. The end-plate joint of Figure 5.2c is likely to be 'semi-rigid' (defined later). The bolts at A are usually designed for tension only, and bolts in the compression zone (B and C) are designed for vertical shear only.

To achieve a 'rigid' connection it may be necessary to use an extended end plate and to stiffen the column web in regions D and E, as shown in Figure 5.2b.

5.3.1.1 Resistance of an end-plate joint

An account of the Eurocode design methods for steel and composite joints is available (Jaspart and Weynand, 2016). The calculations needed for a partial-strength end-plate joint, as in Figure 5.2c, are extensive. These are explained, with an example, in a *Designers' Guide* to EN 1994-1-1 (Johnson, 2012). An outline of the method is now given, assuming beams of the same depth on opposite sides of the column. Tabulated results for joints with typical dimensions could be provided, but were not available in 2017.

The rotation capacity of the joint is ensured by using a thin end plate, so that yield lines form in it before the bolts at A fracture in tension. Plastic bending of the column flange and yielding of the column web at D may also occur. The check on bolt fracture may need to allow for prying action. This is the increase of bolt tension caused by compressive force where the edges of the end plate bear against the column flange, as shown in Figure 5.25. The tensile resistance of the top bolts, $F_{T,Rd}$, is given by the weakest of these types of deformation.

The longitudinal reinforcement in the slab is assumed to be at yield in tension, so the force $F_{t,s,Rd}$ is known. Assuming that any axial force in the beam is negligible, the compressive force at the bottom of the joint cannot exceed $F_{T,Rd} + F_{t,s,Rd}$. Failure could occur by buckling of the column web at E, so this resistance is found next, allowing for the axial compression in the column. If buckling governs, a stiffener can be added, but this is rarely necessary. The compressive force to cause yielding of the bottom flange, $F_{a,fl,Rd}$, is then found. If it is less than the total tensile force, an area of web is assumed also to yield, such that

$$F_{a,fl,Rd} + F_{a,web,Rd} = F_{T,Rd} + F_{t,s,Rd}$$

The lines of action of these four forces are known, so the bending resistance of the joint, $M_{j,Rd}$, is found.

For beams of unequal depth, or at an external column, checks are also needed on the resistance of the column web to shear and the transfer to the column of the unbalanced tensile force in the slab reinforcement. The methods can also allow for concrete encasement to the column and/or the webs of the beams.

The resistance of the joint to vertical shear is normally provided by the bolts at B and C, and may be limited by the bearing strength of the end plate or column flange. The allocation of some of the shear to the bolts at A would reduce their design resistance to tension.

5.3.1.2 Moment-rotation curve for an end-plate joint, and stiffness coefficients

The other information needed for design is a curve of hogging bending moment against rotation of the joint, ϕ. This is defined as the rotation additional to that which would occur if the joint were rigid and the beam continued to its intersection with the centre-line of the column, as shown in Figure 5.4.

For steel connections, methods are given in EN 1993-1-1 (BSI, 2014b) for the prediction of this curve. They are applicable also to composite joints, with modifications given in Annex A of EN 1994-1-1. These allow for the effect of slip of the shear connection on the longitudinal stiffness of the top reinforcement, and are explained in Johnson (2012). The shorter account here is limited to beams of equal depth and with equal hogging bending moments, so that no in-plane shear is applied to the column web. The steel bottom flange is assumed to be strong enough for force $F_{a,web}$ in Figure 5.2 to be zero. The connections between the column and the two beams are identical, with pairs of bolts at levels A, B and C.

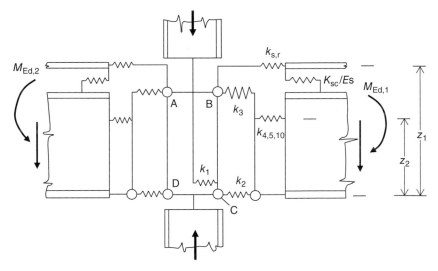

Figure 5.3 Model for elastic analysis of end-plate connections between composite beams and a column

Figure 5.4 Rotation of a joint

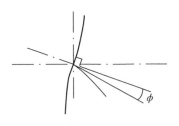

The elasticity of their components is represented by springs (Figure 5.3) with stiffness coefficients that are given in EN 1993-1-8 as follows:

k_1 shear deformation of the column web panel, ABCD

k_2 compression of the web of the column, and of encasement, if any, at the level of the bottom flange

k_3 extension of the web of the column

k_4 bending of the column flange caused by the tension in bolt group A

k_5 bending of the end plate caused by the tension in bolt group A

k_{10} extension of bolt row A

The additional coefficients given in EN 1994-1-1 are:

$k_{s,r}$ extension of longitudinal reinforcement in the slab

K_{sc}/E slip of the shear connection, with E taken as the value for steel, E_a

The stiffness of the connection in compression is given by k_2. The overall stiffness in tension is found from all the other coefficients, taking account of their lever arm for bending, z. This is their distance above the centre of compression, which is aligned with stiffness k_2. The distances z_1 (for slab reinforcement and slip) and z_2 (for bolts in tension) are shown in Figure 5.3.

Combination of these stiffnesses gives the initial elastic rotational stiffness (moment/rotation) of the connection, $S_{j,ini}$, which is defined by

$$S_{j,ini} = \frac{M}{\phi} = Ez^2 \Big/ \sum_i \left(\frac{1}{k_i} \right) \tag{5.1}$$

where ϕ is the rotation for a bending moment M, and z is the effective lever arm for bending.

The coefficients are found to have a dimension of length. This is illustrated for the simple case where there is no reinforcement and stiffness in compression is infinite. Then, Equation 5.1 is

$$S_{j,ini} = Ez_2^2 k_{10} = \frac{M}{\phi} \tag{5.2}$$

The bolts, of total net area $2A_b$ and effective length L_b, resist tension M/z_2. Their elongation is

$$e = (M/z_2)(L_b/2A_b E) \quad \text{and} \quad \phi = e/z_2 = ML_b/(2A_b Ez_2^2)$$

Equation 5.2 then gives the result, with a dimension of length: $k_{10} = 2A_b/L_b$.

5.3.1.3   Stiffness in tension

For tension in the steel beam, the flexibilities $1/k$ are in series, so the total stiffness k_t is given by

$$1/k_t = 1/k_3 + 1/k_4 + 1/k_5 + 1/k_{10}$$

Slip of the shear connection causes a flexibility $1/K_{sc}$ which is found as a ratio of displacement to force, so for consistency with $k_{s,r}$, the stiffness coefficient is K_{sc}/E, which is in series with $k_{s,r}$. The combined stiffness for reinforcement is given by

$$1/k_{s,red} = 1/k_{s,r} + E/K_{sc} \tag{5.3}$$

The equivalent stiffness in tension, $k_{t,eq}$, and the effective lever arm z are defined by

$$k_{t,eq}z = k_{s,red}z_1 + k_t z_2 \quad \text{and} \quad M/\phi = Ek_{t,eq}z^2 \tag{5.4}$$

by analogy with Equation 5.1.

The calculation of $k_{t,eq}$ and z is illustrated by assuming that when bending moment M causes rotation ϕ, there is tension T_1 at level z_1 and tension T_2 at level z_2, so that

$$M = T_1 z_1 + T_2 z_2$$

The elongations at these levels are:

$$e_1 = T_1/Ek_{s,red} \quad \text{and} \quad e_2 = T_2/Ek_t$$

and for rotation ϕ about the bottom flange,

$$e_1 = \phi z_1 \quad \text{and} \quad e_2 = \phi z_2$$

Elimination of e_1, e_2, T_1 and T_2 from the last five equations gives

$$M/E\phi = k_{s,red}z_1^2 + k_t z_2^2 \tag{5.5}$$

Equations (5.4) and (5.5) can be solved for z and $k_{t,eq}$, giving:
for the effective lever arm:

$$z = (k_{s,red}z_1^2 + k_t z_2^2)/(k_{s,red}z_1 + k_t z_2) \tag{5.6}$$

for the equivalent stiffness in tension:

$$k_{t,eq} = (k_{s,red}z_1 + k_t z_2)/z \tag{5.7}$$

5.3.1.4 Elastic stiffness of the joint

Including the stiffness in compression, from Equation 5.1,

$$S_{j,ini} = Ez^2/(1/k_{t,eq} + 1/k_2) \tag{5.8}$$

This stiffness is assumed to be applicable for bending moments $M_{j,Ed}$ up to $2M_{j,Rd}/3$, where $M_{j,Rd}$ is the bending resistance of the joint.

At higher bending moments, EN 1993-1-8 (BSI, 2005a) gives the stiffness as

$$S_{j,ini}/S_j = (1.5M_{j,Ed}/M_{j,Rd})^\psi \tag{5.9}$$

where ψ depends on the type of joint, and is 2.7 for a welded or bolted end-plate joint. The moment–rotation curve for $\psi = 2.7$ is shown as 0ABC in Figure 5.5, in which $\phi_{0.67}$ is the rotation for $M_{j,Ed} = 2M_{j,Rd}/3$.

It is inconvenient for analysis to have a joint stiffness that depends on the bending moment. For beam-to-column joints, EN 1993-1-8 permits the simplification that for all values of $M_{j,Ed}$, $S_j = S_{j,ini}/2$. This is line OB in Figure 5.5.

An end-plate connection of this type is designed in Section 5.10.2, where results of the main calculations are given. It is found to be *rigid* and *partial-strength*, terms which are defined in Section 5.3.2.

Figure 5.5 Moment–rotation curve for an end-plate joint

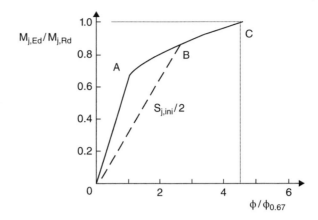

Figure 5.6 Classification of joints by initial stiffness

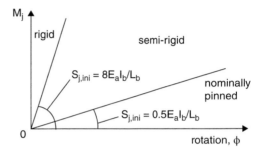

5.3.2 Classification of joints

As shown in Table 4.1 in Section 4.1, beam-to-column joints are classified in Eurocode 4, as in Eurocode 3, by rotational stiffness, which is relevant to elastic global analysis, and by resistance to bending moment, which is relevant to the resistance of a frame to ultimate loads. The three stiffness classes are shown in Figure 5.6.

A *nominally pinned* joint has

$$S_{j,ini} \leq 0.5E_aI_b/L_b \tag{5.10}$$

where E_aI_b is the rotational stiffness of the connected beam, of length L_b. The value of E_aI_b should be consistent with that taken for a cross-section adjacent to the joint in global analysis of the frame. The significance of this limit to $S_{j,ini}$ can be illustrated by considering a beam of span L_b and uniform section that is connected at each end to rigid columns, by connections with $S_{j,ini} = 0.5E_aI_b/L_b$. It can be shown by elastic analysis that for a uniformly distributed load w per unit length, the restraining (hogging) moments at each end of the beam are low, even for a pinned joint of maximum stiffness:

$$M_{el} = (wL_b{}^2/8)/7.5 \tag{5.11}$$

These end moments act also on the columns, the flexibility of which would in practice reduce the moments below M_{el}.

In practice, the joint stiffness is often taken as zero, and the bending moment applied by a beam to a supporting column defined by assuming the location of the zero-moment 'pin' (Section 5.4.4.2).

A joint is *rigid* if

$$S_{j,ini} \geq k_b E_a I_b / L_b$$

where $k_b = 8$ for a braced frame, a term defined in EN 1993-1-8 (BSI, 2005a).

The amount of redistribution of elastic moments caused by the flexibility of a connection that is just 'rigid' can be quite significant. As an example, we consider the same beam as before, with properties $E_a I_b$ and L_b, supported at each end by rigid columns, with uniform load such that both end moments are $2M_{j,Rd}/3$, when the joints have stiffness $S_{j,ini} = 8E_a I_b / L_b$.

Elastic analysis for a uniform beam shows that the end moments are $wL^2/15$. They would be $wL^2/12$ if the joints were truly rigid, so their flexibility causes a 20% redistribution of hogging moment. The situation for a composite beam in practice is more complex because $E_a I_b$ is not uniform along the span, and the columns are not rigid.

A *semi-rigid* joint has an initial rotational stiffness between these two limits (Figure 5.6).

The classification of joints by strength is as follows.

A joint with design resistance $M_{j,Rd}$ is classified as *nominally pinned* if $M_{j,Rd}$ is less than 25% of the bending resistance of the weaker of the members joined, and if it has sufficient rotation capacity. It is not difficult to design connections that satisfy these conditions. An example is given in Section 5.10.1.

A *full-strength* joint has a design resistance (to bending, taking account of co-existing shear) at least equal to $M_{pl,Rd}$ of the connected beam, or to the sum of the plastic resistances of the column lengths connected at the joint. There is a separate requirement to check that the rotation capacity of the connection is sufficient. This can be difficult. It is waived if

$$M_{j,Rd} \geq 1.2 M_{pl,Rd} \tag{5.12}$$

In practice, a 'full-strength' beam-to-column connection can be designed to satisfy Condition (5.12). It can then be assumed that inelastic rotation occurs in the beam, not in the connection. The rotation capacity is then assured by the classification system for steel elements in compression.

A *partial-strength* joint has a resistance less than that of the members joined, but must have sufficient rotation capacity, if it is at the location of a plastic hinge, to enable all the necessary plastic hinges to develop under the design loads.

5.4 Design of non-sway composite frames

5.4.1 Imperfections

The scope of this Section is limited to multistorey structures of the type shown in Figure 5.1, modelled as two sets of plane frames as explained in Section 5.1. It is assumed that the layout of the beams and columns and the design ultimate gravity loads on the beams are known, and that these members do not contribute to resistance to wind loading.

The first step is to define the imperfections of the frame. These arise mainly from lack of verticality of columns, but also have to take account of lack of fit between members, effects of residual stresses in steelwork, and other minor influences, such as non-uniform temperature of the structure. The term 'column' is here used to mean a member that may extend over the whole height of the building. A part of it with a length equal to a storey height is referred to as a 'column length', where this is necessary to avoid ambiguity.

Imperfections within a storey-height column of length L are represented by an assumed initial bow, e_0. This ranges from $L/100$ to $L/300$, depending on the type of column and the axis of bending. For major-axis bending of the columns shown in Figure 5.1, EN 1994-1-1 specifies a bow of $L/200$, or 20 mm in 4.0 m. This has to be allowed for in verification of the member, but not in first-order global analysis. The condition of EN 1994-1-1 for neglecting member imperfections in second-order analysis is, essentially, that $N_{Ed} \le 0.25\,N_{cr}$, where N_{cr} is the elastic critical axial compression for the member, allowing for creep, given by Equation 5.29.

A bow of 20 mm in 4 m exceeds the tolerance that would be acceptable in construction, because it allows also for other effects, such as residual stresses in steel. The out-of-straightness e_0 is assumed to occur at mid-length. No assumption is made about the shape of the imperfection.

Imperfections in beams are allowed for in the classification system for steel elements in compression, and in design for lateral buckling.

Frame imperfections are represented by an initial angle of sidesway, ϕ, as shown in Figure 5.7a for a single column length of height h, subjected to an axial load N. The action effects in the column are the same as if it were vertical and subjected to horizontal forces $N\phi$, as shown.

It is assumed that the angle ϕ for a composite frame is the same as for the corresponding steel frame. This is given in EN 1993-1-1 as a function of the height of the structure

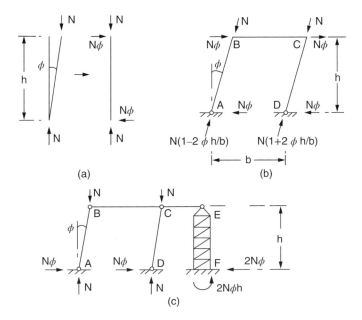

Figure 5.7 Unbraced and braced frames

in metres, h, and the number of columns in the plane frame considered, m, as follows:

$$\phi = \alpha_h \alpha_m / 200 \tag{5.13}$$

where

$$\alpha_h = 2/\sqrt{h} \quad \text{with} \quad 2/3 \le \alpha_h \le 1 \tag{5.14}$$

and

$$\alpha_m = (0.5 + 1/m)^{1/2} \tag{5.15}$$

Thus, in the frame in Figure 5.1b, $h = 36$ m, $\alpha_h = 2/3$, $m = 3$, $\alpha_m = 0.816$, and $\phi = 0.67 \times 0.816/200 = 1/366$. This initial sway applies in all horizontal directions, and is uniform over the height of the frame. In this example, the overall out-of-plumb of each column is assumed to be $36\,000/366 = 98$ mm.

Let the total design ultimate gravity load on the frame, for a particular combination of actions, be $G + Q$ per storey. The imperfections can then be represented by a notional horizontal force $\phi(G + Q)$ at each floor level – but there may or may not be an equal and opposite reaction at foundation level.

To illustrate this, we consider the single-bay, single-storey, unbraced frame ABCD shown in Figure 5.7b, with pin joints at A and D and loads N as shown, and assume $\sin \phi = \phi$, $\cos \phi = 1$. The use of additional forces $N\phi$ at B and C is associated with the assumption that the loads N still act along the columns. There are obviously horizontal reactions $N\phi$ at A and D; but the vertical reactions N are replaced by reactions $N(1 \pm 2\phi h/b)$ at angle ϕ to the vertical. The total horizontal reaction at A is therefore

$$N\phi - N\phi(1 - 2\phi h/b) = 2N\phi^2 h/b \approx 0$$

The maximum first-order bending moment in the perfect frame is zero. The imperfection ϕ increases it to $N\phi h$, at corners B and C, which may not be negligible.

If there are pin joints at these corners, the frame has to be braced against sidesway, for example by connection to the top of a stiff vertical cantilever EF (Figure 5.7c). The external reactions now do include horizontal forces $N\phi$ at A and D, with an opposite reaction $2N\phi$ at F; and the vertical reactions at A and D are independent of ϕ.

These simple analyses are *first order*: that is, they neglect any increase in the assumed sway ϕ caused by the deformations of the structure under load. Analyses that take account of this effect are referred to as *second order*. A simple example is the elastic theory for the lateral deformation of an initially crooked pin-ended strut.

5.4.2 Elastic stiffnesses of members

The determination of these properties requires consideration of the behaviour of joints and of creep and cracking of concrete. Creep in beams will be allowed for by using a modular ratio $n = 2n_0 = 20.2$, as before. For columns, EN 1994-1-1 gives the effective modulus for concrete as

$$E_c = E_{cm} / [1 + (N_{G,Ed}/N_{Ed})\varphi_t] \tag{5.16}$$

where N_{Ed} is the design axial force, $N_{G,Ed}$ is the part of N_{Ed} that is permanent, and φ_t is the creep coefficient. Expressions for the stiffnesses of composite columns are given in Section 5.6.3 in terms of E_c. Typical values for the short-term elastic modulus E_{cm} are given in Section 3.2.

For cracking it will be assumed, from Section 4.3.2, that the 'cracked reinforced' section of each beam is used for a length of 15% of the span on each side of the central column. The joints are assumed to be rigid at the central column and nominally-pinned at the external columns, as explained earlier.

5.4.3 Methods of global analysis

The condition of EN 1993-1-1 for the use of first-order analysis is $\alpha_{cr} \geq 10$, where α_{cr} is the factor by which the design loading would have to be increased to cause elastic instability in a sway mode.

The factor α_{cr} used to be calculated for simple frames by a hand method involving s and c functions, which have been tabulated (Coates et al., 1988); but if the approximation that follows is not satisfied, computer analysis is now used. The frame is modelled without imperfections, and repeated elastic analyses are done for the design load multiplied by an increasing factor, until an analysis fails to converge. The loading then is the *elastic critical buckling load*.

For beam-and-column plane frames in buildings, EN 1993-1-1 gives the approximation

$$\alpha_{cr} = (H_{Ed}/V_{Ed})(h/\delta_{H,Ed}) \tag{5.17}$$

to be satisfied separately for each storey of height h. In this expression, V_{Ed} and H_{Ed} are the total vertical and horizontal reactions at the bottom of the storey, with forces $N\phi$ included in H_{Ed}, and $\delta_{H,Ed}$ is the change in lateral deflection over the height h.

It will be shown later that, for the frame in Figure 5.1, the lateral stiffness of the floor slabs is so much greater than that of the columns that almost the whole of the lateral load is transferred to the core and end walls. These are stiff enough for $\delta_{H,Ed}$ to be very small, so α_{cr} for each storey far exceeds 10. Hence, first-order global analysis can be used for each frame. The shear walls require a separate check.

In its clause on columns, Eurocode 4 gives in outline a 'General method' of design which involves *elasto-plastic* non-linear global analysis, and includes second-order effects, where needed. This analysis will involve members adjacent to the column considered, if not the whole frame. (Elasto-plastic analysis is defined in EN 1990 as a method that uses stress–strain or moment–curvature relationships that consist of a linear-elastic part followed by a plastic part that may or may not include strain-hardening.)

This method does not give sufficiently clear guidance on the use of partial factors, and is rarely used. It will probably be revised in the next Eurocode 4.

The 'General method' is followed in Eurocode 4 by a 'Simplified method of design', applicable to a length of a column between lateral restraints. It is described in Section 5.6 and used in examples in Sections 5.7 and 5.8.

5.4.4 First-order global analysis of braced frames

5.4.4.1 Actions

This Section should be read with reference to Section 4.3, on global analysis of continuous beams, much of which is applicable. Braced frames do not have to be designed for horizontal actions, so the load cases are similar to those for beams. Imposed load is the leading variable action, and neither wind nor the $N\phi$ horizontal forces need be included.

No serviceability checks are normally required for braced frames, or for composite columns. The columns are designed using elastic global analysis and plastic section analysis; that is, as if their cross-sections were in Class 2. The method of construction, propped or unpropped, is therefore not relevant.

For most types of imposed load, the probability of the occurrence of the factored design load becomes less as the loaded area increases. EN 1991-1-1 recommends that the imposed load on a floor or roof with a loaded area of A m^2 may be reduced by a factor α_A, given by:

$$\alpha_A = 5\psi_0/7 + A_0/A \leq 1.0 \tag{5.18}$$

where $A_0 = 10.0$ m^2 and ψ_0 is the combination factor for the relevant type of imposed load. For the loading used in the example, $\psi_0 = 0.7$, and $\alpha_A = 1.0$ (no reduction) when the loaded floor area A is 20 m^2. In the example here, with a column grid 9.5 m by 4 m, a span of a beam carries 38 m^2 of floor, and Equation 5.18 gives $\alpha_A = 0.763$. This reduction of the imposed loading, almost 24%, was not done in the example in Section 4.6.

The characteristic imposed load on a column that carries load from n storeys may be reduced by applying the recommended factor

$$\alpha_n = [2 + (n - 2)\psi_0]/n \tag{5.19}$$

For $\psi_0 = 0.7$, this gives $\alpha_n < 1$ for three or more storeys.

Maximum bending moments in columns in rigid-jointed frames occur when some of the nearby floors do not carry imposed load. For a column length AB in a frame with many similar storeys, the most adverse combination of axial load and bending moment is likely to occur when the imposed load is applied as in Figure 5.8a, for an external column, or Figure 5.8b, for an internal column. The bending-moment distributions for these columns are likely to be as shown.

5.4.4.2 Eccentricity of loading, for columns

The use of 'nominally pinned' beam-to-column joints reduces bending moments in the columns, with corresponding increases in the sagging moments in the beams. For the beams, it is on the safe side to assume that the moments in the connections are zero. If this were true, the load from each beam would be applied to the column at an eccentricity slightly greater than half the depth h_a of the steel column section (Figure 5.2a) for major-axis connections.

An elastic analysis that modelled the real (non-zero) stiffness of the connection would give an equivalent eccentricity greater than this. The real behaviour is more complex. Initially, the end moments increase the tendency of each column length to buckle; but as the load increases, the column becomes less stiff, the end moments change sign, and the greater the stiffness of the connections, the more beneficial is restraint from the beams for the stability of the column.

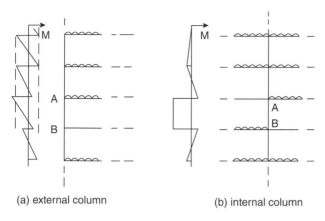

(a) external column   (b) internal column

Figure 5.8 Arrangements of imposed load, for column design

Typically, British codes of practice for steel columns have allowed for this stabilizing effect by modelling each storey-height column with an 'effective length' of between 70% and 85% of its actual length (e.g. $L_e/L \approx 0.7$), but have specified an 'equivalent eccentricity of loading', $e_c \approx 0.5 h_a + 100$ mm, for the calculation of end moments. For the 206×204 UC section to be used here, it will be assumed that the load from a supported major-axis beam acts at 0.2 m from the axis of the column. This applies a bending moment to the column that is independent of its stiffness, and reduces the design span of the beam by 0.2 m at each connection.

In some other European countries, the practice has been to assume $L_e \approx L$, which makes buckling more critical, and $e_c = 0$, which eliminates bending moments from columns. One justification for using $e_c = 0$ is that the bending moments in the beams are calculated using the span between column centres, rather than the smaller span between the centres of the 'pin' connections. Eurocodes 3 and 4 at present give no guidance on this subject.

In the following example, it will be assumed that $L_e = L$ and that the load from a nominally pinned connection acts at 100 mm from the face of the steel column section, so that $e_c = 0.5 h_a + 100$ mm. The beam is assumed to be simply-supported at the pin, so its span is less than that to the column centre-line.

5.4.4.3 Elastic global analysis

First-order elastic analysis is applicable to braced composite frames that satisfy the condition in Section 5.4.3. The flexural stiffness of hogging moment regions of beams is treated as in Section 4.3.2. For columns, concrete is assumed to be uncracked, and the stiffness of the longitudinal reinforcement is usually included, as it may not be negligible.

Bending moments in beams may be redistributed as in Section 4.3.2, but end moments found for composite columns may not be reduced, because the extent of redistribution to be allowed would depend on the rotation capacity of the columns, and there is insufficient knowledge of this subject.

Where the beam-to-column joints are nominally pinned, as in the external columns in the example in Section 5.7, the bending moments in a column are easily found by moment distribution for that member alone.

5.4.4.4 Rigid-plastic global analysis

The use of this method for a braced frame is not excluded by EN 1994-1-1, but there are several conditions that make it unattractive in practice. In addition to the conditions that apply to beams (Section 4.3.3), these include the following.

(1) All connections must be shown to have sufficient rotation capacity, or must be full-strength connections with $M_{j,Rd} \geq 1.2M_{pl,Rd}$, as explained in Section 5.3.2.
(2) Unless verified otherwise, it should be assumed that composite columns do not have rotation capacity.

5.4.5 Outline sequence for design of a composite braced frame

This overview is intended to provide an introduction to the subject; it is not comprehensive. Its scope is limited to regular multistorey braced frames of the type used in the examples in Chapters 3, 4 and 5. It is assumed that the detailing will provide the required resistance to fire, and that the following decisions are made at the outset:

- number of storeys, storey heights, column positions, layout and spans of beams;
- use of floors to transfer lateral forces to bracing elements;
- type and location of bracing elements;
- imposed floor loadings and wind loading;
- type(s) of beam-to-column joints (pinned, semi-rigid or rigid);
- nominal eccentricity for any nominally-pinned joints to columns;
- strengths of materials and densities of concretes to be used.

5.4.5.1 Ultimate limit states

(1) Design the floor slabs (concrete or composite), spanning between the beams.
(2) Find imposed-load reductions (if any) for the beams.
(3) Do preliminary designs for beams, neglecting interaction with internal columns, as these have little influence, even if joints are rigid. Use of nominally-pinned joints to external columns enables their influence on beam design also to be ignored. Include the flexibility of any semi-rigid joints in analyses of continuous beams.
(4) Find imposed-load reductions for columns, and do preliminary designs. Check that columns are not so slender that their imperfections should be included in global analysis.
(5) Consider creep and cracking of concrete, and find elastic stiffnesses for all beams and columns.
(6) Do elastic first-order global analyses for all frames, for gravity loads only, to find action effects in beams and columns. Moments in beams may be redistributed, within permitted limits. Neglect frame and member imperfections.
(7) Check beam designs. Revise if necessary.
(8) Increase bending moments in each column to allow for second-order effects within the column length and for column imperfections. Check column designs and revise if necessary.
(9) Allow for frame imperfections by notional horizontal forces. Compare these with forces from wind, and decide whether, for horizontal loading, the leading variable action should be imposed gravity loading or wind loading.

(10) Do preliminary designs for bracing elements, to find their stiffnesses.
(11) Do elastic first-order global analyses for the complete structure, for horizontal loads plus appropriate gravity loads, to find lateral deflections. The arrangement of imposed loading should be that which gives maximum sidesway.
(12) Repeat (11) with all beam–column intersections displaced laterally by the amounts found in step (11). Check that the increases in the action effects that govern design of members are all less than 10%. (This condition for the use of first-order global analyses is now assumed to be satisfied.)
(13) Check the design of the bracing elements, taking account of imperfections and second-order effects. (The example in this chapter does not include this.)

5.4.5.2 Design for serviceability limit states
Re-analyse the frames for unfactored loads to check deflections and susceptibility to vibration. Detail reinforcement to control crack widths, as necessary.

5.5 Example: composite frame

5.5.1 Data

To enable previous calculations to be used, the structure to be designed has a composite slab floor that spans 4.0 m between two-span continuous composite beams with spans of 9.5 m. There are nine storeys, each with floor-to-floor height of 4.0 m, as shown in Figure 5.1. The outer columns are assumed to be nominally pinned at ground level, and the internal columns to be nominally pinned at basement level, 4.0 m below the beam at ground level. For simplicity, it is assumed that the roof has the same loading and structure as the floors, though this would not be so in practice. The building stands alone, and the horizontal span of its floors is 28 m between walls that provide lateral restraint, as explained earlier.

The materials and loadings are as used previously (Sections 3.2, 3.11 and 4.6), and the composite floor is as designed in Section 3.4. The two-span composite beams are as designed in Section 4.6, with nominally pinned connections to the external columns (Section 5.10), except that they are not continuous over a central point support. There is instead a composite column to which each span is connected by a joint assumed at first to be 'rigid' and 'full-strength'. These terms are defined in Section 5.3.2.

The only gravity loads additional to those carried by the beams are the weight of the columns and the external walls. The characteristic values are assumed to be as follows:

- for each column, $g_k = 3.0 \, \text{kN/m} = 12.0 \, \text{kN}$ per storey.
- for each external wall, $g_k = 60 \, \text{kN}$ per bay per storey.

The 60-kN load is for $4 \times 4 = 16 \, \text{m}^2$ of wall, which is assumed to be supported at each floor level by a beam spanning 4.0 m between adjacent columns.

The design ultimate gravity load per storey for each column is therefore the load from one main beam plus:

$$
\left.
\begin{array}{ll}
\text{- for the internal column,} & 12 \times 1.35 = 16.2 \, \text{kN} \\
\text{- for an external column,} & 72 \times 1.35 = 97.2 \, \text{kN}
\end{array}
\right\}
\qquad (5.20)
$$

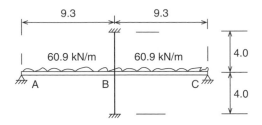

Figure 5.9 Beam loading for maximum hogging bending moment at B

The characteristic wind load is based on wind in a direction parallel to the longitudinal axes of the main beams. It causes pressure on the windward wall, and suction (i.e. pressure below atmospheric) on the leeward wall. The sum of these two effects is assumed to be:

$$q_{k,wind} = 1.5 \text{ kN/m}^2 \text{ of windward wall} \tag{5.21}$$

The effects of wind blowing along the building are not considered.

The properties of materials are as summarized in Section 4.6.1, except that the concrete in the composite columns is of normal density, with properties

$$f_{ck} = 25 \text{ N/mm}^2, \quad E_{cm} = 31.0 \text{ kN/mm}^2 \tag{5.22}$$

The design initial sidesway angle ϕ of a frame such as DEF in Figure 5.1 was found in Section 5.4.1 to be $\phi = 1/366$.

5.5.2 Design action effects and load arrangements

The whole of the design variable load for a typical frame is transferred to its three columns by the major-axis beams. Permanent loading is symmetrical about the plane of the frame, so the beams that support the external walls apply negligible minor-axis bending moments to the external columns. The additional gravity loads (Expressions 5.20) are assumed to cause no major-axis bending moments. These are caused in the columns only by the loads from the major-axis beams.

Load arrangements for the two-span beams   For each action effect in a member, the appropriate arrangement of imposed load is that which gives the most adverse value (usually, the highest value). For the beams, the limited frame shown in Figure 5.9 can be used, because at the joints 4 m above and below joint B, any difference from the fixed-end condition assumed will have negligible effect on the action effects in beam ABC.

For hogging bending in a beam, and for axial force in columns, imposed load should be applied to both spans of every beam. From Table 4.4, the design loads are as shown in Figure 5.9. The loaded floor area is $18.6 \times 4 = 72.4 \text{ m}^2$, so from Equation 5.18 a reduction factor $\alpha_A = 0.5 + 10/72.4 = 0.64$ could be applied. This is not done, for simplicity, so that earlier results can be used.

From symmetry, this loading causes no bending in the internal column. From Section 4.6.2, the bending moment in the beam at B, after 30% redistribution of moments, is

$$M_{Ed,B} = 461 \text{ kN m} \tag{5.23}$$

From equilibrium of length AB, the shear force in each beam at B is

$$V_{Ed,B} = 60.9 \times 9.3/2 + 461/9.3 = 333 \text{ kN} \tag{5.24}$$

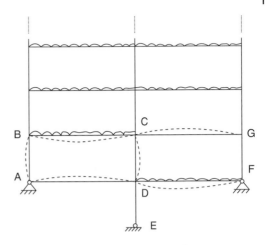

Figure 5.10 Arrangement of variable load for maximum bending in columns AB and CD

To obtain the shear forces at the pin joints at A and C, each span is increased to 9.5 m, so that the loads on the external columns include the correct width of floor. Resolving vertically for span AB,

$$V_{A,Ed} = V_{C,Ed} = 60.9 \times 9.5 - 333 = \mathbf{246\ kN} \tag{5.25}$$

For maximum sagging bending moment in one span of the beam, the other span should have minimum load. It is accurate enough to analyse the limited frame of Figure 5.9 with the loading on the other span reduced to 23.7 kN/m, the design dead load (Table 4.4). It will be noted that this value includes the partial factor 1.35, even though omitting it would slightly increase the sagging moment. This simplification is part of the Eurocode system (Table 1.1).

For the columns, the arrangement of variable load shown in Figure 5.10, with full variable load on all upper floors, will provide an adverse combination of axial force and single-curvature bending in both the column lengths AB and CD. It is assumed that column length DE can have an increased cross-section, if necessary.

The example is continued after the design methods for composite columns have been explained.

5.6 Simplified design method of EN 1994-1-1, for columns

5.6.1 Introduction

This method is preceded in Eurocode 4 by a 'General method', which is referred to in Section 5.4.3. Background information for the simplified method is provided in Sections 5.1 and 5.2. Global analysis provides for each column length in a plane frame a design axial force, N_{Ed}, and applied end moments $M_{1,Ed}$ and $M_{2,Ed}$. By convention, M_1 is the greater of the two end moments, and they are both of the same sign where they cause single-curvature bending.

Initially, concrete-encased H or I sections are considered (Figure 5.11a, b). Where methods for concrete-filled steel tubes (Figure 5.11c) are different, this is explained in Section 5.6.7. The encased cross-sections are assumed to have biaxial symmetry, and to

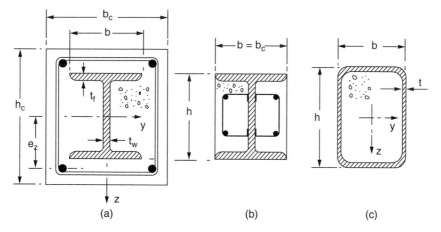

Figure 5.11 Typical cross-sections of composite columns

be uniform along each column length. Applied moments are resolved into the planes of major-axis and minor-axis bending of the column, and their symbols have additional subscripts (y and z, respectively) where necessary.

The two ends of a column length are assumed each to be connected to one or more beams and to be braced laterally at these points, distance L apart. The effective length of each column length is here assumed to be L, as explained in Section 5.4.4.2. Lateral loads on columns are assumed to be applied only at the ends of each column length.

The methods explained below are applied separately for each plane of bending. It often happens that all significant bending occurs in one plane only. If this is minor-axis bending, no major-axis verification is needed. If it is major-axis bending, minor-axis buckling must be checked, as explained in Section 5.6.5.2, because of interaction between the axial load and the minor-axis imperfections.

These methods were aligned with those for concrete columns (Eurocode 2) as far as practicable. There are important differences from the methods for steel columns in Eurocode 3.

5.6.2 Detailing rules, and resistance to fire

Before doing calculations based on an assumed cross-section for a composite column, it is wise to check that the section satisfies relevant limits to its dimensions.

The resistance to fire of a concrete-encased I-section column is determined by the thickness of the concrete cover to the steel section and the reinforcement. For a 90-minute period of resistance and a cross-section with dimensions h_c and b_c of at least 250 mm (for example), the limits to cover given by EN 1994-1-2 are 40 mm to the steel section and 20 mm to the reinforcement. There is a worked example for the fire resistance of a concrete-filled tube column in Section 6.4.3.

The rules of EN 1992-1-1 for minimum cover and reinforcement, and for maximum and minimum spacing of bars, should be followed. These ensure resistance to corrosion, safe transmission of bond forces, and avoidance of spalling of concrete and buckling of longitudinal bars. The ratio of area of reinforcement to area of concrete allowed for in calculating resistances should satisfy

$$0.003 \leq A_s/A_c \leq 0.06 \tag{5.26}$$

The upper limit is to ensure that the bars are not too congested at overlaps.

The thickness of concrete cover, c, to the steel section that may be used in calculations has upper limits $c_y = 0.4b$, and $c_z = 0.3h$, with b and h as in Figure 5.11. These limits arise from the proportions of columns for which this design method has been validated. The *steel contribution ratio* δ and the *slenderness* $\bar{\lambda}$ (Section 5.6.3.1) are limited for the same reason. Slenderness limits are given for steel flanges (Figure 5.11b) and for tube walls (Figure 5.11c), to avoid failure by local buckling.

The steel contribution ratio is defined by

$$\delta = A_a f_{yd}/N_{pl,Rd} \tag{5.27}$$

where f_{yd} is the design yield strength of the structural steel, with the condition $0.2 \leq \delta \leq 0.9$.

If $\delta < 0.2$, the column should be treated as reinforced concrete; and if $\delta > 0.9$, as structural steel. The term $A_a f_{yd}$ is the contribution of the structural steel section to the plastic resistance $N_{pl,Rd}$, given by Equation 5.33.

5.6.3 Properties of column lengths

The characteristic elastic flexural stiffness of a column cross-section about a principal axis (y or z) is the sum of contributions from the structural steel (subscript a), the reinforcement (subscript s) and the concrete (subscript c), and so has the format:

$$(EI)_{eff} = E_a I_a + E_s I_s + K_c E_{c,eff} I_c \tag{5.28}$$

where E is the elastic modulus of the material and I the second moment of area of the relevant cross-section.

The elastic critical axial load is found from

$$N_{cr} = \pi^2 (EI)_{eff}/L^2 \tag{5.29}$$

where L should be taken as the length between the lateral restraints in the plane of buckling considered. The 'concrete' term in Equation 5.28 is based on calibration of results from this method against test data. It was found that $K_c = 0.6$ and that creep should be allowed for by reducing the mean short-term elastic modulus for concrete, E_{cm}, as follows:

$$E_{c,eff} = E_{cm}/[1 + (N_{G,Ed}/N_{Ed})\varphi_t] \tag{5.30}$$

where the symbols are defined after Equation 5.16.

As N_{cr} does not depend on strengths of materials, no partial factors are involved. It is treated as a 'characteristic' value. The use of the characteristic value of EI, denoted $(EI)_{eff}$, is therefore appropriate. A lower 'design' value is also needed, for analysis of second-order effects within a column length. This is given in EN 1994-1-1 as

$$(EI)_{eff,II} = 0.9(E_a I_a + E_s I_s + 0.5 E_{c,eff} I_c) \tag{5.31}$$

with $E_{c,eff}$ from Equation 5.30. The factor 0.5 allows for cracking, and the 0.9 is based on calibration work.

5.6.3.1 Relative slenderness

The non-dimensional relative slenderness of a column length for buckling about a particular axis is defined by

$$\bar{\lambda} = \sqrt{N_{pl,Rk}/N_{cr}} \qquad (5.32)$$

The design resistance to axial load of a straight column too short to buckle, known as the 'squash load', is given by

$$N_{pl,Rd} = A_a f_{yd} + A_s f_{sd} + 0.85 A_c f_{cd} \qquad (5.33)$$

where the design strengths of the materials are:

- for structural steel, $f_{yd} = f_y/\gamma_A$ (symbol f_{yk} is not used because f_y is a nominal value)
- for reinforcement, $f_{sd} = f_{sk}/\gamma_S$
- for concrete in compression, $f_{cd} = f_{ck}/\gamma_C$

and the γ's are the usual partial factors for ultimate limit states. The area A_c for a concrete-encased section is conveniently calculated from

$$A_c = b_c h_c - A_a - A_s \qquad (5.34)$$

For calculating $\bar{\lambda}$, $N_{pl,Rd}$ is replaced by the characteristic squash load,

$$N_{pl,Rk} = A_a f_y + A_s f_{sk} + 0.85 A_c f_{ck} \qquad (5.35)$$

because N_{cr} is a characteristic value.

The following method of column design, from EN 1994-1-1, is limited to column lengths with $\bar{\lambda} \le 2$. This limit is rarely exceeded in practice.

5.6.4 Resistance of a cross-section to combined compression and uniaxial bending

Design for a combination of load along the x-axis and bending about the y- or z-axis is based on an interaction curve between resistance to compression alone, $N_{pl,Rd}$, and resistance to bending about the relevant axis, $M_{pl,Rd}$. The method is explained with reference to Figure 5.12. The plastic resistance $N_{pl,Rd}$ is given above.

The complexity of hand methods of calculation for $M_{pl,Rd}$ and other points on the curve has been a disincentive to the use of composite columns. It is quite easy to prepare a spreadsheet to do this. However, it should be noted that when a rolled I- or H-section is represented by three rectangles, as in the algebra given by Johnson (2012) and outlined below, results will differ slightly from those by hand calculation based on section tables, unless the corner fillets are allowed for.

The assumptions are those used for calculating $M_{pl,Rd}$ for beams: rectangular stress blocks with structural steel at a stress $\pm f_{yd}$, reinforcement at $\pm f_{sd}$, and concrete at $0.85 f_{cd}$ in compression or cracked in tension. The difference is that the plastic neutral axis can be outside or at any depth within the cross-section. Full shear connection is assumed.

The complexity appears in the algebra. For major-axis bending of the section shown in Figure 5.11a, there are five possible locations of the plastic neutral axis, each leading

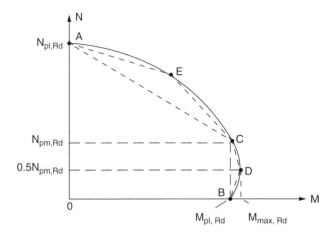

Figure 5.12 Polygonal approximation to *M–N* interaction curve

to different expressions for N_{Rd} and M_{Rd}. A practicable method is to guess a position for the neutral axis, and calculate N_{Rd} by summing the forces in the stress blocks, and M_{Rd} by taking moments of these forces about the centroid of the uncracked section. This gives one point on the interaction curve in Figure 5.12. Other points, and hence the curve, are found by repeating the process.

The simplification made in EN 1994-1-1 is to replace the curve by a polygonal diagram, AECDB in Figure 5.12. An ingenious and fairly simple method of calculating the coordinates of points B, C and D was given in the ENV version of EN 1994-1-1 (CEN, 1992), and is explained in an Appendix in Johnson (2012). It is used in Section 5.7.2.

For major-axis bending of encased I-sections, AC may be taken as a straight line, but for other situations an intermediate point, E, should be found, as line AC can be too conservative. For point E, the first guessed neutral-axis position is usually good enough. A similar method is used for the interaction polygon for axial load with minor-axis bending.

Transverse shear force may be assumed to be resisted by the steel section alone. The design method for moment–shear interaction in beams (Section 4.2.2) may be used. In columns, V_{Ed} is usually less than $0.5V_{pl,Rd}$, and then no reduction in bending resistance need be made. None is assumed here.

5.6.5 Verification of a column length

5.6.5.1 Design action effects for uniaxial bending

It is assumed that the interaction curve or polygon, Figure 5.12, has been determined, and that the design axial force N_{Ed} and the end moments $M_{1,Ed}$ and $M_{2,Ed}$ have been found by global analysis. It is rare for a vertical column length in a building to be subjected to significant transverse load within its length, and none is assumed here.

If, as is likely, member imperfections (Section 5.4.1) were omitted from the global analysis of the frame, the initial bow, of amplitude e_0, is allowed for now. Its first-order effect is to increase the bending moment at mid-length of the column by $N_{Ed}e_0$.

The condition in EN 1994-1-1 for neglect of second-order effects is

$$N_{cr,eff} \geq 10N_{Ed} \tag{5.36}$$

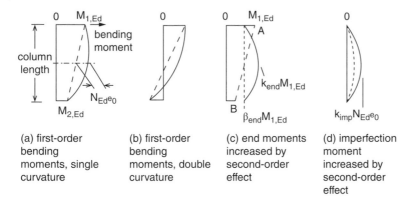

(a) first-order bending moments, single curvature

(b) first-order bending moments, double curvature

(c) end moments increased by second-order effect

(d) imperfection moment increased by second-order effect

Figure 5.13 First-order and second-order bending moments in a column length

where $N_{cr,eff}$ is found from Equation 5.29 with $(EI)_{eff}$ replaced by $(EI)_{eff,II}$ from Equation 5.31. If this condition applies, the design bending moment M_{Ed} for the column is the greatest value given by the curve in Figure 5.13a or b. Otherwise, a second-order analysis is required, or the following simplified methods from EN 1994-1-1 can be used.

Second-order effects in a column length Second-order effects of the end moments and from the $N_{Ed}e_0$ moment are found separately, and can be superimposed. This is possible because they both result from the same axial force. They are always added, because the imperfection e_0 can occur in any lateral direction. Subscripts 'end' and 'imp' are now used, respectively, for these two sets of moments.

The distribution of first-order end moments, for example, AB in Figure 5.13c, is replaced by an equivalent uniform moment $\beta_{end}M_{1,Ed}$, which is increased to $k_{end}M_{1,Ed}$ at mid-length to allow for second-order effects, as shown. The factor k for both sets of moments is

$$k = \beta/[1 - (N_{Ed}/N_{cr,eff})] \tag{5.37}$$

with

$$\beta_{end} = 0.66 + 0.44(M_2/M_1) \geq 0.44 \tag{5.38}$$

Coefficient β_{end} allows for the more adverse effect of single-curvature bending than of double-curvature bending (Figure 5.13b). For the most adverse distribution, $M_2 = M_1$, Equation 5.38 gives $\beta_{end} = 1.1$. It reduces to 0.44 for $M_2 \leq -0.5M_1$, double-curvature bending. The bending moment $k_{end}M_{1,Ed}$ always exceeds $\beta_{end}M_{1,Ed}$, from Equation 5.37.

In Equation 5.37, $N_{cr,eff}$ is found as above, and Eurocode 4 states that here, the effective length should be taken as the column length, which is correct for braced frames. The assumption to be made for a column in an unbraced frame may be clearer in the next Eurocode 4.

In EN 1994-1-1, Equation 5.37 appears with the further condition $k \geq 1.0$. It has been found to be over-conservative to apply this when combining two sets of second-order effects. For example, if $\beta_{end} = 0.66$ and $N_{Ed} = 0.2N_{cr,eff}$, then Equation 5.37 gives $k_{end} = 0.82$. It is now recommended that this need not be increased to 1.0.

For the bending moment from the member imperfection, EN 1994-1-1 specifies $\beta_{\text{imp}} = 1.0$, so from Equation 5.37, k_{imp} always exceeds 1.0, and the imperfection gives the additional bending moment shown in Figure 5.13d, which is assumed to occur about the axis where its effect is the more adverse.

The design bending moment for the column length is usually

$$M_{\text{Ed}} = k_{\text{end}} M_{1,\text{Ed}} + k_{\text{imp}} N_{\text{Ed}} e_0 \qquad (5.39)$$

but is $M_{1,\text{Ed}}$, if that is greater.

Verification, for uniaxial bending The column is strong enough if its cross-section can resist the combination of M_{Ed} with N_{Ed}. The bending resistance M_{Rd} in the presence of axial compression N_{Ed} is found from the interaction diagram, explained in Section 5.6.4.

A correction is required for the unconservative assumption that the rectangular stress block for concrete extends to the plastic neutral axis (Section 3.5.3.1). It is made by reducing the bending resistance, so that the verification condition is

$$M_{\text{Ed}} \leq \alpha_M M_{\text{Rd}} \qquad (5.40)$$

where $\alpha_M = 0.9$ for steel grades up to S355, for which $f_y = 355 \, \text{N/mm}^2$ for sections of thickness up to 40 mm. It is reduced to 0.8 for stronger steels, to take account of the adverse effect of their higher yield strain on the load at which concrete begins to crush.

It may be noticed that over the length BD of the curve in Figure 5.12, increase in axial compression increases resistance to bending. This could be unsafe where the compression and the bending result from independent actions. Eurocode 4 allows for this in a rule that requires the partial factor applied to the action causing compression to be reduced by 20%.

5.6.5.2 Biaxial bending

It has to be decided in which plane of bending failure is expected to occur. This is usually obvious. The bending moment $N_{\text{Ed}} e_0$ is included only for this plane. The axial load N_{Ed} and the maximum design bending moments about both axes, $M_{y,\text{Ed}}$ and $M_{z,\text{Ed}}$, are found, as in Section 5.6.5.1. The verification consists of checking Expression 5.40 separately for each axis, and in addition, satisfying Expression 5.41:

$$M_{y,\text{Ed}}/M_{y,\text{Rd}} + M_{z,\text{Ed}}/M_{z,\text{Rd}} \leq 1.0 \qquad (5.41)$$

5.6.6 Transverse and longitudinal shear

For applied end moments M_1 and M_2 as defined in Section 5.6.1, the transverse shear in a column length without lateral loading is $(M_1 - M_2)/L$. An estimate can be made of the longitudinal shear stress at the interface between steel and concrete, by elastic analysis of the uncracked composite section. This is rarely necessary in multistorey structures, where these stresses are usually very low.

Higher stresses may occur near joints at a floor level where the axial load added to the column is a high proportion of the total axial load. Load added after the column has become composite, N_{Ed} say, is assumed to be transferred initially to the steel section, with cross-section of area A_a. It is then shared between the steel section and its encasement or infill on a transformed area basis:

$$N_{Ed,c} = N_{Ed}(1 - A_a/A) \tag{5.42}$$

where $N_{Ed,c}$ is the force that causes shear at the surface of the steel section and A is the transformed area of the column in 'steel' units. There must be a '*clearly defined load path ... that does not involve an amount of slip at this interface that would invalidate the assumptions made in design*' (from EN 1994-1-1).

There is no well-established method for calculating longitudinal shear stress at the surface of the steel section, τ_{Ed}. Design is usually based on mean values, found by dividing the force by the perimeter of the section, u_a, and an assumed 'load introduction length', ℓ_v:

$$\tau_{Ed} = N_{Ed,c}/u_a\ell_v \tag{5.43}$$

It is recommended in EN 1994-1-1 that ℓ_v should not exceed the least of $2b_c$, $2h_c$ (Figure 5.11) and $L/3$, where L is the column length.

Design shear strengths τ_{Rd} due to bond and friction are given in EN 1994-1-1 for several situations. For completely encased sections,

$$\tau_{Rd} = 0.3 \text{ N/mm}^2 \tag{5.44}$$

This is a low value, to take account of the approximate nature of τ_{Ed}.

Where τ_{Ed} is less than τ_{Rd}, no account need be taken of the further transfer of force by shear between steel and concrete as failure is approached. The best protection against local failure is provided by the transverse reinforcement (links), which are required by EN 1992-1-1 to be more closely spaced near beam–column intersections than elsewhere.

In regions where τ_{Rd} is exceeded, shear connectors should be provided for the whole of the shear. These are best attached to the web of a steel H or I section, because their resistance is enhanced by the confinement provided by the steel flanges. Design rules are given in EN 1994-1-1.

5.6.7 Concrete-filled steel tubes

A cross-section of a concrete-filled steel tube (CFST) column of rectangular section is shown in Figure 5.11c. Section 5.11 gives a design for a column of circular cross-section, including load introduction from the beams supported. Circular cross-sections are widely used, although their curved surfaces make detailing of the joints to beams more difficult. Design rules for columns of elliptical cross-section have been developed, and may be included in the next Eurocode 4.

To reduce size and improve resistance to fire, some CFST columns have a massive steel bar at their centre as well as the outer tube.

To avoid local buckling, the slendernesses of the walls of the tube are limited. For a rectangular cross-section the limit is

$$h/t \leq 52\varepsilon \tag{5.45}$$

The limit is more generous for a circular section of diameter d:

$$d/t \leq 90\varepsilon^2 \tag{5.46}$$

where $\varepsilon = (235/f_y)^{0.5}$, and f_y is the yield strength in N/mm^2.

Design is essentially as for encased H-sections, except as follows. In calculating the squash load $N_{pl,Rd}$, account is taken of the higher resistance of the concrete, caused by lateral restraint from the steel tube. The factor 0.85 in Equations 5.33 and 5.35 is replaced by 1.0. Also, for circular sections only, f_{cd} is increased to an extent that depends on the ratios t/d, f_y/f_{ck}, $\bar{\lambda}$ and $M_{Ed}/(N_{Ed}d)$, provided that the relative slenderness $\bar{\lambda}$ does not exceed 0.5.

For a circular section, there is also a reduction in the effective yield strength of the steel wall used in calculating $N_{pl,Rd}$ to take account of the circumferential tensile stress in the wall. This stress provides restraint to lateral expansion of the concrete caused by the axial load on the column. These rules are based on extensive testing.

Evaporation of water from an exposed concrete surface contributes to both shrinkage and creep strains. It is allowed for in the rules of Eurocode 2, which are applied in the example in Section 5.7.2. This drying does not occur in concrete-filled tubes. It has been found that Eurocode 2 best predicts their behaviour when the drying components of shrinkage and creep are ignored. This would enable a reduced value for the creep coefficient φ_t to be used in Equation 5.30 for $E_{c,eff}$. In the simplified design method of Eurocode 4, shrinkage is usually neglected, as it has little effect on stiffness in flexure.

5.7 Example (continued): external column

5.7.1 Action effects

The use of nominally-pinned joints and a braced frame enables the design of an external column to be completed without further global analysis of the frame of Figure 5.1. From Section 5.5, the design ultimate shear force from a fully loaded beam is 246 kN. The size of the external column is governed by the length 0–1 in Figure 5.14a, and only this length will be designed. It supports load from nine floors, so from Equation 5.19 the reduction factor for imposed load, with $\psi_0 = 0.7$ as before, is

$$\alpha_n = (2 + 7 \times 0.7)/9 = 0.767$$

From Table 4.4, the variable load provides $37.2/60.9 = 61\%$ of the shear force. Hence, the shear force at each pin joint is

$$V_{Ed} = 246 \times 0.61 \times 0.767 + 246 \times 0.39 = 115 + 96 = \mathbf{211 \ kN} \tag{5.47}$$

It is assumed that the steel section for the column will be from the 203×203 UC serial size, so the eccentricity at the pin joint is $0.203/2 + 0.1 = 0.20$ m, and the major-axis bending moment applied to the column at each loaded floor level is

$$M_{Ed} = 211 \times 0.2 = 42.2 \ \text{kN m} \tag{5.48}$$

The bending moment in length 0–1 is determined mainly by the load from level 1, so factor α_n is taken as 1.0 at that level, giving the applied moments and shears shown in Figure 5.14a. The bending moments in the lower part of the column, found by moment distribution, are shown in Figure 5.14b.

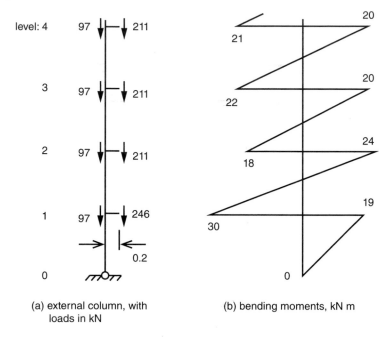

(a) external column, with
loads in kN

(b) bending moments, kN m

Figure 5.14 Dimensions and action effects for external column

For minor-axis bending, all the loading is permanent, and equal on the two sides of the column, so bending moment arises only from the initial bow of the member. Including the permanent load from Equation 5.20, the axial load for length 0–1 is:

$$N_{\text{Ed}} = 9 \times 97 + 8 \times 211 + 246 = \textbf{2807 kN} \tag{5.49}$$

5.7.2 Properties of the cross-section, and y-axis slenderness

A cross-section for the column must now be assumed, and is shown in Figure 5.15. Applying the usual partial factors to the properties of the materials given in Section 4.6.1 and Equation 5.22, the design properties are:

$$f_{\text{yd}} = 355 \, \text{N/mm}^2 \qquad\qquad f_{\text{sd}} = 435 \, \text{N/mm}^2$$
$$0.85 f_{\text{cd}} = 14.2 \, \text{N/mm}^2 \qquad\qquad E_{\text{cm}} = 31.0 \, \text{kN/mm}^2$$

The assumed concrete cover to the reinforcement, 30 mm, and to the structural steel, 57 mm, satisfy the requirements for 90 minutes' fire resistance. From EN 1992-1-1, 30 mm cover should be sufficient if the external face of the column is protected; but if it is exposed to rain and/or freeze/thaw, it would be necessary to increase either the cover or the grade of the concrete.

The cross-sectional areas of the three materials are:

$$A_{\text{a}} = 6640 \, \text{mm}^2 \qquad\quad A_{\text{s}} = 804 \, \text{mm}^2 \qquad\quad A_{\text{c}} = 94\,950 \, \text{mm}^2$$

The ratio $A_{\text{s}}/A_{\text{c}}$ is 0.0085, which is within the range permitted by Expression 5.26.

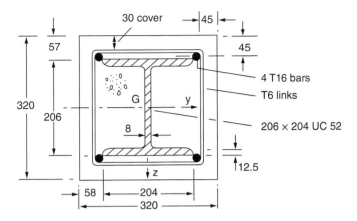

Figure 5.15 Assumed cross-section for external column length 0–1

From Equation 5.33, the design plastic resistance to axial load is

$$N_{pl,Rd} = 6640 \times 0.355 + 804 \times 0.435 + 94.95 \times 14.2$$
$$= 2357 + 350 + 1345 = \textbf{4052 kN} \tag{5.50}$$

With the partial factors taken as 1.0, from Equation 5.35,

$$N_{pl,Rk} = 2357 + 350 \times 1.15 + 1345 \times 1.5 = \textbf{4776 kN} \tag{5.51}$$

The steel contribution ratio (Equation 5.27) is $\delta = 2357/4052 = 0.582$, which is within the permitted range.

Second moments of area of the uncracked section are needed for the calculation of the elastic critical load, N_{cr}.

For the steel section, from tables, $10^{-6}I_a = 52.6\,\text{mm}^4$

For the reinforcement, $10^{-6}I_s = 804 \times 0.115^2 = 10.6\,\text{mm}^4$

For the concrete, $10^{-6}I_c = 320^2 \times 0.32^2/12 - 52.6 - 10.6 = 811\,\text{mm}^4$

The long-term creep coefficient for the column, $\varphi(t_0,\infty)$, is needed for Equation 5.30. It depends on the relative humidity, taken as 50% for a centrally-heated building, on the cross-section of the concrete, and on the 'age at first loading', t_0. There is, of course, no single age for the bottom length of a column, but the result is not sensitive to ages exceeding 28 days. Assuming a (conservative) mean age at first loading of 40 days, EN 1992-1-1 gives $\varphi(t_0,\infty) = 3.0$.

From Equations 5.47 and 5.49,

$$N_{G,Ed}/N_{Ed} = (9 \times 96 + 9 \times 97)/2807 = 0.62$$

From Equation 5.30,

$$E_{c,eff} = 31/(1 + 0.62 \times 3) = 10.8\ \text{kN}/\text{mm}^2$$

From Equation 5.28,

$$10^{-12}(EI)_{eff} = 0.21 \times 52.6 + 0.20 \times 10.6 + 0.6 \times 0.0108 \times 811$$
$$= 11.05 + 2.12 + 5.26 = 18.4\ \text{N mm}^2 \tag{5.52}$$

It is notable that the allowance made in design for creep and cracking of the concrete encasement, including second-order effects, reduces the contribution from the concrete to the effective flexural stiffness of this cross-section to 29%. The initial unfactored uncracked value, using $E_{cm}I_c$, is 66%.

From Equation 5.31,

$$10^{-12}(EI)_{\text{eff,II}} = 0.9(11.05 + 2.12 + 0.5 \times 0.0108 \times 811) = 15.8 \text{ N mm}^2 \qquad (5.53)$$

From Equation 5.29, the elastic critical buckling load is

$$N_{cr} = \pi^2 \times 18.4 \times 1000/4^2 = \mathbf{11\ 350\ kN} \qquad (5.54)$$

From Equations 5.32, 5.51 and 5.54

$$\overline{\lambda} = \sqrt{4776/11350} = 0.65$$

This is less than 2.0, so the design method of Section 5.6 is applicable.

Interaction polygon for major-axis bending Coordinates for the polygonal interaction diagram for major-axis bending are now calculated, using the notation shown in Figure 5.16 and dimensions from Figure 5.15. It is assumed that the plastic neutral axis for pure bending, line B–B, lies between the steel flanges, as shown, with the region above B–B in compression. The results are shown in Figure 5.17.

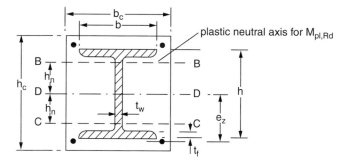

Figure 5.16 Plastic neutral axes for encased I-section

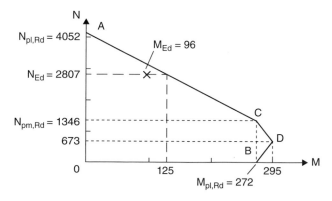

Figure 5.17 Interaction diagram for major-axis bending of external column

The plastic section moduli for the three materials, assuming that concrete is as strong in tension as in compression, are:

$$10^{-6} W_{pa} = 0.568 \text{ mm}^3 \text{ (from tables)}$$
$$10^{-6} W_{ps} = A_s e_z = 0.804 \times 0.115 = 0.0925 \text{ mm}^3$$
$$10^{-6} W_{pc} = b_c h_c^2/4 - 0.568 - 0.0925 = 7.53 \text{ mm}^3$$

For rectangular stress blocks with stresses $f_{yd} = \pm355 \text{ N/mm}^2$ in steel, $f_{sd} = \pm435$ N/mm^2 in reinforcement, and $0.85 f_{cd} = 14.2 \text{ N/mm}^2$ in concrete, in compression only, it is found from longitudinal equilibrium with $N_{Ed} = 0$ that $h_n = 67$ mm.

Plastic section moduli for the region of depth $2h_n$ between lines B–B and C–C in Figure 5.16 are now found:

$$10^{-6} W_{pa,n} = t_w h_n^2 = 8 \times 0.067^2 = 0.036 \text{ mm}^3$$
$$10^{-6} W_{pc,n} = (b_c - t_w)h_n^2 = (320 - 8) \times 0.067^2 = 1.40 \text{ mm}^3$$

At point D on the interaction polygon of Figure 5.12, the neutral axis is line D–D in Figure 5.16. The longitudinal forces in the steel section and the reinforcement sum to zero, from symmetry. Exactly half of the concrete area is assumed to be cracked, so the axial compression is equal to the compression in the uncracked concrete, $N_{pm,Rd}/2$, where

$$N_{pm,Rd} = 0.85 A_c f_{cd} \tag{5.55}$$

Hence,

$$\mathbf{0.5 N_{pm,Rd}} = 0.5 \times 94.95 \times 14.2 = \mathbf{673 \text{ kN}}$$

The bending resistance at point D, with W_{pc} halved to allow for cracking, is

$$M_{max,Rd} = W_{pa} f_{yd} + W_{ps} f_{sd} + 0.85 W_{pc} f_{cd}/2$$
$$= 0.568 \times 355 + 0.0925 \times 435 + 7.53 \times 14.2/2 = \mathbf{295 \text{ kN m}} \tag{5.56}$$

When the plastic neutral axis moves from D–D to C–C, the axial compression changes from $N_{pm,Rd}/2$ to $N_{pm,Rd}$, because the changes in axial force are of the same size (but of opposite sign) as when it moves from D–D to B–B.

When the plastic neutral axis moves from B–B to C–C, the resultant of all the changes in axial force passes through G (from symmetry), so that the bending resistances at points B and C are the same, and are

$$M_{pl,Rd} = M_{max,Rd} - W_{pa,n} f_{yd} - W_{pc,n} f_{cd}/2$$
$$= 295 - 0.036 \times 355 - 1.4 \times 14.2/2 = \mathbf{272 \text{ kN m}} \tag{5.57}$$

The axial force at point C is

$$N_{pm,Rd} = \mathbf{1346 \text{ kN}} \tag{5.58}$$

which gives the position of line AC in Figure 5.17.

5.7.3 Resistance of the column length, for major-axis bending

The design axial compression is $N_{Ed} = 2807$ kN from Equation 5.49. Using $(EI)_{eff,II}$ from Equation 5.53 gives:

$$N_{cr,eff} = \pi^2 \times 15.8 \times 1000/4^2 = 9750 \text{ kN}$$

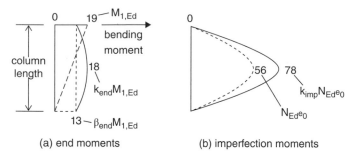

Figure 5.18 Major-axis bending-moment diagrams for external column

The condition $N_{cr,eff} \geq 10N_{Ed}$ (Equation 5.36) is not satisfied. Second-order effects will be allowed for by the method of Section 5.6.5.1.

From Figure 5.14b, $M_{1,Ed} = 19$ kN m and $M_{2,Ed} = 0$.

From Equation 5.38, $\beta_{end} = 0.66$.

From Equation 5.37, $k_{end} = 0.66/(1 - 2807/9750) = 0.66 \times 1.404 = 0.93$.

For the initial bow $e_0 = 20$ mm (Section 5.4.1), $\beta_{imp} = 1.0$, and its second-order bending moment is

$$N_{Ed}e_0 = 2807 \times 0.02 = 56 \text{ kN m}$$

From Equation 5.37, $k_{imp} = 1.404$.

From Equation 5.39 and as shown in Figure 5.18,

$$M_{y,Ed} = 0.93 \times 19 + 1.4 \times 56 = 18 + 78 = 96 \text{ kN m} \tag{5.59}$$

With $N_{Ed} = 2807$ kN m, the interaction diagram in Figure 5.17 gives

$$M_{y,Rd} = 125 \text{ kN m} \tag{5.60}$$

From Equation 5.40,

$$\alpha_M M_{y,Rd} = 0.9 \times 125 = 113 \text{ kN m} \tag{5.61}$$

This exceeds $M_{y,Ed}$, so this column length has sufficient major-axis resistance. It can be shown that although the column length above has a higher end moment, 30 kN m, it also is strong enough, because it is in double-curvature bending.

5.7.4   Resistance of the column length, for minor-axis bending

The margin of resistance to major-axis bending (above) is quite low, so two more T16 reinforcing bars were added to the cross-section, as shown in Figure 5.19a, to increase minor-axis resistance.

The cross-sectional areas given in Section 5.7.2 now become:

$$A_a = 6640 \text{ mm}^2 \qquad A_s = 1206 \text{ mm}^2 \qquad A_c = 94\,550 \text{ mm}^2$$

and $N_{pl,Rd}$ is increased from 4052 kN to 4224 kN.

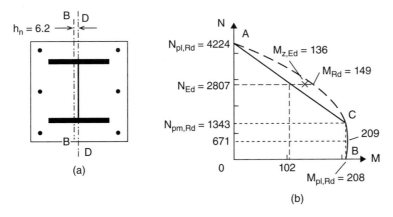

Figure 5.19 Minor-axis bending of external column. (a) Position of neutral axis. (b) Interaction diagram

The minor-axis second moments of area are:

For the steel section, from tables, $10^{-6}I_a = 17.7\ \text{mm}^4$

For the reinforcement, $\qquad 10^{-6}I_s = 1206 \times 0.115^2 = 15.9\ \text{mm}^4$

For the concrete, $\qquad 10^{-6}I_c = 320^2 \times 0.32^2/12 - 17.7 - 15.9 = 840\ \text{mm}^4$

With $E_{c,eff} = 10.8\ \text{kN/mm}^2$ (Section 5.7.2), and from Equation 5.31,

$$10^{-12}(EI)_{eff,II} = 0.9(0.21 \times 17.7 + 0.20 \times 15.9 + 0.5 \times 0.0108 \times 840)$$
$$= 10.3\ \text{N mm}^2$$

Using this value in Equation 5.29,

$$N_{cr,eff} = \pi^2 \times 10.3 \times 1000/42 = 6353\ \text{kN}$$

This is less than $10N_{Ed}$, so the second-order effect of the initial bow must be included. With $\beta_{imp} = 1.0$ as before, and from Equation 5.37,

$$k_{imp} = 1/(1 - 2807/6353) = 1.79$$

From EN 1994-1-1 and Section 5.4.1, the initial bow for the minor axis is $L/150 = 27$ mm, so from Equation 5.39,

$$M_{z,Ed} = 1.79 \times 2807 \times 0.027 = \textbf{136 kN m} \tag{5.62}$$

5.7.4.1 Interaction diagram for minor-axis bending

The method of calculation is similar to that used in Section 5.7.2. The plastic neutral axis for pure bending usually intersects the steel flanges, but not the web, and was found for this cross-section to be as shown in Figure 5.19a, with $h_n = 6.2$ mm.

The required plastic section moduli are:

$10^{-6}W_{pa} = 0.263\ \text{mm}^3$ (from tables)
$10^{-6}W_{ps} = 1.206 \times 0.115 = 0.139\ \text{mm}^3$

$$10^{-6} W_{\text{pc}} = 3.2^3/4 - 0.263 - 0.139 = 7.79 \text{ mm}^3$$
$$10^{-6} W_{\text{pa,n}} = 0.004 \text{ mm}^3$$
$$10^{-6} W_{\text{pc,n}} = 0.010 \text{ mm}^3$$

As in Equation 5.56,

$$M_{\text{max, Rd}} = 0.263 \times 355 + 0.139 \times 435 + 7.79 \times 14.2/2 = \textbf{209 kN m}$$

As in Equation 5.57

$$M_{\text{pl, Rd}} = 209 - 0.004 \times 355 - 0.010 \times 14.2/2 = \textbf{208 kN m}$$

From Equation 5.55,

$$N_{\text{pm, Rd}} = 94.55 \times 14.2 = \textbf{1343 kN}$$

The interaction polygon, shown in Figure 5.19b, gives

$$M_{\text{z, Rd}} = \textbf{102 kN m} \tag{5.63}$$

From Equation 5.40,

$$\alpha_{\text{M}} M_{\text{z, Rd}} = 0.9 \times 102 = \textbf{92 kN m} \tag{5.64}$$

This is less than $M_{\text{z,Ed}}$, 136 kN m, so this column length is found to be too weak in minor-axis bending, when the polygonal approximation for curve AC in Figure 5.19b is used.

Using the computed curve (the dashed curve), $M_{\text{z,Rd}}$ is found to be 149 kN m, reduced by α_{M} to 134 kN m, which is still not quite enough, and resistance to biaxial bending remains to be checked.

5.7.4.2 Biaxial bending

For the biaxial check to Expression 5.41, the initial bow is assumed to be in the more adverse plane, so either $M_{\text{y,Ed}}$ or $M_{\text{z,Ed}}$ is reduced. Here, $N_{\text{Ed}}e_0$ is greater for minor-axis bending, so $M_{\text{y,Ed}}$ is reduced from 96 kN m to 18 kN m (Figure 5.18). From Equations 5.60 and 5.62, and using $M_{\text{z,Rd}} = 149$ kN m,

$$18/125 + 136/149 = 0.14 + 0.91 = 1.05$$

but this value should not exceed 1.0. To satisfy both checks, $M_{\text{z,Rd}}$ would have to be increased to $136/(0.91 - 0.05) = 158$ kN m. Provision of more minor-axis reinforcement would do this.

5.7.5 Checks on shear, and closing comment

These checks are described in Section 5.6.6. The major-axis design transverse shear is greatest in column length 1–2, and is

$$V_{\text{Ed}} = (M_1 - M_2)/L = (24 + 30)/4 = 13.5 \text{ kN}$$

This is obviously negligible; $V_{\text{pl,Rd}}$ for the web of the steel section is over 300 kN.

From Figure 5.14b, the total vertical load applied to the column at level 1 is 246 + 97 kN, but for longitudinal shear the self-weight of the column can be deducted, giving $N_{\text{Ed}} = 327$ kN. Any load transferred to the concrete encasement by direct bearing of the three steel beams connected to the column at this floor level is conservatively neglected. Poor compaction of the concrete under their flanges is possible. Creep reduces the load transfer, so the short-term modular ratio, $n_0 = 10.1$, is used for the transformed area of the column cross-section A. With areas from Section 5.7.4,

$$10^{-3}A = 6.64 + 1.206 + 94.55/10.1 = 17.2 \text{ mm}^2$$

From Equation 5.42, the load to be transferred from the steel section to the concrete casing is

$$N_{Ed,c} = 327(1 - 6.64/17.2) = 201 \text{ kN}$$

The perimeter of the steel section is $u_a = 1140$ mm. Assuming a transmission length of 640 mm, twice the least lateral dimension, Equation 5.43 gives

$$\tau_{Ed} = N_{Ed,c}/u_a \ell_v = 201/(1.14 \times 640) = 0.27 \text{ N/mm}^2$$

This is less than τ_{Rd} (Equation 5.44), so local bond stress is not excessive, and shear connection is not required.

This completes the validation for this column length, provided that analysis for lateral loading (Section 5.9) confirms the assumption that it is all transferred by the floors to the end walls and the central core.

The initial cross-section for this column was chosen to be barely strong enough, so that some checks would fail. The second-order moments are significant, so that although the axial compression higher up the building is much lower, there would be little scope for reducing the cross-section much below 320 mm^2.

In practice, the cross-section chosen would be such that it would be known from experience that some of the rules would not govern, so that fewer checks would be needed. Calculations for interaction diagrams are straightforward on a spreadsheet, although tedious by hand calculation.

5.8   Example (continued): internal column

A typical internal column between level 0 and level 1 is now designed, for the arrangement of variable loading shown in Figure 5.10. Full permanent load acts on all of the beams. Variable load acts on all beams at levels 2 to 9, but not on beams AD and CG, as this increases the single-curvature bending moment in length CD of the column. Any rotational restraint from point E, at basement level, is neglected.

In Section 5.7.1, the live-load reduction factor was found to be $\alpha_n = 0.767$. This reduces the design ultimate load on each beam from 60.9 kN/m (Table 4.4) to 52.2 kN/m. This reduction is not made for the beams that cause bending in the column, so the beam loadings for the global analysis are as shown in Figure 5.20.

5.8.1   Global analysis

It is assumed that the major-axis joints between the beams and the internal column are rigid and full-strength, and that the frame is braced against sidesway. The uncracked

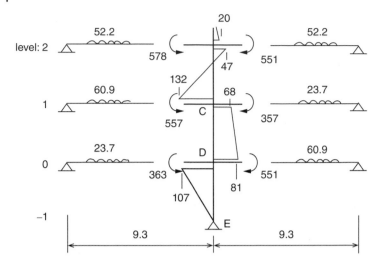

Figure 5.20 Loadings and major-axis bending moments for an internal column

unreinforced second moment of area of the beam, with $n = 20.2$, is given in Table 4.5 as $I = 636 \times 10^6$ mm^4, so

$$(EI)_{\text{beam}} = 636 \times 210 \times 10^6 = 1.34 \times 10^{11} \text{ kN mm}^2 \tag{5.65}$$

The column was at first assumed to have a 254×254 UC73 steel section, encased to 400 mm square, and with longitudinal reinforcement of six T20 bars. To obtain its stiffness, an effective modulus for the concrete is required. A low value, relative to that for the beams, reduces the bending moments in the column, so the creep coefficient $\varphi = 3.0$, used in Section 5.7.2 for resistance, may be too high. The design is relatively insensitive to this assumption, so the value $n = 2n_0$ is now used for the whole loading, as for the beams (Section 3.2). For the Grade C25/30 normal-density concrete used, $E_{\text{cm}} = 31$ kN/mm^2, so $n = 2 \times 210/31 = 13.6$. This leads to

$$(EI)_{\text{column}} = 0.60 \times 10^{11} \text{ kN mm}^2 \tag{5.66}$$

The beam members in Figure 5.20 are over twice as long as the column members, so the stiffnesses $(EI)/L$ of beams and columns at nodes C, D, and so on, are similar. For global analysis by moment distribution it is accurate enough to consider only the limited frame in Figure 5.20, which also shows the bending moments found. These are plotted on the tension side of each member.

For the beams, from Section 4.2.1.2, the bending resistance at the internal column is $M_{\text{pl,Rd}} = 510$ kN m. Cracking and inelastic behaviour are assumed (Section 4.6.2) to reduce the beam moments that exceed 510 kNm to that value, by redistribution of moments to mid-span, without altering the column moments. This enables the shear force in each beam at nodes C, D, and so on, to be found, and hence the total axial load in the column just above node D, including its weight. The result is

$$N_{\text{Ed}} = \textbf{5401 kN} \tag{5.67}$$

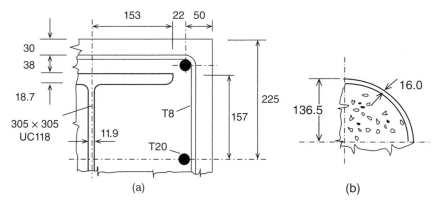

Figure 5.21 Part cross-sections for length 0–1 of internal columns. (a) Encased I-section. (b) Concrete-filled steel tube with high-strength materials

5.8.2 Resistance of an internal column

Approximate calculations then showed that the initial column cross-section was too weak. The larger doubly-symmetric cross-section partly shown in Figure 5.21a was assumed. Its resistance was checked by the methods used in Section 5.7 for external columns, taking account of the single-curvature bending moments $M_{y,Ed,1} = 81\,\text{kN m}$ and $M_{y,Ed,2} = 68\,\text{kN m}$ (Figure 5.20), the $N_{Ed}e_0$ moments, and the second-order moments about both axes; and checking uniaxial and biaxial bending. The cross-section shown in Figure 5.21b is for an alternative design, given in Section 5.11.

Major-axis bending was found to be the most critical, with $M_{y,Ed} = 216\,\text{kN m}$ and $0.9M_{y,Rd} = 443\,\text{kN m}$, when $N_{Ed} = 5401\,\text{kN}$. The margins on bending resistance so found were assumed to be sufficient to cover the increase in the end moments caused by the change in column cross-section, and the small increase in N_{Ed} from the heavier column.

5.8.3 Comment on column design

It is evident, above, that response to the uncertainties of loading, cracking, creep, and inelastic behaviour involve some judgement and approximation. A small increase in the cross-section of a column reduces its slenderness (and hence the secondary bending moments) as well as increasing all its resistances (N_{Rd}, $M_{y,Rd}$, etc.); so there is little saving in cost from seeking an 'only-just-adequate' design.

5.9 Example (continued): design of frame for horizontal forces

As explained in Section 5.1, horizontal loads in the plane of a typical frame, such as DEF in Figure 5.1a, are transferred by the floor slabs to a central core and to two shear walls at the ends of the building (Figure 5.22). It is now shown by approximate calculation that the system is so stiff, and the relevant stresses are so low, that rigorous verification is unnecessary.

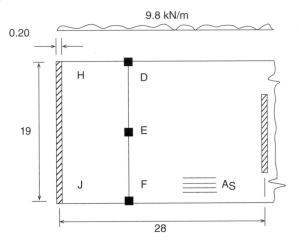

Figure 5.22 Part plan of typical floor slab

It is shown in Section 5.4.1 that allowance for the frame imperfections is made by applying at each floor level of each frame a notional horizontal force $H_{Ed} = (G + Q)/366$, where G and Q are the total design ultimate dead and imposed loads for the relevant storey.

From EN 1994-1-1, second-order effects may be neglected in the global analysis if the deformations from first-order global analysis increase the relevant internal action effects by less than 10%. It will be found that this exemption applies.

It is assumed that the concrete above the profiled sheeting in each floor slab acts as a reinforced concrete beam of breadth 80 mm (its thickness) and depth 19 m, spanning 28 m. For simplicity, this span is assumed to be simply-supported. The lateral stiffnesses of these deep 'beams' and of the shear walls are so much higher than that of each frame, such as DEF in Figure 5.22, that the presence of the frames can be ignored.

5.9.1 Design loadings, ultimate limit state

The force H_{Ed} is greatest when live load is applied to all floors, so the reduction factor $\alpha_n = 0.767$ (Section 5.7.1) is applicable. From Table 4.4, the imposed beam loading is

$$0.767 \times 37.2 = 28.5 \text{ kN/m}$$

Dead loads are as in Section 5.5.1, except that the design weight of a 4-m length of internal column has increased from 16.2 kN to 30.0 kN.

The total permanent load per storey from a 4-m length of the building is

$$G = 23.7 \times 19 + 2 \times 97.2 + 30 = 675 \text{ kN}$$

The imposed load is

$$Q = 28.5 \times 19 = 542 \text{ kN}$$

From Section 5.5.1, with wind pressure 1.5 kN/m² and $\gamma_F = 1.5$, the design wind load on a wall area 4 m² is

$$W = 1.5 \times 1.5 \times 4 \times 4 = 36 \text{ kN}$$

These values are such that wind should be taken as the leading variable action. For imposed load, the combination factor $\psi_0 = 0.7$, and $\phi = 1/366$, so

$$H_{Ed} = W + (G + \psi_0 Q)\phi = 36 + (675 + 0.7 \times 542)/366 = 39 \text{ kN}$$

The lateral load applied to one edge of each floor is

$$h_{Ed} = 39/4 = 9.8 \text{ kN/m}$$

and for the 28-m span of the floor,

$$M_{Ed} = h_{Ed}L^2/8 = 9.8 \times 28^2/8 = 960 \text{ kN m}.$$

5.9.2 Stresses and stiffness

Assuming a lever arm of about $0.8 \times 19 = 15.2$ m, the area of reinforcement needed near each edge of each floor slab (Figure 5.22) is tiny:

$$A_s = (960/15.2)/0.435 = 145 \text{ mm}^2$$

The shear force applied to each shear wall is $9.8 \times 28/2 = 137$ kN per storey.

The reinforced concrete wall HJ in Figure 5.22 is a cantilever 36 m high. For storeys 4 m high, the horizontal load in its plane is $137/4 = 34.3$ kN/m. The bending moment at its base is

$$M_{Ed} = 34.3 \times 36^2/2 = 22\,230 \text{ kN m}$$

Elastic analysis for a beam 19 m deep and 200 mm wide gives the maximum bending stress as less than 2 N/mm^2. The in-plane deflection at the top of the wall, including shear deformation, and with $n = 2n_0 = 13.6$ (as for the columns), is less than 6 mm. This is an additional sidesway of $6/36\,000 = 1/6000$, which is less than 10% of the frame imperfection of $1/366$. The resulting increase in action effects is less than 10%, so the use of first-order global analysis is confirmed.

5.10 Example (continued): joints between beams and columns

5.10.1 Nominally-pinned joint at external column

The design vertical shear for this joint is 246 kN, from Equation 5.25. A worked example of this design method is available (Steel Construction Institute and British Constructional Steelwork Association, 2014). From standard details for partial-depth end-plate joints (Malik, 2009) the initial design of the joint is as shown in Figure 5.23, with six M20 8:8 bolts. The 220×150 mm end plate is of S275 steel and only 8 mm thick, so that its plastic deformation can provide the necessary end rotation for the beam while transmitting very little bending moment to the column.

Preliminary calculation showed that a 4-bolt joint is just adequate for the vertical shear. However, it may be necessary to resist a tensile force of about 75 kN, depending on how the robustness of the structure is to be assured. Also, the greater depth of a 6-bolt joint provides better torsional restraint to the beam during erection.

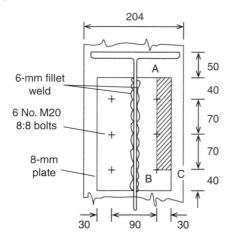

Figure 5.23 End-plate joint to major axis of external column

Detailed calculations to EN 1993-1-1 and EN 1993-1-8 are not given, as this is not a composite joint; but the results may be of interest. The calculated design resistances are as follows:

- shear of six M20 bolts, $V_{Rd} = 565$ kN;
- bearing of bolts on end plate, $\Sigma F_{b,Rd} = 471$ kN;
- shear resistance of 6-mm fillet welds to beam web, $V_{w,Rd} = 377$ kN;
- shear resistance of 220-mm depth of beam web, $V_{Rd} = 767$ kN;
- block tearing failure of end plate, both sides on surfaces ABC (Figure 5.23), $V_{Rd} = 412$ kN.

Thus, the resistance of the joint to vertical shear, neglecting bending moment and axial tension, is 377 kN, which provides sufficient margin for these other effects.

5.10.2 End-plate joint at internal column

The internal support for the two-span beam designed in Section 4.6 was assumed to be a wall, so full continuity was assumed at that cross-section. Here, the beam is interrupted by a supporting column, so its spans are referred to as 'beams'. The options for its joint with the beams are explained in Section 5.3. The local bending moments will depend on both the stiffness and strength of the joint and the stiffness of the column lengths above and below. Rigid and full-strength connections to the column would minimize the deflections of the beams, but are expensive.

A suitable design would be a full-depth end-plate joint, with the local reinforcement in the slab continuous past the column. In the earlier design for the continuous beam, this reinforcement was six 12-mm bars at 200 mm spacing, within the effective width of 1.225 m. It was found in Section 4.6.6 that to control crack widths to 0.4 mm, they should be replaced by bars of smaller diameter. However, at ultimate load, there is high local tensile strain in this reinforcement. Tests have shown that small-diameter bars can fracture. It is now recommended (Couchman and Way, 1998) that the bar size at a joint should be at least 16 mm. The maximum stress in 16-mm bars (from serviceability loading) for cracks to be controlled to 0.4 m is about 280 MPa.

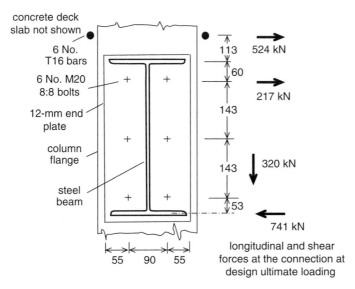

concrete deck
slab not shown

6 No.
T16 bars

6 No. M20
8:8 bolts

12-mm end
plate

column
flange

steel
beam

113   524 kN

60

143

143   320 kN

53

741 kN

55   90   55

217 kN

longitudinal and shear
forces at the connection at
design ultimate loading

Figure 5.24 Details of end-plate connection of composite beam to internal column

The trial use of six 16-mm bars at 200 mm spacing is now studied. It is assumed that for serviceability checks, the hogging bending moment at joint B (Figure 5.9) is 275 kN m, from Section 4.6.6, allowing for the use of unpropped construction. The dimension 35 mm in Figure 4.14b is increased to 37 mm to allow for the increase in bar diameter from 12 mm to 16 mm. With $A_s = 1206$ mm^2 and other values as before, the second moment of area of the cracked hogging cross-section of the beam is $I = 319 \times 10^6$ mm^4. The bars are 273 mm above the elastic neutral axis, and the tensile stress in them is

$$\sigma_s = My/I = 275 \times 273/319 = 236 \text{ MPa}$$

This is below 280 MPa, so crack-width control is provided.

5.10.2.1   Resistances and strength class of a full-depth end-plate connection

A full-depth end-plate connection is proposed, with six M20 8:8 bolts and the dimensions shown in Figure 5.24. The plate is 12 mm thick, of S275 steel.

The design tensile strength of the reinforcement, at yield, with notation as in Figure 5.2, is

$$F_{t,s,Rd} = A_s f_{yd} = 1206 \times 0.435 = 524 \text{ kN}$$

From Figure 5.24, its distance above the centre of the bottom flange is

$$z_1 = 406 + 150 - 37 - 7 = 512 \text{ mm}$$

The end plate and column flange are each modelled as T-stubs. The end plate is assumed to develop yield-line mechanisms around the top pair of bolts. The thicker column flange is assumed to remain elastic. The adjacent concrete is ignored. The top pair of bolts each have a design tensile strength of 141 kN, but $F_{T,Rd}$ may be lower, because of prying action.

Figure 5.25 represents a horizontal cross-section through the bolts in tension, shown by dashed lines, the beam web, the column flange and the thin end plate, shown as a

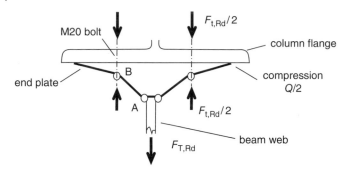

Figure 5.25 Prying forces in a T-stub at an end-plate connection

single line. Its three-dimensional yield-line mechanism is shown as plastic hinges at A and B. The tensile force in the bolts at failure, $F_{t,Rd}$, acts on the end plate and column flange as shown. The edges of the end plate apply compressive force Q to the column (the prying force), so the tension transferred from the steel top flange to the column, $F_{T,Rd}$, is less than the resistance of the bolts.

The ratio $Q/F_{t,Rd}$ depends on the assumed yield-line pattern. These are of two main types. A third possibility is that the end plate is too thick to fail; then $Q = 0$. The force $F_{T,Rd}$ is the lowest found from these alternatives. Here, with $F_{t,Rd} = 282$ kN, the force Q is found to be 65 kN. The force needed for calculating the resistance moment is therefore

$$F_{T,Rd} = 282 - 65 = 217 \text{ kN}$$

From Figure 5.24, its distance above the centre of the bottom flange is

$$z_2 = h_b - t_f/2 = 406 - 60 - 7 = 339 \text{ mm}$$

Taking moments about the bottom flange, $M_{j,Rd} = 524 \times 0.512 + 217 \times 0.339 = \textbf{342 kN m.}$

For the beam with six T12 bars, $M_{pl,Rd} = 510$ kN m (Section 4.2.1.2) without including concrete encasement, if any. The beam with six T16 bars will be stronger, so this connection is clearly *partial strength*.

The design vertical shear for the composite beam was 354 kN (Section 4.6.2). With this partial-strength connection it will be lower. Assuming $M_{Ed} = M_{j,Rd}$, the new design ultimate vertical shear is

$$V_{j,Ed} = 60.9 \times 9.3/2 + 342/9.3 = \textbf{320 kN}$$

From EN 1993-1-8, the design shear resistance of these M20 bolts is 94.1 kN/bolt, so four can resist 376 kN, which is sufficient.

The new design ultimate bending moment at mid-span is

$$M_{Ed} = wL^2/8 - M_{j,Rd}/2 = 60.9 \times 9.3^2/8 - 171 = 487 \text{ kN m}$$

The sagging resistance moment is much higher, at 709 kN m from Equation 3.115.

5.10.2.2 Flexural stiffness

The stiffness coefficients given for a steelwork joint, from EN 1993-1-8, are defined in Section 5.3.1 and shown in Figure 5.3. Their values for this example are as follows. Here, the moments $M_{Ed,1}$ and $M_{Ed,2}$ are equal, so there is no shear deformation of the column web.

For extension of the web of the column at level A (Figure 5.2), $k_3 = 4.9$ mm.
Bending of the column flange at level A is negligible, so $1/k_4$ is taken as zero.
For bending of the end plate (the mechanism), $k_5 = 8.5$ mm.
For the extension of row A of bolts, $k_{10} = 8.9$ mm.
For compression of the column web, if uncased, $k_2 = 2.2$ mm, but the concrete encasement increases it to $k_{2,c} = 19.6$ mm.

Formulae for these quite complex calculations are available (BSI, 2005a; Couchman and Way, 1998). Calculations are given for these coefficients for a steel structure with the same end plate and bolts, but different beam and column sections, in Johnson (2012).

Calculations are now given for the two additional coefficients defined in Eurocode 4 for a composite structure. For a double-sided joint with equal end moments, the coefficient for the extension of the slab reinforcement is

$$k_{s,r} = 2A_s/h = 2 \times 1206/314 = 7.7 \text{ mm}$$

where h is the depth of the steel section for the column. Slip of the shear connection is allowed for by reducing $k_{s,r}$ using the coefficient k_{slip}. The formulae for it in Annex A of Eurocode 4 are too complex to repeat here. They involve several geometric parameters and the number, strength and stiffness of the shear connectors. For 19-mm studs, the 'approximate' stiffness is given as 100 kN/mm, but only where the reduction factor k_t (for resistance of a stud in a trough of sheeting) is unity. Where $k_t < 1$, as here, the code refers to the use of push testing, which is unhelpful in practice. Here, a guessed value, 80 kN/mm, was used, leading to the result that the stiffness of the shear connection K_{sc} is 463 kN/mm, and the stiffness coefficient is

$$K_{sc}/E = 463/210 = 2.2 \text{ mm}$$

This is in series with $k_{s,r}$, so from Equation 5.3 in Section 5.3.1,

$$1/k_{s,red} = 1/7.7 + 1/2.2 = 1/1.71$$

This stiffness is at distance z_1 ($=512$ mm) from the bottom flange.

For tension in the steel beam, k_t is found from the flexibilities $1/k$ numbered 3, 4, 5 and 10 in Figure 5.3, which are in series:

$$1/k_t = 1/4.9 + (\text{neg.}) + 1/8.5 + 1/8.9 = 1/2.31 \text{ mm}^{-1}$$

The effective lever arm for the two tensile stiffnesses is given by Equation 5.6 as

$$z = (1.71 \times 0.512^2 + 2.31 \times 0.339^2)/(1.71 \times 0.512 + 2.31 \times 0.339) = 0.43 \text{ m}$$

The equivalent stiffness in tension is given by Equation 5.7 as

$$k_{t,eq} = (1.71 \times 0.512 + 2.31 \times 0.339)/0.43 = 3.85 \text{ mm}$$

Equation 5.1 is now used to combine this with the stiffness in compression, $k_2 = 19.6$ mm:

$$S_{j,ini} = 210 \times 0.43^2/(1/3.85 + 1/19.6) = \textbf{125 kN m/mrad}$$

It is notable that the only one of the seven flexibilities used here that is clearly negligible is the bending of the column flange, and that the largest single contribution to stiffness comes from compression in the encasement of the column web. If k_2 is reduced to the uncased value, 2.2 mm, $S_{j,ini}$ is reduced from 125 to 54 kN m/mrad.

The condition for a connection to be classed as rigid is

$$S_{j,ini} \geq 8EI_b/L$$

where EI_b is the value used for the global analysis of the connected beam. Using the value at mid-span, $636 \times 10^6 \text{ mm}^4$,

$$8EI_b/L = 8 \times 210 \times 636/9.3 = 115 \text{ kN m/mrad}$$

so the connection is *rigid* (just). The use of stiffness lower than $S_{j,ini}$ at high ratios $M_{Ed}/M_{j,Rd}$ does not alter its stiffness class.

5.10.2.3 Deflection of a beam with an end-plate connection at one end

Deflections of this beam, treated as continuous, are calculated in Section 4.6.5. The addition of semi-rigid connections at the internal support adds rotations ϕ, which reduce the local bending moment and increase the deflection of the spans. Global re-analysis including these rotations will, in general, find changes in bending moments is all nearby members. For simplicity, it is assumed for this example that the characteristic imposed loads act on both spans. From symmetry, one span can then be analysed in isolation, treating it as the propped cantilever of uniform section shown in Figure 5.26.

Elastic analysis of this beam, with $\phi = M_j/S_j$, leads to

$$M_j = (wL^2/8)/(1+J) \tag{5.68}$$

with

$$J = 3(EI)/S_jL \tag{5.69}$$

and (EI) taken as the uncracked unreinforced stiffness of the beam. If $M_j < 2M_{j,Rd}/3$, S_j can be taken as $S_{j,ini}$. Otherwise, and also where there is doubt, EN 1993-1-8 gives the simplification $S_j = S_{j,ini}/2$.

Further analysis gives the deflection at mid-span as

$$\delta = wL^4/(192EI) + 3\phi L/16 \tag{5.70}$$

These equations give the usual results for a propped cantilever when $\phi = 0$.

Eurocode 4 gives correction factors for both cracking of concrete (f_1) and yield of the steel section (f_2), which are used in Section 4.6.5. They reduce the 'uncracked' bending moment at the support, in that case by 58%. Eurocode 4 gives no guidance on how to proceed where this bending moment is also reduced (to M_j) by flexibility of a connection. It is assumed here that, because this flexibility is present before cracking or yielding can occur, these factors are determined using the moments M_j.

Separate calculations are needed for the steel beam, with loading $w_a = 12.4 \text{ kN/m}$ and for the composite beam, with loading $w_c = 30 \text{ kN/m}$ (from Section 3.11.3.1). The values used and the results are given in Table 5.1.

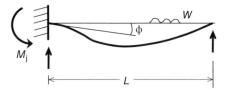

Figure 5.26 Model for elastic analysis of a span with an end-plate connection at one end

The stiffness of the steel connection, without reinforcement, is found from coefficients k_t and k_2, found above, with lever arm z_2:

$$S_{j,ini,a} = 210 \times 0.339^2/(1/2.31 + 1/19.6) = \textbf{49.8 kN m/mrad}$$

The second moment of area of the composite beam is taken as the mid-span value because the analysis assumes a uniform cross-section. Assuming $S_j = S_{j,ini}$ for the steel beam, Equation 5.69 gives

$$J_a = (3 \times 210 \times 215)/(49.8 \times 9.3) = 0.292$$

For the composite beam it is not yet clear whether M_j exceeds $2M_{j,Rd}/3$, so $S_{j,ini}$ is halved:

$$J_c = (3 \times 210 \times 636)/(62.5 \times 9.3) = 0.689$$

From Equation 5.68,

$$M_{j,k,a} = (12.4 \times 9.3^2/8)/1.292 = 104 \text{ kN m}$$

$$M_{j,k,c} = (30 \times 9.3^2/8)/1.689 = 192 \text{ kN m}$$

The total $M_{j,k}$, 296 kN m, is 85% of $M_{j,Rd}$, so inelastic behaviour at this load level is likely, and the use of $S_{j,ini}/2$ for the composite connection is appropriate.

To find whether cracking of concrete should be allowed for, the maximum 'uncracked' tensile stress is found:

$$\sigma_c = M_{j,k,c} y/I = 121 \times 231/(508 \times 20.2) = 4.3 \text{ MPa}$$

with y and I from Table 4.5 for the uncracked cross-section with $b_{eff} = 1.225$ m. The stress exceeds $1.5f_{ltcm}$ (3.5 MPa), so as in Section 4.6.5, $f_1 = 0.6$.

A similar calculation for the stress in the steel beam finds that with $M_{j,k,c}$ reduced to $192 \times 0.6 = 115$ kN m, its bottom flange does not yield.

The rotations at the connection are given by $\phi = M/S_j$; hence

$$\phi_a = 104/49.8 = 2.1 \text{ mrad} \quad \phi_c = 115/62.5 = 1.8 \text{ mrad}$$

From Equation 5.70 the mid-span deflections are:

$$\delta_a = 12.4 \times 9.3^4/(192 \times 210 \times 0.215) + 3 \times 2.1 \times 9.3/16 = 10.7 + 3.7 \text{ mm}$$

$$\delta_c = 30 \times 9.3^4/(192 \times 210 \times 0.636) + 3 \times 1.8 \times 9.3/16 = 8.7 + 3.1 \text{ mm}$$

The total deflection is 26.2 mm (span/355), and would be higher with variable load on one span only. It is similar to the result for the fully-continuous beam, 29.4 mm in Section 4.6.5, which was for the more adverse non-uniform loading.

Table 5.1 Data and results for deflection of a beam with an end-plate connection

Connection	Loading kN/m	$S_{j,ini}$ kN m/mrad	I m²mm²	J	M_j kN m	ϕ mrad	δ mm
Steel	12.4	49.8	215	0.292	104	2.1	14
Composite	30.0	125	636	0.689	192	1.8	12

The methods used here include many assumptions and simplifications and some step functions, such as that for S_j. The reliability of the results is lower than the use of three significant figures implies, and the methods need more input from experience in practice.

5.11   Example: concrete-filled steel tube with high-strength materials

Reference is made in Section 1.4 to the increasing use in practice of columns with much stronger materials than those used so far in this worked example. The compressive strain at which concrete crushes reduces with increasing strength, but encasement in a steel tube counteracts this effect. For this reason, in high-rise buildings concrete-filled steel tube (CFST) columns are often preferred to concrete-encased steel members. The stronger materials enable smaller columns to be used.

This is illustrated by replacement of the internal column designed in Section 5.8 by a CFST column using C90/105 concrete and S550 steel. These are the strongest materials within the scope of a design guide published as an extension to Eurocode 4 (Liew and Xiang, 2015). The *Guide* shows that strong steel requires the use of strong concrete to avoid crushing of the concrete core before the steel yields. The 'simplified method' of clause 6.7.3 of Eurocode 4 is found to be applicable with a minor change. It is that a reduction factor, $\eta = 0.8$, is applied to the compressive strength of the concrete. In Eurocode 2, the elastic modulus E_{cm} is given in terms of f_{ck}, and factor η is also applied here. In MPa, the revised expression (in MPa units) is:

$$E_{cm} = 22\ 000[(0.8 f_{ck} + 8)/10]^{0.3} \tag{5.71}$$

5.11.1   Loading

The column length to be designed is CD in Figure 5.20, with the loadings on the beams at levels 0 and 1 chosen for maximum single-curvature bending moment in the column at joints C and D, and the beams at higher levels loaded for maximum compression at level 1. The design compression, N_{Ed}, is taken as 5401 kN at joint D (Equation 5.67) to enable the two designs to be compared. This revision does not include the joints. The new column is more slender, so the bending moments M_y at its ends will be lower than those shown in Figure 5.20. It is assumed, as before, that the loading does not apply bending moment about the z-axis.

For simplicity, a cross-section without internal reinforcement is assumed, so the cross-section is defined by its external diameter d and wall thickness t. As all the steel is exposed, resistance to fire must be checked. There is a worked example in Chapter 6.

5.11.2   Action effects for the column length

The sequence of calculation here follows that in the flow chart in Appendix A2 of Liew and Xiang (2015). The trial and error involved is quite complex, because as slenderness increases, the reduction in end moments is more than offset by the second-order effects, and the ratio of area of steel to area of concrete has a strong influence. The penalties applied to the stiffness of the concrete suggest the use of a quite thick steel tube. The previous design led to a section 450 mm square. The first guess here was a steel tube with external diameter $d = 406$ mm and $t = 12.5$ mm. It was found that greater slenderness is possible. The calculations here are for $d = 273$ mm and $t = 12.5$ mm, so $d/t = 21.8$ and $A_c = 0.0483$ m^2. Later, t is increased to 16.0 mm, and that cross-section is compared in Figure 5.21 with the one designed in Section 5.8.

For the steel, $\gamma_A = 1.0$, so $f_{yd} = f_y = 550$ MPa, for thicknesses up to 40 mm. For the concrete, $f_{ck} = 90$ MPa, reduced by factor η to 72 MPa. With $\gamma_C = 1.5$, $f_{cd} = 48$ MPa. From Equation 5.71,

$$E_{cm} = 22[(72 + 8)/10]^{0.3} = 41.0 \text{ GPa}$$

From tables for steel sections, the properties of the steel tube are:

$$A_a = 0.0102 \text{ m}^2, \qquad 10^6 I = 87.0 \text{ m}^4, \qquad 10^6 W = 637 \text{ m}^3$$

The method is not applicable to Class 4 cross-sections. The limit for Class 3 in Eurocode 3-1-1 (BSI, 2014b) is $d/t \leq 38$ for S550 steel, easily satisfied here. The shear force is the column is low and need not be checked.

From Equation 5.33 with factor 0.85 omitted, the plastic resistance to compression is

$$N_{pl,Rd} = A_a f_{yd} + A_c f_{cd} = 5626 + 2319 = 7945 \text{ kN}$$

Similarly,

$$N_{pl,Rk} = 5626 + 2319 \times 1.5 = 9104 \text{ kN}$$

The steel contribution ratio is $\delta = 5626/9104 = 0.62$, a value within the scope of Eurocode 4.

5.11.3 Effect of creep

Allowance for the effect of creep is explained in Section 5.6.3. In Section 5.7.2 the long-term creep coefficient for the external column was estimated to be 3.0, with 62% of the design loading being permanent. Equation 5.30 then gave $E_{c,eff} = 0.35 E_{cm}$, for an assumed age at first loading of 40 days. For this internal column, about 40% of the loading is permanent, and 40 days is conservative for a column length with eight storeys to be built above it, so 60 days is now assumed. It is then found that $E_{c,eff}$ exceeds $0.5 E_{cm}$. The approximation $E_{c,eff} = 0.5 E_{cm}$ is used here.

From Equation 5.28,

$$(EI)_{eff} = E_a I_a + 0.6 \times 0.5 E_{cm} I_c = 18\,260 + 2290 = 20\,550 \text{ kN m}^2$$

Similarly, from Equation 5.31,

$$(EI)_{eff,II} = 18\,150 \text{ kN m}^2$$

The steel tube provides a high proportion of these stiffnesses, so the results are not sensitive to the uncertain nature of creep.

5.11.4 Slenderness

The effective length of the column is the length between lateral restraints, 4.0 m. From Equation 5.29,

$$N_{cr} = \pi^2 (EI)_{eff}/L^2 = \pi^2 \times 20\,550/4^2 = 12\,670 \text{ kN}$$

Similarly, using $(EI)_{eff,II}$,

$$N_{cr,eff} = 11\,200 \text{ kN}$$

The slenderness can now be found. From Equation 5.32,

$$\bar{\lambda} = \sqrt{N_{pl,Rk}/N_{cr}} = \sqrt{(9104/12\,670)} = 0.847$$

This is below the upper limit for this method, 2.0.

5.11.5 Bending moment

From clause 6.7.3.4, internal forces are found using $(EI)_{eff,II}$, so the end stiffness EI/L for this column is 18 150 / 4 = 4537 kN m. From Table 4.5 for the beams, $10^6 I = 636 \text{ m}^4$ and $L = 9.3$ m, so EI / L is is much greater, at 14 360 kN m. The bending of the column caused by the unbalanced loading on the beams is lower than before, and is found to be:

$$M_{1,Ed} = M_{2,Ed} = 32 \text{ kN m}$$

Second-order effects are allowed for by the method given in Section 5.6.5. For the end moments, $\beta = 1.1$ from Equation 5.38, and Equation 5.37 gives

$$k_{end} = 1.1/(1 - 5401/11\,200) = 2.12$$

The equivalent member imperfection (Section 5.4.1) is $L/300$, so $e = 13.3$ mm here, and $\beta = 1.0$. From the result for k_{end},

$$k_{imp} = 2.12/1.1 = 1.93$$

The design bending moment for the column is

$$M_{y,Ed} = k_{end}M_{1,Ed} + k_{imp}N_{Ed}e = 2.12 \times 32 + 1.93 \times 5401 \times 0.0133 = 206 \text{ kN m}$$

5.11.6 Interaction polygon, and resistance

The calculation and use of this diagram for resistance to combined axial compression and bending have been described earlier (Figures 5.12, 5.17 and 5.19). The method needs values for $N_{pm,Rd}$, $M_{pl,Rd}$ and $M_{max,Rd}$. The equations for them are usually presented for rectangular cross-sections, but the modifications needed for circular cross-sections are available (Johnson, 2012, Appendix C; Liew and Xiang, 2015, Table 3.3). Their use leads to these results, which are shown in Figure 5.27:

$$N_{pl,Rd} = 7945 \text{ kN}, \ N_{pm,Rd} = 2319 \text{ kN},$$
$$M_{pl,Rd} = 511 \text{ kN m}, \ M_{max,Rd} = 528 \text{ kN m}$$

From the Figure,

$$M_{pl,N,Rd} = 511(7945 - 5401)/(7945 - 2319) = 231 \text{ kN m}$$

Figure 5.27 Interaction diagram for concrete-filled steel tube internal column, with $t = 12.5$ mm

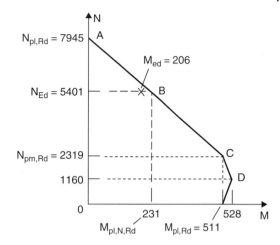

Eurocode 4 requires the ratio $M_{Ed}/M_{pl,N,Rd}$ to be less than a value α_M, which it gives as 0.8 for steel grades S420 and S460. Liew and Xiang (2015) reported that tests have shown that this limit is applicable for steel grades up to S550. Here, the ratio is

$$M_{Ed}/M_{pl,N,Rd} = 206/231 = 0.892$$

so the resistance of the column with this cross-section is too low.

The wall thickness was then increased from 12.5 mm to 16 mm, with the same external diameter. This reduced the ratio from 0.892 to 0.54, well below 0.8, and illustrates the non-linear nature of these verifications.

5.11.7 Discussion

The calculation finds that the first cross-section assumed is not strong enough. When this problem arose in Section 5.7.4, the resistance was checked against the curve for which line AC in Figure 5.19 is a conservative approximation. This would be a difficult calculation here, because the algebra for a circular cross-section is more complicated.

The example in Section 5.7.4 was completed by a check on biaxial bending, which was found to be more adverse than uniaxial bending. The present column has the same resistances about both axes, and no z-axis bending from the loading. If the imperfection is assumed to cause z-axis bending rather than contributing to $M_{y,Ed}$, the design value of $M_{y,Ed}$ is reduced by an amount equal to the new $M_{z,Ed}$. The check on biaxial resistance, Expression 5.41, would then give the same ratio, 0.892, as found above.

This example illustrates the important influence of imperfections and slenderness. For the column with $t = 12.5$ mm, they increased to 206 kN m the design bending moment of 32 kN m found at the ends of the column length by first-order analysis.

As shown in Figure 5.21, the use of high-strength materials has enabled the cross-sectional area of the CFST column to be less than 30% of that of the encased I-section, and it uses 40% less structural steel. However, it will be more difficult to ensure resistance to fire because the steel component is more exposed and provides a higher proportion of the total resistance. It will need more complex joints to the beams, and steel tubes of this strength may be difficult to procure. This design is outside the scope of the current Eurocode 4.

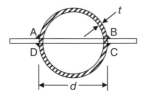

Figure 5.28 Cross-section through CFST column showing gusset plate for load introduction

Load introduction In Section 5.5.1, rigid full-strength joints were used between the floor beams and the central column. These are practicable, although quite expensive, for the concrete-encased column designed in Section 5.8. They would be so difficult to provide for this smaller CFST column that it would probably not be used unless the joints to the beams could be designed as nominally pinned. The depth of the beams would need to be increased. It is now assumed that this is done, to provide an example of the use of the clause in Eurocode 4 on load introduction.

The problem, explained in Section 5.6.6, is to find a load path that enables the vertical shear from each beam to be shared between the steel tube and the concrete core. A method given in Eurocode 4 is now used. A steel gusset plate is threaded through slots in the steel tube, as shown in Figure 5.28, and bolted to the webs of the beams (not shown here). These joints can be treated as nominal pins.

The load to be transferred is a maximum when both beams are fully loaded. From Figure 5.20, with simply-supported beams, it is

$$N_{Ed} \approx 60.9 \times 9.3 = \textbf{566 kN}$$

as the extra weight of the redesigned steel beams will be small.

The method of Eurocode 4 is based on tests (Porsch and Hanswille, 2006). It assumes that the gusset plate loads the column by a bearing stress $\sigma_{c,Rd}$ on an area which, for this symmetrical loading, is ABCD in Figure 5.28, denoted A_ℓ. For a CFST column of circular cross-section, $\sigma_{c,Rd}$ is given by

$$\sigma_{c,Rd} = f_{cd}[1 + 4.9(t/d)(f_y/f_{ck})](A_c/A_\ell)^{1/2}, \quad \leq A_c f_{cd}/A_\ell, \leq f_{yd} \tag{5.72}$$

where A_c is the area of the concrete cross-section of the column. Symbols t and d are shown in Figure 5.28. Here,

$$A_c = \pi(273 - 32)^2/4 = 45\ 620 \text{ mm}^2$$

The web thickness of the beams is 7.8 mm (Figure 3.31). Assuming a gusset plate 10 mm thick, $A_\ell = 273 \times 10 = 2730 \text{ mm}^2$, and $(A_c/A_\ell)^{1/2} = 4.09$.

With $f_{cd} = 48$ MPa, $f_{ck} = 72$ MPa, $t = 16$ mm, $d = 273$ mm and $f_{yd} = 355$ MPa for the gusset plate, Equation 5.72 gives

$$\sigma_{c,Rd} = 48(1 + 1.42)(4.09) = 474, \quad \leq 802, \leq 355 \text{ MPa}$$

so $\sigma_{c,Rd} = 355$ MPa and the load introduction is verified, because

$$N_{Rd} = \sigma_{c,Rd}A_\ell = 0.355 \times 2730 = \textbf{969 kN}$$

Eurocode 4 refers to $\sigma_{c,Rd}$ as the 'local design strength of the concrete', even though in this example it exceeds $7f_{cd}$. This high value appears to allow for the proportion of the total load that is transferred from the plate to the tube by the welds shown in Figure 5.28. Eurocode 4 does not give this proportion, which is needed for the design of the welds.

It can be estimated using transformed areas. Allowing for creep:

- the modular ratio is $n = 210 \times 2/41.0 = 10.2$;
- the cross-section of the steel tube is $\pi(273^2 - 241^2)/4 = 12\,920\,\text{mm}^2$;
- the transformed area in terms of steel is $12\,920 + 45\,620/10.2 = 17\,400\,\text{mm}^2$;
- so the load to be transferred to the steel tube is $566 \times 12.9/17.4 = 420\,\text{kN}$.

The details of the bolted joints to the beam webs will determine the depth of gusset plate required.

6

Fire Resistance

Yong C. Wang

6.1   General introduction and additional symbols

The first five chapters of this book cover structural design at ambient temperature. This chapter deals with the accidental limit state of exposure to fire, which is important for buildings. The general fire safety requirements for buildings (Building Regulations 2013, 2013) are:

- means of escape to ensure safe evacuation of occupants of the building,
- control of surface lining materials to prevent and delay further ignition, and to reduce the rate of heat release and generation of smoke and toxic gases,
- containment of fire to prevent fire spread from compartment to compartment,
- prevention of external fire spread and
- fire-fighting to control and to extinguish fire as quickly as possible.

This chapter deals with fire safety of the structure, which is part of fire containment, often referred to as fire resistance.

Wherever possible, the same symbols as for chapters 1–5 of this book are used. The additional fire-related symbols are listed below.

A	inner surface of the fire protection material per unit length of the protected steel element
B	external width (shorter dimension) of rectangular or elliptical hollow section
c	specific heat
d	thickness
D	diameter of circular hollow section
H	external depth (longer dimension) of rectangular or elliptical hollow section
K	coefficient
$R30$	30-minute fire resistance time (and $R60$, etc)
t	fire resistance time
u	axial distance of reinforcement bar from concrete surface
V	volume per unit length of steel member enclosed by fire protection material
λ	thermal conductivity

Composite Structures of Steel and Concrete: Beams, Slabs, Columns and Frames for Buildings,
Fourth Edition. Roger P. Johnson.
© 2019 John Wiley & Sons Ltd. Published 2019 by John Wiley & Sons Ltd.

η   load level factor

θ   temperature, degrees C

φ   stiffness reduction factors due to thermal stress in composite column

Subscripts

ay	strength of steel
fi	fire situation
i	ith part of a steel section
m	solid section of concrete-filled tube
p	fire protection
sec	secant
t	fire exposure time
δ	eccentricity
θ_{cr}	critical temperature
∞	ambient

6.1.1 Fire resistance requirements

Fire resistance is determined by the ability of the construction to achieve adequate *insulation, integrity* and *load-bearing performance.* These three requirements are to prevent the three modes of fire spread from one compartment to another compartment. The insulation requirement is concerned with prevention of excessive temperature on the unexposed surface of the construction to cause further ignition. This is quantified by limiting the temperature increase on the unexposed surface to be no more than 140°C on average, and 180°C at the maximum. Integrity failure occurs when fire burns through the construction. Load-bearing capacity is necessary to ensure that the structure does not fail, leading to fire spread from one compartment to another.

Depending on the usage, a structural member may not need to meet all of the above three fire resistance criteria (insulation, integrity, load-bearing). For example, a composite floor that separates one fire-resistant compartment from another would need to achieve all three requirements. However, the composite beams and composite columns that support the composite floor would only need to have sufficient load-bearing capacity.

This chapter focuses on the assessment of the load-bearing capacity of composite structural members in fire. It should be appreciated that because the objective of a fire-resistant construction is to prevent fire spread from one compartment to another, it may not be necessary for every structural member to maintain its load-bearing capacity in fire, provided the loss of load-bearing capacity of this structural member does not lead to fire spread. For example, application of tensile membrane action in composite floor slabs, which will be introduced in Section 6.2.3, is based on the premise that the loss of load-bearing capacity of some of the internal secondary beams does not lead to fire spread. On the other hand, a member-based approach, in which every structural member of the building has been demonstrated to have sufficient load-bearing capacity in fire, does not guarantee overall fire resistance of the structure. Failure of the World Trade Center buildings on September 11, 2001, tragically illustrates this point.

It is important that structural interactions of members in fire are adequately considered. However, these topics are beyond the scope of this chapter on fire resistance of composite structural members. Interested readers may wish to consult more specialist literature on the subject, such as Wang et al. (2012).

6.1.2 Fire resistance design procedure

When evaluating the load-bearing capacity of structural members in fire, the assessment procedure follows three generic steps:

(1) Quantifying the fire exposure condition to the structure, in the form of a fire temperature–time relationship. EN 1991-1-2 (BSI, 2002d) can be used to specify the fire temperature–time relationship, which may be the nominal standard fire temperature–time curve or, more realistically, parametric fire curves. In the standard fire temperature–time curve, the fire temperature increases indefinitely with time. Therefore, it is necessary to stop at a specific time. This time is referred to as the fire resistance time or *fire rating*. It is usually specified in multiples of 30 minutes.
(2) Calculating the temperature field in the structural member under the above fire condition. Heat transfer analysis is carried out. However, some analytical equations are available in EN 1994-1-2 (BSI, 2014c) for calculating temperatures in composite structural elements.
(3) Calculation of the remaining load-carrying capacity of the structural member at elevated temperatures and comparison with the applied load.

This chapter will only deal with evaluation of load-carrying capacity at given elevated temperatures. Such calculations can be performed in either the temperature domain or the load domain.

In the temperature domain, the applied load on the structure is provided and the aim is to find the critical or limiting temperature of the structure at which the residual load-carrying capacity of the structure equals the applied load. This is usually the flow of information in fire resistance design because, in practice, the applied load is known, and the purpose of fire design is to find the critical temperature of the structure which is then used to determine the required fire protection thickness.

Calculation in the load domain is more familiar to structural engineers. In this calculation procedure, the structural temperatures are known and the task is to check whether the structure has sufficient load-carrying capacity at the elevated temperatures to resist the applied load. In EN 1994-1-2 terminology, the design requirement is:

$$E_{fi,d,t} \leq R_{fi,d,t} \tag{6.1}$$

where $E_{fi,d,t}$ is the design effect of actions for the fire situation, and $R_{fi,d,t}$ is the corresponding design resistance in the fire situation.

If operating in the load domain, but the critical temperature (which is the maximum temperature that the critical element of a structural member can be allowed to attain) is sought, an iterative process will be necessary.

This chapter will consider composite slabs, composite beams and composite columns. Structural fire resistance design may be performed by using tabulated data or calculations. Because tabulated data are based on fire tests, their scope of application is limited. However, composite slabs are usually proprietary systems for which manufacturers of the systems have to carry out fire testing to demonstrate their fire resistance. The designer has little flexibility in choosing details of a proprietary system. Therefore, fire resistance design of composite slabs is primarily based on using tabulated data.

On the other hand, for composite beams and columns, calculation methods offer great advantage in design flexibility. Therefore, in this chapter, the tabulated method will be described for composite slabs, and the calculation methods will be provided for composite beams and composite columns.

6.1.3 Partial safety factors and material properties

Because fire is an accidental condition, the applied mechanical loads on the structure are similar to those for serviceability limit state design. The design mechanical load is calculated for the following combination:

$$G_k + \psi_{fi} Q_{k,1} \tag{6.2}$$

where G_k is the characteristic value of the permanent action, $Q_{k,1}$ is the characteristic value of the leading variable action, and ψ_{fi} is the combination factor for fire design, usually taken as the frequent value ($\psi_{1,1}$).

Also, the material safety factors are reduced for fire design, being 1.0 for steel, concrete and reinforcement.

Thermal properties are required for heat transfer calculations, and mechanical properties are necessary for load-carrying capacity calculations. For composite construction, temperature-dependent mechanical properties of steel and concrete are provided in EN 1994-1-2. This code also provides thermal properties of steel and concrete. However, because assumptions about fire protection are often necessary when calculating temperatures of steel sections in composite construction, it is also necessary to obtain thermal properties of fire protection materials. At present, there is a lack of reliable and accurate data of temperature-dependent thermal properties of fire protection materials. Therefore, even though methods (Section 6.3.2) exist for calculating structural temperatures in fire, when fire protection is required one has to resort to manufacturers' test data.

6.2 Composite slabs

6.2.1 General calculation method

Composite slabs are flexural members. Therefore, the fire resistance of composite slabs is governed by the bending moment resistance of the slab. The calculation method for bending resistance of composite slabs is the same as that at ambient temperature, except that the mechanical properties of the materials (steel sheeting, reinforcement, concrete) at ambient temperature are replaced by those at elevated temperatures. Because the temperature distribution in composite slabs is non-uniform, the main calculation effort is to obtain equivalent uniform temperatures for the different components of the slab.

In EN 1994-1-2, a temperature calculation method is proposed, based on regression analysis of fire testing and numerical heat transfer simulation results. The allowed field of application of this method may render this method not applicable to many composite slabs.

Table 6.1 An example of tabulated data for fire resistance of a proprietary composite slab (not for general use)

Props	Span	Fire rating	Slab depth (mm)	Mesh	Maximum span (m) with no additional reinforcements Deck thickness (mm)								
					0.9			1			1.2		
					Total applied load (kN/m²)								
					3.5	5.0	10.0	3.5	5.0	10.0	3.5	5.0	10.0
No temporary props	Single span	1 hr	120	A142	2.8	2.8	2.8	3	3	2.9	3.4	3.4	3.2
		1.5hr	130	A142	2.8	2.8	2.3	3	3	2.3	3.3	3.3	2.6
		2 hr	140	A193	2.5	2.2	1.8	2.7	2.4	1.9	2.9	2.6	2
	Double span	1 hr	120	A142	3.4	3.4	2.9	3.5	3.5	3	4	4	3.2
		1.5 hr	130	A142	3.3	3.3	2.6	3.4	3.4	2.6	3.7	3.7	2.8
		2 hr	200	A393	2.8	2.8	2.8	3	3	3	3.4	3.4	3.4

6.2.2 Tabulated data

Selection of composite slabs for fire resistance is usually based on tabulated data. The slabs are proprietary systems developed by manufacturers who provide fire resistance rating data for their products. Therefore, the calculation method is rarely used by the engineer who will simply refer to tabulated data from manufacturers. An example is shown in Table 6.1. Should modification become necessary, for example to use additional reinforcement to increase the maximum span, or to sustain additional load, advice should be sought from the manufacturer of the profiled sheeting for the slab.

6.2.3 Tensile membrane action

In the mid-1990s, a series of large-scale structural fire experiments were carried out in an eight-storey composite structure at Cardington, in Bedfordshire in the UK (Newman et al., 2006). In the so-called compartment fire tests, the internal secondary beams were not fire-protected. These beams experienced very high temperatures and distorted badly, therefore clearly contributing very little to fire resistance of the structure. Yet the composite floors maintained their load-bearing capacity. The load-carrying mechanism that enabled this performance was identified as tensile membrane action (TMA) in the composite slab at very large deflections. TMA can be used to safely eliminate fire protection to interior secondary beams, as illustrated in Figure 6.1.

At ambient temperature, composite slabs are considered as one-way spanning between secondary beams. However, in the fire situation, if there is no fire protection to the interior secondary beams, the composite slab becomes two-way spanning and is supported by the edge beams. At small deflections, bending behaviour governs, and the load-carrying capacity of two-way spanning slabs is quantified based on yield-line theory. However, at very large slab deflections that can be generated in fire, tensile membrane action can develop, which can provide the composite slab with load-carrying capacities many times the yield line capacity. It is this increased load-carrying capacity in tensile membrane action that enables the composite floor slab to achieve sufficient resistance without the interior secondary beams.

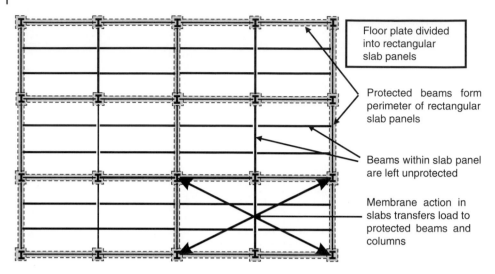

Figure 6.1 Scheme to use tensile membrane action to eliminate fire protection to interior secondary beams (Wang et al., 2012, with permission from CRC Press)

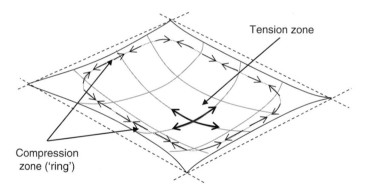

Figure 6.2 Membrane action in a horizontally unrestrained concrete slab (Wang et al., 2012, with permission from CRC Press)

Tensile membrane action, as the name suggests, is mobilized when the tensile reinforcement mesh of the composite slab acts as a tensile net. The in-plane tensile force of the tensile net is balanced by the compressive force that can develop in a concrete ring around the edges of the slab, as illustrated in Figure 6.2. Therefore, there is no requirement for external horizontal (in-plane) restraint to anchor the tensile force in the reinforcement mesh.

The load-carrying capacity of the slab under tensile membrane action increases with increasing deflection until eventually the reinforcement fractures, or large through-thickness cracks develop, which signals integrity failure of the slab. Predicting reinforcement fracture or slab cracking is a major challenge that has not been resolved. Therefore, to allow the benefit of tensile membrane action to be utilized in fire resistance design, the maximum allowable slab deflection is limited. Detailed practical design guidance is provided in Newman et al. (2006).

There are two additional requirements for the use of tensile membrane action. It develops because the lateral (vertical) restraints along the slab edges prevent folding of the slab. If restraint along any edge of the slab is missing, the slab folding mechanism can occur freely and TMA does not develop. Therefore, if TMA is used in fire design, the edge beams must be fire-protected to ensure they provide the necessary lateral (vertical) support to the slab. Furthermore, the load transfer path from the slab to the edge beams in TMA is different from that at ambient temperature. At ambient temperature, the applied load on the slab is transferred to the secondary beams, and whence to the primary beams. Therefore, the applied load on the edge secondary beams can be very low. When using TMA in fire design (Figure 6.1), the interior secondary beams are not fire-protected, so they have very little resistance. Therefore, the applied load on the slab is transferred to the supporting edge beams directly. This can result in forces and bending moments in the edge secondary beams that are much greater than those calculated assuming the ambient temperature load path. It is important that this additional load is included when determining fire protection to the edge secondary beams. Because the loads on the edge secondary beams are increased, those on the edge primary beams are reduced. However, it would not be wise to take this benefit into consideration. Therefore, when determining fire protection to the edge primary beams, using the ambient temperature load transfer path is recommended.

6.3   Composite beams

As explained in Section 6.1.2, structural fire resistance assessment can be carried out either in the temperature domain (to find out the critical or limiting temperature, given the design load) or in the load domain (to calculate the structural load-carrying capacity, knowing the structural temperatures). Accordingly, EN 1994-1-2 provides two calculation methods. These two methods should give very similar results, but there are some differences due to different assumptions. This section will present these two calculation methods, and then use worked examples, based on the composite beam design example of Section 3.11.1, to illustrate these calculation methods and also to give guidance on how to select the appropriate calculation method to use.

EN 1994-1-2 covers conventional composite beams with a concrete slab on top of the steel section, and composite beams with a partially encased steel section in which concrete fills the space between the flanges. The load-bearing capacity calculation method is the same for both types of beam. Due to a complex distribution of temperatures in composite beams with partially encased steel section, an entire Annex in EN 1994-1-2 is devoted to calculating equivalent uniform temperatures for the different components of this type of construction. This chapter will not repeat such information and will only focus on the calculation method for elevated temperature load-carrying capacity of conventional composite beams.

6.3.1   Critical temperature method

Although prominence in EN 1994-1-2 is given to calculating the load-carrying capacity of a composite beam at elevated temperature (load domain), the target of fire resistance design is often to find the critical temperature that can be survived by the beam at a given loading condition (temperature domain). This is the logical flow of design information: the loading condition will have been determined from ambient temperature considerations, but the structural temperatures are unknown due to unknown fire protection. The load-bearing capacity calculation method can be used to find the critical temperature, but an iterative process is necessary.

To find the critical temperature without iteration, the critical temperature method may be used. The critical temperature is the maximum temperature in the critical element of the member that can be survived by the structure without losing load-bearing capacity. For composite beams, the critical element is the lower flange of the steel section. The critical temperature is a function of the level of applied load of the structural member.

In EN 1994-1-2, the critical temperature is found in the following way:

(1) *Calculate the load level*:

$$\eta_{fi,t} = E_{fi,d,t}/R_d \qquad (6.3)$$

Since bending moment resistance usually governs fire design, the design effect of actions for fire design ($E_{fi,d,t}$) is the maximum bending moment in the beam for fire design, and R_d is the composite beam bending resistance for ambient temperature design.

(2) *Calculate the maximum yield stress ratio of steel*:

The load level is then converted to the minimum value to which the elevated temperature yield stress of steel can be reduced, in the following way:

$$\frac{f_{ay,\theta cr}}{f_y} = \eta_{fi,t} \qquad (6.4)$$

for fire resistance rating greater than 30 minutes, or

$$\frac{f_{ay,\theta cr}}{f_y} = 0.9\eta_{fi,t} \qquad (6.5)$$

for 30 minutes fire resistance rating (R30).

(3) *Find the critical temperature*:

Using this steel yield stress reduction factor, the critical temperature can be found based on the yield stress reduction factor–temperature relationship in EN 1994-1-2.

The 0.9 multiplier in Equation 6.5 gives a yield stress ratio for R30 lower than that for higher fire resistance ratings. Therefore, at the same $\eta_{fi,t}$, the critical temperature for R30 is higher than that for higher fire resistance ratings. This slight increase in critical temperature (compared with that for higher fire ratings at the same load level) is to take advantage of non-uniform temperature distribution in the steel section, as explained below.

In fire, the temperature of the steel top flange at first rises more slowly than that of the bottom flange, because the top flange is in contact with the cooler concrete slab. The critical temperature method assumes that the temperature distribution on the steel section is everywhere that of the exposed bottom flange – a valid assumption for long fire resistance periods when a longer heat conduction time allows the upper flange temperature to reach almost the same temperature as elsewhere. It is conservative for 30-minute fire resistance, because the upper flange maintains a much lower temperature than elsewhere due to a shorter heat conduction time. The use of a higher critical temperature compensates for this.

Worked examples The worked examples in this section are based on the composite slab in Section 3.4 and the composite beam in Section 3.11. The standard fire resistance rating is 60 minutes (R60).

(i) *Complete shear interaction*

From Equation 3.99, the full-interaction bending resistance at ambient temperature is

$$M_{pl,Rd} = \textbf{829 kN m}$$

Assuming the combination factor for variable actions is 0.7, then the applied load in fire is:

$$1.0 \times 17.6 + 0.7 \times 14.8 = 27.96 \ \text{kN/m}$$

The maximum bending moment for fire design is:

$$27.96 \times \frac{8.6^2}{8} = \textbf{258.5 kN m}$$

From Equation 6.4,

$$\frac{f_{ay,\theta cr}}{f_y} = \eta_{fi,t} = 258.5/829 = 0.312$$

The steel yield strength is reduced to 0.47 and 0.23 of the ambient temperature value at 600°C and 700°C respectively (BSI, 2014c). Linear interpolation gives a critical temperature of **666°C**.

For comparison, if the fire resistance rating were R30, and the beneficial effect of non-uniform temperature distribution in the steel section were taken into account, then

$$\frac{f_{ay,\theta cr}}{f_y} = 0.9\eta_{fi,t} = 0.9 \times 0.312 = 0.281$$

for which the critical temperature would be 679°C.

(ii) *Partial shear interaction*

From Section 3.11.2, the degree of shear connection is 0.46 and the bending moment resistance is 693 kN m, so the load ratio is:

$$\frac{f_{ay,\theta cr}}{f_y} = \eta_{fi,t} = 258.5/693 = 0.373$$

Interpolation using the results above gives the critical temperature as **640°C**.

When exposed to fire from beneath, the shear studs are shielded by the concrete slab. Their temperature is much lower than the steel section temperature. Therefore, shear studs suffer much lower reduction in strength compared with the steel section. This means that the degree of shear connection tends to increase in fire compared with that at ambient temperature (example (iii) of Section 6.3.3). Therefore, the critical temperature for partial shear interaction tends to be higher than that for complete shear interaction, if the load ratio is the same. However, EN 1994-1-2 does not take advantage of this benefit.

6.3.2 Temperature of protected steel

To limit the temperature rise of the steel section to the critical temperature, fire protection is usually required. The required fire protection thickness depends on the fire protection method as well as the fire protection material. Usually, specific fire protection material manufacturers should be consulted. If reliable thermal properties of fire protection materials are available, the following temperature calculation equation may be used to find the required fire protection thickness.

$$\Delta\theta_{a,t} = \left[\left(\frac{\lambda_p/d_p}{c_a\rho_a}\right)\left(\frac{A_{p,i}}{V_i}\right)\left(\frac{1}{1+w/3}\right)(\theta_{fi,t}-\theta_{a,t})\Delta t\right] - \left[(e^{w/10}-1)\Delta\theta_{fi,t}\right]$$

(6.6)

with $w = \left(\frac{c_p\rho_p}{c_a\rho_a}\right)d_p\left(\frac{A_{p,i}}{V_i}\right)$
where:

λ_p is the thermal conductivity of fire protection material (W/m K)
d_p is the thickness of fire protection material (m)
$A_{p,i}$ is the area of the inner surface of the fire protection material per unit length of part i of the protected steel member
V_i is the volume of part i of the steel member enclosed by fire protection material
c_a, ρ_a are the temperature-dependent specific heat (kJ/kg K) and density (kg/m³) of steel
c_p, ρ_p are the temperature-dependent specific heat (kJ/kg K) and density (kg/m³) of fire protection material
$\theta_{fi,t}$ is the fire temperature (°C) at time t
$\Delta\theta_{fi,t}$ is the increase of fire gas temperature (°C) during the time interval Δt (s).

For standard fire resistance design, the fire temperature–time relationship (BSI, 2014c), is:

$$\theta_{fi,t} = 345 \times \log_{10}(8t+1) + \theta_\infty$$

(6.7)

in which t is time (in minutes) and θ_∞ is the temperature of ambient air in °C.

Equation 6.6 should be applied incrementally, with the time interval Δt not exceeding 30 seconds. Due to the presence of the second part on the right-hand side in Equation 6.6, the calculated steel temperature increase may be negative at the start of the calculation. If this is the case, the negative values should be replaced by zero.

The value $\frac{A_{p,i}}{V_i}$ is referred to as the section factor. It expresses the ratio of the surface area of the steel member exposed to fire to the volume of the steel member being heated up. Depending on how the fire protection is applied, the section factor is calculated differently.

Figure 6.3 shows the cross-section profile of a steel section with a concrete slab on top. If box protection is used, the entire steel member is assumed to have the same section

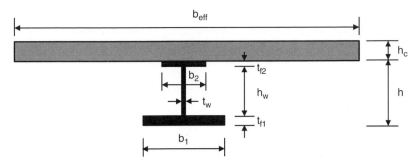

Figure 6.3 Geometry of a composite beam with concrete slab on top of steel section

factor. It is the ratio of the inner surface area of the box protection per unit length to the cross-sectional area (A_a) of the steel member per unit length:

$$\frac{A_p}{V} = \frac{2h + b_1}{A_a} \tag{6.8}$$

If profiled fire protection is used, EN 1994-1-2 allows the steel section to be divided into the top flange, the web and the lower flange, and their temperatures to be calculated individually.

For the lower flange, the section factor is $\frac{2(b_1+t_{f1})}{b_1 \times t_{f1}}$. For the example in Section 3.11.1, this gives

$$2 \times \frac{12.8 + 178}{12.8 \times 178} \times 1000 = 167.5 \text{ m}^{-1}$$

For the upper flange, it is possible to take advantage of the contact between the concrete slab and the top of the flange to reduce the section factor. Therefore, if at least 85% of the upper flange is in contact with the concrete slab or any void between the upper flange and the steel deck is filled with non-combustible material, the upper flange section factor is $\frac{(b_2+2t_{f2})}{(b_2 \times t_{f2})}$. If the upper flange of the steel section in the example has all the voids filled, its section factor is

$$\frac{2 \times 12.8 + 178}{12.8 \times 178} \times 1000 = 89.4 \text{ m}^{-1}$$

This value is roughly half of the value for the lower flange because the exposed surface area (mainly the lower side) of the upper flange is about half of the exposed surface area (both upper and lower sides) of the lower flange.

However, if less than 85% of the upper flange is in contact with concrete, the upper surface of the upper flange is assumed to be exposed to fire and the section factor of the upper flange is $\frac{2(b_2+t_{f2})}{(b_2 \times t_{f2})}$.

EN 1994-1-2 states that the web temperature can be taken the same as the lower flange temperature if the steel section height does not exceed 500 mm. However, it does not provide any guidance for steel sections with height greater than 500 mm. In a protected steel section, it is likely that heat conduction inside the steel section will be very rapid (due to high thermal conductivity of steel) compared with heat transfer from the fire to the steel section (due to fire protection). Therefore, it can be assumed that the steel web temperature is the same as the lower flange temperature for any steel section height.

Worked example For the steel section in the example in Section 3.11.1, box fire protection using vermiculite is assumed. What is the required fire protection thickness so that the maximum temperature in the steel section is 666°C as calculated in the previous worked example?

The thermal properties of materials are as follows.

For structural steel (BSI, 2014c): density $\rho_a = 7850$ kg/m³; and for temperatures up to 600°C, specific heat (J/(kg°C)):

$$c_a = 425 + 0.773 \times \theta_a - 1.69 \times 10^{-3}\theta_a^2 + 2.22 \times 10^{-6}\theta_a^3 \tag{6.9}$$

where θ_a is the temperature of the steel in °C.

The fire protection material is assumed to be vermiculite. Its thermal properties are (Wang et al., 2012):

$$\text{density}: \quad \rho_p = 600\,\text{kg/m}^3;$$

$$\text{specific heat}: \quad c_p = 900\,\text{J/(kg.°C)}$$

$$\text{thermal conductivity}: \quad \lambda_p = 0.162 + 0.072\left(\frac{\theta_p + 273}{1000}\right)^3 \tag{6.10}$$

where θ_p is the fire protection temperature, which is the mean of θ_a and the fire temperature in °C.

Using values from Figure 3.31, the section factor for the entire steel section is:

$$\frac{A_p}{V} = \frac{2 \times 406 + 178}{7600} \times 1000 = 130.3\ \text{m}^{-1} \tag{6.11}$$

The required fire protection thickness for complete shear interaction is 12.5 mm and that for partial shear interaction is 13.6 mm. Figure 6.4 shows the steel temperature–time relationships, calculated using Equation 6.6. The steel temperatures at 60 minutes are 665°C and 640°C respectively, not exceeding the critical temperatures of 666°C and 640°C for complete and partial shear interaction.

6.3.3 Load-carrying capacity calculation method

The load-carrying capacity method for fire design is the same as that for ambient temperature design, except that the elevated temperature mechanical properties of steel and concrete, and material partial safety factors for fire design, should be used.

Worked examples For comparison with the critical temperature method presented in the previous section, it is assumed that the lower flange temperature in the load-carrying capacity method is the same as the critical temperature obtained in the previous worked example.

(i) *Complete shear interaction with uniform temperature in the steel section*
   For 60 minutes of fire resistance, the uniform temperature in the steel section is 666°C. The yield strength of the steel is $0.312 \times 355 = 111$ N/mm².

$$N_{fi,a,pl} = 0.312 \times 2698 = 842\ \text{kN}$$

From rules in EN 1994-1-2, the effective thickness of the composite slab is 125 mm. The 60-minute thermal insulation criterion requires an effective thickness of 80 mm, reduced by a screed layer, if any. No strength reduction is required

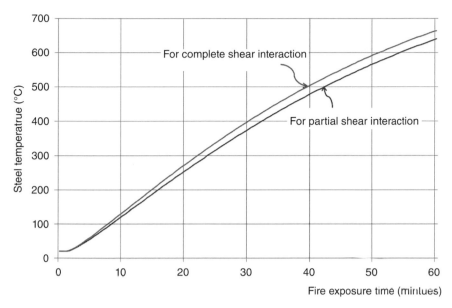

Figure 6.4 Steel temperature–time relationship

for concrete temperatures below 250°C, and the insulation rules show that it will be less than 180°C. Hence, the concrete design strength need not be reduced. In fire, it is $25/1.0 = 25\,\text{N/mm}^2$ (because the material partial safety factor for concrete, 1.5 for ambient temperature, is 1.0 for fire). From Equation 3.98 in Section 3.11.1:

$$N_{\text{fi,c}} = 2555 \times 1.5 = 3832\text{ kN}$$

Thus, the whole of the steel section is at yield in tension, and the depth of concrete in compression is $842/3832 \times 80 = 17.6\,\text{mm}$. Hence,

$$M_{\text{fi,pl,Rd}} = 842 \times \frac{\left(\frac{406}{2} + 70 + 80 - \frac{17.6}{2}\right)}{1000} = \textbf{290 kN m}$$

This value is 12% higher than the bending moment of 258.5 kN m used to calculate the critical temperature in the previous section. However, in most cases, the difference would be much smaller. The relatively large increase in this particular example is due to a large depth of the concrete slab in compression at ambient temperature. In most cases at ambient temperature, a smaller depth of the concrete slab in compression would be required. In the fire condition, the depth of concrete in compression is reduced due to the reduced steel yield strength. This results in an increase in the lever arm. However, the increase in lever arm would be small compared with the total lever arm. Therefore, because the critical temperature method and the moment capacity method use the same reduced yield stress of steel in fire, these two methods give similar results for the bending resistance of the beam. Furthermore, the critical temperature method is simple to use, gives results similar to those from the bending moment capacity method, and on the safe side. In conclusion, because such composite beams normally would require fire protection, and the starting point in determining the required fire protection thickness is the critical temperature, the critical temperature method should be the first choice in fire resistance design of composite beams.

(ii) *Complete shear interaction with non-uniform temperature in the steel section*

To illustrate the effects of a non-uniform temperature distribution in the steel section, the temperature in the web and lower flange is 679°C, the same as the critical temperature for R30 in the worked example in Section 6.3.1, for which the steel yield stress reduction factor was found to be 0.281.

The upper flange temperature is usually about 150–250°C lower than the lower flange/web temperatures. Assume the temperature difference is 200°C, so the upper flange temperature is 479°C. Because the upper flange is close to the plastic neutral axis, the bending resistance is not very sensitive to values within this temperature difference range.

The steel yield strength reduction factors are 1.0 and 0.78 at 400°C and 500°C respectively (BSI, 2014c), giving a value of 0.827 at 479°C, by linear interpolation. From Equation 3.98, $N_{a,pl} = 2698$ kN, so:

$$N_{fi,a,pl} = 0.281 \times 2698 + (0.827 - 0.281) \times 0.355 \times (12.8 \times 178)$$
$$= 758.1 + 441.6$$
$$= 1200 < 3832 \text{ kN}$$

The depth of concrete in compression is $1200/3832 \times 80 = 25$ mm.

$$M_{fi,pl,Rd} = 758.1 \times \frac{\left(\frac{406}{2} + 150 - \frac{25}{2}\right)}{1000} + 441.6 \times \frac{\left(\frac{12.8}{2} + 150 - \frac{25}{2}\right)}{1000}$$
$$= 248.1 + 63.5$$
$$= \mathbf{322 \text{ kN m}}$$

This value is 24.4% higher than the bending moment value of 258.5 kN m in the previous section which was used to calculate the critical temperature. Compared with the calculation results for uniform steel temperature distribution, this indicates that it is worthwhile exploring the advantage of the bending moment capacity method when the steel temperature distribution is non-uniform. This is particularly the case for slim floor/asymmetric beam/shelf angle construction where the concrete slab is placed between the top and bottom flanges, thus shielding much of the steel section from fire exposure and creating severe non-uniform temperature distribution in the steel section.

(iii) *Partial shear interaction*

To enable this method to be compared with the critical temperature method, the lower flange temperature is assumed to be the critical temperature given by that method, 640°C for partial shear interaction, with a steel yield strength reduction factor of 0.373.

For calculating shear connector resistance, the rules for composite slabs give the concrete temperature as $0.4 \times 640 = 256$°C. The concrete strength reduction factor is $k_{c,\theta} = 0.894$, based on concrete strength reduction factors of 0.95 and 0.85 at 200°C and 300°C respectively (BSI, 2014c). The temperature of the stud shear connectors is $0.8 \times 640 = 512$°C. Their strength reduction factor is $k_{u,\theta} = 0.742$.

Using values from Section 3.2, the design shear connector resistance in a solid slab is the lesser of $60.2 \times 1.25 \times 0.894 = 67.3$ kN, for concrete failure, and $0.8 \times 91 \times 1.25 \times 0.742 = 67.5$ kN, for shank failure, and so is 67.3 kN/stud. The increase above the resistance at ambient temperature results from the reduction in the partial safety factor from 1.25 to 1.0. As before,

$$k_t = 0.70, \quad \text{so} \quad P_{fi,Rd} = 67.3 \times 0.7 = 47.1 \text{ kN}$$

From here on, the calculation procedure is the same as at ambient temperature, except for the changes in the design strengths of the steel section, the concrete slab and the shear connectors.

The resistance of the steel section to axial load is

$$N_{fi,a,pl} = 0.373 \times 2698 = 1006 \text{ kN}$$

The resistance of the shear connectors in a half span is $47.1 \times 28 = 1319$ kN > 1006 kN, so shear connection is 100% for fire design.

The depth of concrete in compression is $1006/(3832 \times 0.894) \times 80 = 23.5$ mm.

$$M_{fi,pl,Rd} = 1006 \times \frac{\left(\frac{406}{2} + 150 - \frac{23.5}{2} \right)}{1000} = \textbf{343 kN m}$$

This value is about 33% higher than the bending moment of 258.5 kN m used in the previous section to obtain the critical temperature. Thus, the critical temperature method is conservative. This is due to the critical temperature method of not taking advantage of the lower reduction in shear stud resistance compared with the steel section. A composite beam with partial shear interaction at ambient temperature may become full shear interaction, as demonstrated in this example, in the fire condition.

With such high conservatism, it is worth exploring the benefit of using the load-carrying capacity method for composite beams with partial shear connection, especially if the degree of shear interaction at ambient temperature is low even though an iterative process may be necessary to obtain the critical temperature. To obtain a more accurate estimate of the critical temperature, the calculated bending resistance can be used to modify the steel yield strength reduction factor to give:

$$\frac{f_{ay,\theta cr}}{f_y} = \frac{\text{applied bending moment in fire}}{\text{calculated bending moment capacity in fire}} \times \eta_{fi,d,t} \tag{6.12}$$

The following calculations are presented to illustrate the iterative process.

$$\frac{f_{ay,\theta cr}}{f_y} = \frac{258.5}{343} \times 0.373 = 0.281$$

Linear interpolation of the steel yield stress reduction factors of 0.78 and 0.47 at 600°C and 700°C respectively gives the critical temperature as 679°C.

To confirm that this new temperature is a very close and safe estimate of the critical temperature, the bending moment capacity method is reapplied to calculate the bending resistance of the composite beam.

$$N_{fi,a,pl} = 0.281 \times 2698 = 758 \text{ kN}$$

The concrete temperature is $0.4 \times 679 = 271.5$°C. From linear interpolation, $k_{c,\theta} = 0.878$. By inspection, the degree of shear connection is 100%.

The depth of concrete in compression is $\frac{758}{(3832.5 \times 0.878)} \times 80 = 18$ mm.

$$M_{\text{fi,pl,Rd}} = 758 \times \frac{\left(\frac{406}{2} + 150 - \frac{18}{2}\right)}{1000} = \mathbf{261 \ kN \ m} > 258.5 \ \text{kN m}$$

6.3.4 Appraisal of different calculation methods for composite beams

While either the critical temperature or the load-carrying capacity calculation method may be used to check fire resistance of composite beams, the following advice is offered to the designer.

For conventional composite beams with concrete/composite slab above a steel section
- If the composite beam has complete shear interaction and the calculation task is to obtain the steel critical temperature, whence to determine the necessary fire protection thickness, the critical temperature method is preferred.
- If the composite beam has a low degree of partial shear connection, the critical temperature method may still be used. However, if the fire design (e.g. cost of fire protection) is sensitive to the critical temperature, then the load-carrying capacity method should be used to reduce the conservatism of the critical temperature method by taking advantage of the increased bending resistance of the composite beam, as a result of non-uniform temperature distribution and higher retention of shear connector strength compared with the steel section in fire.
- If the design target is to check whether the beam has sufficient load-carrying capacity at given elevated temperatures, then the load-carrying capacity method should be used.

For composite beams with steel section encased in concrete
- The load-carrying capacity method should be used because temperatures of the web and the upper flange are much lower than that of the lower flange.

6.3.5 Shear resistance

EN 1994-1-2 requires vertical shear resistance of the steel section and longitudinal shear resistance to be checked. The calculation procedures are identical to those at ambient temperature, except that the design strengths of steel and shear connectors are reduced at elevated temperatures.

As long as the bending resistance of the composite beam is sufficient, the longitudinal shear resistance in fire will always be sufficient because the shear connectors are at lower temperatures and hence retain higher strengths, and also because of the lower partial safety factor for fire design compared with that for ambient temperature design.

The steel web temperature is usually taken to be the same as the lower flange temperature. Therefore, the ratio of beam shear resistance in fire to that at ambient temperature is the same as the ratio of the beam bending resistances. Therefore, if the beam shear resistance does not govern ambient temperature design, which is usually the case, it will not govern fire design.

6.4 Composite columns

When a composite structural member is heated in fire, temperatures in different parts of the member are different, due to the low thermal conductivity of concrete. The different temperatures induce different thermal strains. In order for plane sections to remain plane, thermal stresses are generated in the composite cross-section. In composite beams, because plastic analysis can be used, the effect of thermal stress has little influence. However, in composite columns, the thermal stresses can strongly influence the stiffness of cross-sections, and hence the buckling resistance of the composite column. This should be considered.

EN 1994-1-2 allows the effective (buckling) length of a composite column in braced construction to be reduced if the column is fully connected to the columns above and below, and separated from the connected columns by fire-resistant construction. This is to take advantage of the increase in the rotational stiffness of the cold columns above and below, compared with the diminishing stiffness of the fire-exposed column. As the stiffness of the fire-exposed column approaches zero, the relative rotational restraint provided by the cold columns becomes very high. Therefore, EN 1994-1-2 allows the effective length of the fire-exposed column on an intermediate floor to be reduced to $L_{e,fi} = 0.5L$ and that on the top floor to be reduced to $L_{e,fi} = 0.7L$, where L is the height of the fire-exposed column. This is illustrated in Figure 6.5.

However, this proposed reduction in column effective length is based on the most optimistic assumption that the relative rotational stiffness at the ends of the fire exposure column is nearly infinite. In realistic situations when the fire exposure column will always retain some rotational stiffness, this is unlikely to be reached. The effective length factors, 0.5 and 0.7, are nationally determined parameters (recommended values), and have been increased to 0.7 and 0.85, respectively, in the UK's national annex to EN 1994-1-2.

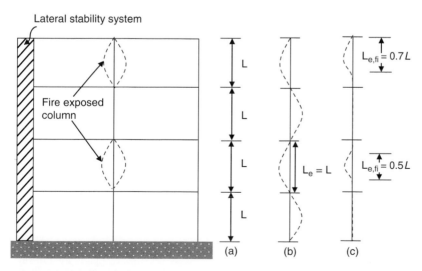

Figure 6.5 Effective length of column in braced frame in fire. (a) Section through the building. (b) Deformation mode and effective length at room temperature. (c) Deformation mode and effective length at elevated temperature

6.4.1 General calculation method and methods for different types of columns

There are significant issues with the fire resistance design methods for composite columns. First, there are different types of composite column, and past research studies that led to the development of the fire resistance design methods for the different types of composite columns were carried out by different research groups. Therefore, there is no consistent approach for fire resistance design of composite columns. For example, in EN 1994-1-2, the design method for columns with a partially encased steel section is completely different from that for concrete-filled tubular (CFT) columns.

Second, the development of fire resistance design methods for composite columns lags behind that for ambient temperature design. Therefore, whilst the ambient temperature design method assumes there are always bending moments in composite columns, and is now firmly rooted in second-order analysis, this is only being gradually phased in for fire resistance design. It is expected that, in due course, the fire resistance and ambient temperature calculation methods for composite columns will have the same basis.

This section outlines the general calculation method of EN 1994-1-2 for composite columns and introduces the methods for different types of column. Section 6.4.2 presents details of a proposed new EN 1994-1-2 design method for concrete-filled tubular columns. Its application is shown in an example in Section 6.4.3. This decision was made because CFT columns are a very efficient and elegant type of composite column and their fire resistance design method is now aligned to the ambient temperature design method. In contrast, there is no calculation method for fully-encased steel sections, and the calculation method for composite columns with partially encased steel section has little structural engineering basis when dealing with eccentric loading.

6.4.1.1 Axial buckling load

The resistance to axial compression of a composite column in fire ($N_{\text{fi,Rd}}$) is:

$$N_{\text{fi,Rd}} = \chi N_{\text{fi,pl,Rd}} \tag{6.13}$$

where χ is the reduction coefficient for buckling curve 'c' of EN 1993-1-1 as a function of the relative slenderness $\overline{\lambda}_\theta$.

The design value of plastic resistance to axial compression of the composite cross-section in the fire situation is:

$$N_{\text{fi,pl,Rd}} = \sum_j (A_a f_{\text{ay},\theta})/\gamma_{\text{M,fi,a}} + \sum_k (A_s f_{\text{sy},\theta})/\gamma_{\text{M,fi,s}} + \sum_m (A_c f_{\text{c},\theta})/\gamma_{\text{M,fi,c}} \tag{6.14}$$

where subscripts 'a', 's' and 'c' refer to steel, reinforcement and concrete, and A_i is the area of part i of the cross-section.

To calculate the elastic critical load of the column, the effective flexural stiffness of the composite section is given by:

$$(EI)_{\text{fi,eff}} = \sum_j (\varphi_{\text{a},\theta} E_{\text{a},\theta} I_a) + \sum_k (\varphi_{\text{s},\theta} E_{\text{s},\theta} I_s) + \sum_m (\varphi_{\text{c},\theta} E_{\text{c,sec},\theta} I_c) \tag{6.15}$$

where I_i is the second moment of area of part i of the cross-section; $E_{\text{c,sec},\theta}$ is the characteristic value of the secant modulus of concrete in the fire situation, given by dividing the peak stress of concrete ($f_{\text{c},\theta}$) by the strain at the peak stress ($\varepsilon_{\text{c1},\theta}$); and φ is the reduction coefficient to account for the effect of thermal stresses.

When calculating the elastic critical load of the column, the reduced buckling length, explained in the previous section, can be used.

In EN 1994-1-2, an entire Annex is devoted to implementing the above calculation procedure for composite columns with partially encased steel section, to deal with non-uniform temperature distribution in the cross-section. This is not repeated in this book.

6.4.1.2 Eccentric loading

For composite columns with partially encased steel section under eccentric loading, the axial resistance of the column with eccentric load ($N_{fi,Rd,\delta}$) is reduced from that without eccentric load using the following equation:

$$N_{fi,Rd,\delta} = N_{fi,Rd}(N_{Rd,\delta}/N_{Rd}) \tag{6.16}$$

where N_{Rd} and $N_{Rd,\delta}$ are the ambient temperature compression resistances of the column without and with eccentric load.

For composite columns made of fully-encased steel sections, EN 1994-1-2 does not give any advice for the stiffness reduction factors in Equation 6.15. Neither does it propose a method for such columns under eccentric loading. Therefore, their design has to follow the tabular method, which specifies minimum dimensions, minimum amount of reinforcement, minimum cover to the steel section and minimum axial distance of reinforcement.

6.4.2 Concrete-filled tubes

Concrete-filled tubular columns are usually constructed without external fire protection. For such columns under the standard fire condition, the design method is aligned to the ambient temperature calculation method, presented in Section 5.6. Modifications are necessary to deal with temperature effects. A method in a recent publication (Espinós et al., 2017) has been proposed for the next Eurocode 4. A shorter presentation of it is given here, and further simplification should be possible.

To simplify design, it is assumed that each part of the CFT cross-section (steel, concrete, reinforcement) has an equivalent uniform temperature.

For the concrete core:

$$\theta_{c,eq} = 81.8 - 5.05t_{fi} + 0.003t_{fi}^2 - 15.07\frac{A_m}{V} +$$
$$0.331\left(\frac{A_m}{V}\right)^2 - 0.875t_{fi}\frac{A_m}{V} + 7.43t_{fi}^{0.842}\left(\frac{A_m}{V}\right)^{0.714} \tag{6.17}$$

for the steel tube:

$$\theta_{a,eq} = -825 - 5.58t_{fi} + 0.007t_{fi}^2 - 0.009t_{fi}\frac{A_m}{V} + 645t_{fi}^{0.269}\left(\frac{A_m}{V}\right)^{0.017} \tag{6.18}$$

and for the reinforcement:

$$\theta_{s,eq} = \beta_3(t_{fi}/u_s^2)^3 + \beta_2(t_{fi}/u_s^2)^2 + \beta_1(t_{fi}/u_s^2) + \beta_0 \tag{6.19}$$

where t_{fi} is the design fire resistance rating (in minutes); A_m/V is the solid section factor, calculated as the circumference of the tube divided by the enclosed area (in m^{-1}); and u_s is the axial distance of the reinforcement bar from the concrete surface.

Table 6.2 Coefficients β_0 to β_3 for different types of steel tubes, depending on u_s

	u_s (mm)	β_3	β_2	β_1	β_0
CHS	20	7236	−10 458	5498	19.38
	30	58 714	−41 328	10 910	11.18
	35	0	−12 732	6518	91.21
	50	0	−55 639	13 768	−19.90
	55	0	−43 201	10 790	24.23
	70	0	0	8858	96.68
SHS	20	8151	−11 323	5595	93.39
	30	85 460	−54 898	12 825	−22.08
	35	0	−18 802	8223	116.34
	50	0	−67 134	15 912	16.12
	55	0	−78 597	14 878	−43.03
	70	0	0	11 922	23.26
RHS	20	7863	−10 978	5456	108.38
	30	82 790	−53 604	12 626	−8.45
	35	0	−20 109	8575	53.01
	50	0	−79 340	17 108	−54.08
EHS	30	79 543	−51 871	12 481	−45.48
	40	304 952	−117 159	18 180	−111.73
	55	0	−100 810	18 531	−35.74
	65	0	−157 800	23 377	−86.43

Values for the constants β_0 to β_3 are given in Table 6.2 for different types of tube shapes. CHS, SHS, RHS and EHS refer to circular, square, rectangular and elliptical hollow sections respectively.

Using the effective constant temperatures, the design strengths of steel, concrete and reinforcement are obtained from the respective reduction factor–temperature relationships. Equations 5.33 and 5.35 are then used to obtain the design and characteristic squash loads, respectively, replacing 0.85 by 1.0 for CFTs. In fact, since the material partial safety factors in fire design are 1.0, these equations give the same value.

To account for the effects of thermal stresses, it is necessary to modify the cross-section effective flexural stiffness, used for calculating the elastic critical load. It is obtained using the same equation as for composite columns with partially encased steel section (Equation 6.15). However, the reduction coefficients are obtained from Table 6.3 for steel and Table 6.4 for reinforcement respectively. In Table 6.3, L_{fi} is the buckling length in the fire situation and B is the shorter external dimension of the RHS/EHS tube.

The stiffness reduction coefficient for concrete is $\phi_{c,\theta} = 1.2$ when using the secant modulus, as in Equation 6.15.

For calculating the effective flexural stiffness of the cross-section as part of second-order analysis, further correction factors are applied, and the calculation

Table 6.3 Stiffness reduction coefficient for steel

CHS (circular hollow section)	$\varphi_{a,\theta} = 0.75 - 0.023 \left(\dfrac{A_m}{V} \right)$
SHS (square hollow section)	$\varphi_{a,\theta} = 0.15 - 0.001 \left(\dfrac{A_m}{V} \right)$
RHS and EHS (rectangular/elliptic hollow section)	$\varphi_{a,\theta} = 0.012 \left(\dfrac{L_{fi}}{B} \right)$

Table 6.4 Stiffness reduction coefficient for reinforcement

CHS & SHS	$\varphi_{s,\theta} = 0.8 - 0.002t_{fi}$
RHS	$\varphi_{s,\theta} = 0.7$
EHS	$\varphi_{s,\theta} = 0.95$

equation is:

$$(EI)_{fi,eff,II} = K_{\theta}K_0[\varphi_{a,\theta}E_{a,\theta}I_a + \varphi_{s,\theta}E_{s,\theta}I_s + K_{\theta,II}\varphi_{c,\theta}E_{c,sec,\theta}I_c] \tag{6.20}$$

where $K_{\theta} = 0.67$, $K_{\theta,II} = 0.5$ and $K_0 = 0.9$.

From here on, the calculations follow the exact procedure as at ambient temperature, as explained in Section 5.6.

The limits of application of this method are:

For CHS columns :
$$5 \le A_m/V \le 30$$
$$10 \le D/t \le 60$$
$$5 \le L_{fi}/D \le 30$$

For SHS columns
$$5 \le A_m/V \le 35$$
$$10 \le B/t \le 40$$
$$5 \le L_{fi}/B \le 30$$

For EHS columns
$$10 \le A_m/V \le 30$$
$$5 \le B/t \le 20$$
$$5 \le L_{fi}/B \le 30$$
$$H/B = 2$$

For RHS columns
$$10 \le A_m/V \le 45$$
$$5 \le B/t \le 20$$
$$5 \le L_{fi}/B \le 30$$
$$1.5 \le H/B \le 3$$

- The percentage of reinforcement should be lower than 5%.
- For concentrically loaded, initially unreinforced CHS and SHS columns with relative slenderness $\bar{\lambda}$ greater than 0.5, a minimum amount of 2.5% of reinforcement is required.

- The relative load eccentricity e/D, e/B or e/H should be lower than 1.
- The method can be used for fire exposure times ranging between 30 to 240 minutes.

6.4.3 Worked example for concrete-filled tubes with eccentric loading

Figure 6.6 shows a concrete-filled tubular section with internal reinforcement. The characteristic yield stress of steel, of reinforcement and cylinder strength of concrete are 355 N/mm², 500 N/mm² and 30 N/mm² respectively. The fire resistance rating is 30 minutes. Calculate the squash load $N_{fi,pl,Rd}$, the effective flexural stiffness $EI_{fi,eff}$, and the effective flexural stiffness $EI_{fi,eff,II}$ for second-order analysis.

The solid section factor is $\frac{A_m}{V} = \frac{4}{0.1937} = 20.65$ m^{-1}
The equivalent constant temperature of concrete (Equation 6.17) is: $\theta_{c,eq} = 352°C$.
The concrete strength reduction factors at 300°C and 400°C are 0.85 and 0.75 respectively (BSI, 2014c). Interpolating,

$$f_{c,352} = 0.798 \times 30 = 23.94 \text{ N/mm}^2$$

The concrete strains at the peak stress at 300°C and 400°C are 0.007 and 0.01 respectively (BSI, 2014c), therefore: $\varepsilon_{c1,352} = 8.56 \times 10^{-3}$.
The secant modulus for concrete is:

$$E_{c,sec,352} = \frac{23.94}{8.56} \times 1000 = 2797 \text{ N/mm}^2$$

The area of longitudinal reinforcement is 1206 mm², whence

$$A_c = \pi \times \frac{177.7^2}{4} - 1206 = 23\,570 \text{ mm}^2$$

Using I_s as calculated below,

$$I_c = \pi \times \left(\frac{177.7^4}{64} \right) / 10\,000 - 188 = 4707 \text{ cm}^4$$

The equivalent constant temperature of steel, Equation 6.18, is $\theta_{a,eq} = 704°C$.
The steel yield stress and Young's modulus reduction factors are 0.23 and 0.11, and 0.13 and 0.09, at 700°C and 800°C respectively (BSI, 2014c). Interpolating for 704°C,

$$f_{a,704} = 0.225 \times 355 = 79.8 \text{ N/mm}^2$$

$$E_{a,704} = 0.128 \times 210\,000 = 26\,900 \text{ N/mm}^2$$

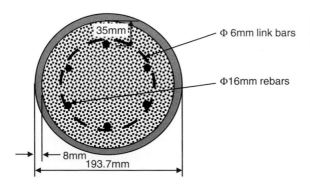

Figure 6.6 Dimensions of concrete-filled tubular column cross-section for the worked example

Φ 6mm link bars

Φ16mm rebars

35mm

8mm
193.7mm

$$A_a = \frac{\pi(193.7^2 - 177.7^2)}{4} = 4667 \ mm^2$$

$$I_a = \left[\frac{\pi(193.7^4 - 177.7^4)}{64} \right] / 10\,000 = 2015 \ cm^4$$

The equivalent constant temperature of reinforcement with $u_s = 35\,mm$, $\beta_3 = 0$, $\beta_2 = -12\,732$, $\beta_1 = 6518$, and $\beta_0 = 91.21$ is given by Equation 6.19 as: $\theta_{s,eq} = 248°C$.

The reinforcing steel yield stress and Young's modulus reduction factors are 1.0 and 1.0, and 0.87 and 0.72, at 200°C and 300°C respectively (BSI, 2014c). Therefore:

$$f_{s,248} = 1.0 \times 500 = 500 \ N/mm^2$$

$$E_{s,248} = 0.805 \times 210\,000 = 169\,100 \ N/mm^2$$

$$A_s = 6 \times \pi \times \frac{16^2}{4} = 1206 \ mm^2$$

The distances of the bars from axes through the centre are:

$$177.7/2 - 35 = 53.85\,mm, \qquad 57.85/2 = 26.92\,mm$$

Their second moment of area about the $y-y$ and $x-x$ axes is

$$I_s = 2 \times \pi \times \frac{16^2}{4} \times \frac{(53.85^2 + 2 \times 26.92^2)}{10\,000} = 188 \ cm^4$$

Results   The squash load of the composite cross-section, Equation 6.14, is

$$N_{fi,pl,Rd} = (23\,570 \times 23.94 + 4667 \times 79.8 + 1206 \times 500)/1000 = \mathbf{1540 \ kN}$$

The effective flexural stiffness of cross-section, Equation 6.15, with

$$\varphi_{a,\theta} = 0.75 - 0.023 \times 20.65 = 0.275 \quad \text{(Table 6.3)}$$

$$\varphi_{s,\theta} = 0.8 - 0.002 \times 30 = 0.74 \quad \text{(Table 6.4)}$$

$$(EI)_{fi,eff} = (0.275 \times 26\,900 \times 2015 + 1.2 \times 2797 \times 4707$$
$$+ 0.74 \times 169\,100 \times 188)/100\,000$$
$$= \mathbf{542 \ kN \ m^2}$$

The effective flexural stiffness of cross-section for second-order analysis, from Equation 6.20, is

$$(EI)_{fi,eff,II} = 0.67 \times 0.9 \times (0.275 \times 26\,900 \times 2015 + 0.5 \times 1.2 \times 2797 \times 4707$$
$$+ 0.74 \times 169\,100 \times 188)/100\,000$$
$$= \mathbf{279.5 \ kN \ m^2}$$

Appendix A

Partial-interaction theory

A.1 Theory for simply-supported beam

This subject is introduced in Section 2.6, which gives the assumptions and notation used in the theory that follows. The tension-positive sign convention is used. On first reading, it may be found helpful to rewrite the algebraic work in a form applicable to a beam with the very simple cross-section shown in Figure 2.2. This can be done by making these substitutions.

- Replace A_c and A_a by bh, h_{cf} and h_c and d_c by h.
- Replace I_a and I_c by $bh^3/12$.
- Ignore creep of concrete and assume $n = 1$, so that $E_{c,eff}$, E_c and E_a are replaced by E.

One end of the beam to be analysed is shown in Figure 2.17a. Figure A.1 shows in elevation part of it, of length dx, and distant x from the mid-span cross-section. Transverse profiled sheeting (not shown) is assumed to be present, and the two components are shown separated. The slip at top-flange level of the steel beam is s at cross-section x, and increases over the length of the element to $s + (ds/dx)dx$, which is written as s^+. This notation is used in Figure A.1 for increments in the other functions of x, which are M_c, M_a, N_c, N_a, V_c and V_a. These are respectively the sagging bending moments, tensile axial forces and anticlockwise vertical shears acting on the two components of the beam, the suffixes c and a indicating concrete and steel. They are shown in the directions taken as positive. From longitudinal equilibrium, $N_c = -N_a$. The interface vertical force r per unit length is unknown.

Let the interface longitudinal shear be v_L per unit length, so the force on each component is $v_L dx$. It must be in the direction shown, to be consistent with the sign assumed for the slip, s. The stiffness of the shear connection, shear force per unit length per unit slip, is k_c, so the load-slip relationship is

$$v_L = k_c s \tag{A.1}$$

The method is to deduce equations from equilibrium, elasticity and compatibility, and by elimination obtain a differential equation relating s to x; then solve it and insert boundary conditions. These are as follows.

(1) Zero slip at mid-span, from symmetry, so $s = 0$ when $x = 0$. (A.2)

(2) At the supports, M and N are zero in both materials, so the difference between the longitudinal strains at the interface is the free shrinkage strain in the concrete, ε_{cs}.

Composite Structures of Steel and Concrete: Beams, Slabs, Columns and Frames for Buildings,
Fourth Edition. Roger P. Johnson.
© 2019 John Wiley & Sons Ltd. Published 2019 by John Wiley & Sons Ltd.

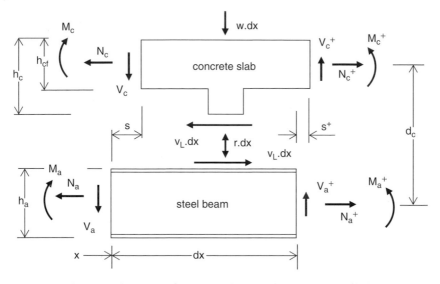

Figure A.1 Elevation of element of composite beam with transverse profiled sheeting

As the tension-positive convention is used, shrinkage strain is a negative number. With this sign convention,

$$ds/dx = \varepsilon_{cs} \quad \text{when} \quad x = \pm L/2 \tag{A.3}$$

Equilibrium   Resolve longitudinally for one component:

$$dN_a/dx = -dN_c/dx = -v_L \tag{A.4}$$

For moment equilibrium,

$$dM_c/dx + V_c = v_L(h_c - h_{cf}/2), \quad dM_a/dx + V_a = v_L h_a/2 \tag{A.5}$$

The vertical shear at cross-section x is wx, so

$$V_c + V_a = wx \tag{A.6}$$

From Figure A.1,

$$d_c = h_a/2 + h_c - h_{cf}/2 \tag{A.7}$$

so from Equations A.5 to A.7,

$$dM_c/dx + dM_a/dx + wx = v_L d_c \tag{A.8}$$

Elasticity   In beams with adequate shear connection, the effects of uplift are negligible in the elastic range. The two components must then have the same curvature, κ, and simple beam theory gives the moment-curvature relations. Using a modular ratio n that allows for creep (Equation 2.19):

$$\kappa = M_a/E_a I_a = nM_c/E_a I_c \tag{A.9}$$

The longitudinal strains in concrete and in steel at the top surface of the steel flange are:

$$\varepsilon_c = \kappa(h_c - h_{cf}/2) + nN_c/(E_a A_c) + \varepsilon_{cs} \tag{A.10}$$

$$\varepsilon_a = -\kappa h_a/2 + N_a/(E_a A_a) \tag{A.11}$$

where ε_{cs} is the free shrinkage strain of the concrete – a negative number.

Compatibility   The difference between ε_c and ε_a is the slip strain, so from Equations A.7, A.10 and A.11,

$$\frac{ds}{dx} = \kappa d_c - \frac{N_a}{E_a A_0} + \varepsilon_{cs} \tag{A.12}$$

where

$$1/A_0 = n/A_c + 1/A_a \tag{A.13}$$

The differential equation for s is found by eliminating M_c and M_a from Equations A.1, A.8 and A.9:

$$E_a I_0 (d\kappa/dx) + wx = v_L d_c = k_c d_c s \tag{A.14}$$

where

$$I_0 = I_c/n + I_a \tag{A.15}$$

Differentiating Equation A.12 and eliminating κ from Equation A.14, N_a and N_c from Equations A.4, and v_L from Equation A.1 leads to:

$$\frac{d^2 s}{dx^2} = \frac{k_c s}{E_a I_0 A'} - \frac{w d_c x}{E_a I_0}$$

where

$$1/A' = d_c^2 + I_0/A_0 \tag{A.16}$$

To present the result in a standard form, these symbols are defined:

$$\alpha^2 = k_c/(E_a I_0 A') \quad \text{and} \quad \beta = A' d_c/k_c \tag{A.17}$$

which gives

$$\frac{d^2 s}{dx^2} - \alpha^2 s = -\alpha^2 \beta wx \tag{A.18}$$

Solving for s:

$$s = K_1 \sinh(\alpha x) + K_2 \cosh(\alpha x) + \beta wx \tag{A.19}$$

Inserting the boundary conditions, Equations A.2 and A.3, gives the expression for longitudinal slip:

$$s = \beta wx - \left(\frac{\beta w - \varepsilon_{cs}}{\alpha}\right) \operatorname{sech}\left(\frac{\alpha L}{2}\right) \sinh \alpha x \tag{A.20}$$

Other results Differentiation of Equation A.20 gives the slip strain at mid-span:

$$\left(\frac{ds}{dx}\right)_{x=0} = \beta w - (\beta w - \varepsilon_{cs})\text{sech}(\alpha L/2) \tag{A.21}$$

Using Equations A.1 and A.4, integration of Equation A.20 gives the longitudinal force N_a, with boundary condition $N_a = 0$ at $x = \pm L/2$:

$$N_a' = (k_c \beta w/8)[L^2 - 4x^2 + (8/\alpha^2)\{\text{sech}(\alpha L/2)\cosh(\alpha x) - 1\}] \tag{A.22}$$

Equating the external and internal bending moments:

$$M_{ext} = (w/8)(L^2 - 4x^2) = M_c + M_a + N_a d_c \tag{A.23}$$

As N_a is known, equations A.22 and A.23 can be solved for the internal bending moments:

$$M_c + M_a = (w/8)(L^2 - 4x^2)(1 - k_c d_c \beta) - \frac{k_c d_c \beta w}{\alpha^2}\{\text{sech}(\alpha L/2)\cosh(\alpha x) - 1\} \tag{A.24}$$

Equation A.9 gives the ratio M_c/M_a, which is independent of x, so M_a can be found, and hence the deflections by integration of the curvature, which is $\kappa = M_a/E_a I_a$.

Figure A.1 shows that the steel beam carries proportion r/w of w, the total load on the composite member. This ratio can be found as follows. For equilibrium of length dx of the steel beam,

$$r = \frac{dV_a}{dx} \quad \text{and} \quad \frac{dM_a}{dx} + V_a = \frac{v_L h_a}{2} \tag{A.25}$$

The use of equations A.1, A.7, A.20 and A.24, ignoring shrinkage, then leads to

$$r/w = I_a/I_o + k_c \beta(h_a/2 - I_a d_c/I_o)\{1 - \text{sech}(\alpha L/2)\cosh(\alpha x)\} \tag{A.26}$$

so the ratio at the ends of this simply-supported beam is I_a/I_o.

Comment This theory is of limited scope, and clearly unsuitable as a design method. The different algebraic route of Ranzi et al. (2003) is better in some situations. The analyses lead to some useful conclusions, which are given in Section A.2.

A.2 Example: partial interaction

Section 3.11 gives design calculations for a simply-supported beam of span 8.6 m, with the cross-section shown in Figure 3.31. It was used in Section 2.7 to illustrate the effect of degree of shear connection on its behaviour under the loading used for serviceability checks. The calculations that gave the results in Table 2.2 are more fully presented here. The shrinkage strain ε_{cs} was taken as zero.

If unpropped construction is assumed, the unfactored loading acting on the composite member is 5.2 kN/m of permanent load and 24.8 kN/m of variable load, so the loading is $w = 30$ kN/m.

There is, in addition, 14.4 kN/m acting on the steel section, which causes bending stresses of ± 109 MPa at mid-span.

For two studs per trough, the design shear resistance is 42 kN per stud. The compressive force in the slab for full interaction at ULS is 2555 kN. The stud layout had to suit the troughs in the sheeting, which led to pairs at 0.3 m spacing, or 28 in a half span. The degree of shear connection is $28 \times 42/2555 = 0.46$.

The design uses 19-mm stud connectors of height 125 mm. Push-out tests give their ultimate shear resistance as about 100 kN, with slips at half this load usually between 0.2 and 0.4 mm. A connector modulus $k = 150$ kN/mm is assumed here, corresponding to slip of 0.33 mm at a shear force of 50 kN per connector. The mean stud spacing is 0.154 m, so the stiffness of the connection is

$$k_c = k/p = 150/154 = 0.974 \text{ kN/mm}^2$$

For ultimate-strength design, the thickness of the slab was taken as 80 mm, which is the height above the top rib of the sheeting (Figure 3.8). Its elastic stiffness corresponds to a greater thickness, so the effective thickness is here taken as 95 mm. The effective flange width is 2.25 m (Figure 3.31).

Elastic analysis for loading on the composite member The modular ratio used for all loading is 20.2 (Section 3.2), so with $E_a = 210$ GPa, the effective modulus for concrete is $210/20.2 = 10.4$ GPa.

Further values for substitution in the equations in Section A.1 are:

$A_a = 0.0076 \text{ m}^2$;　$A_c = 0.214 \text{ m}^2$;　$I_a = 2.15 \times 10^{-4} \text{ m}^4$;　$I_c = 1.608 \times 10^{-4} \text{ m}^4$
From Equation A.13,　$1/A_0 = 226 \text{ m}^{-2}$
From Equation A.15,　$I_0 = 2.33 \times 10^{-4} \text{ m}^4$
From Equation A.16,　$1/A' = 0.1437 \text{ m}^2$
From Equation A.17,　$\alpha^2 = 2.99 \text{ m}^{-2}$,　and　$\beta = 2.18 \times 10^{-6} \text{ m/kN}$
From Equation A.20, the slip (m units) is

$$s = 6.54 \times 10^{-5}x - 4.46 \times 10^{-8} \sinh(1.729x) \tag{A.27}$$

which gives the slip at the ends of the span as 0.244 mm. The 'x' term in this expression gives 0.281 mm, so the slip is closely proportional to x. Its distribution, and that of the shear flow (from Equation A.1) are therefore similar to that of the vertical shear force. This is true generally where the maximum slip and maximum shear force occur at the same cross-section.

For this simply-supported beam, the maximum slip strain, found by differentiating Equation A.25, is 66×10^{-6} at mid-span. This strain distribution is shown in Figure 2.17b.

From Equation A.22, the longitudinal force N_a is, in kN m:

$$N_a = 568 - 31.9x^2 + 0.0251 \cosh(1.729x)$$

with a distribution similar to that of bending moment, given next.

From Equation A.24, $M_a + M_c = 103.9 - 5.27x^2 - 0.0077 \cosh(1.729x)$

These results at mid-span, for degree of shear connection $\eta = 0.46$, are compared in Section 2.6 and Table 2.2 with results for $\eta = 0$ and $\eta = 1.0$.

Equation A.26 above gives the ratio r/w. In this example, it is 0.923 at the ends of the beam, reducing gradually to 0.75 at mid-span. This agrees well with the assumption, made in design, that the whole of the vertical shear is resisted by the steel member.

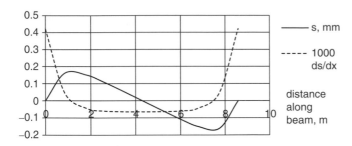

Figure A.2 Distributions of slip and slip strain for a fixed-ended beam

Fixed-ended and continuous beams It was found above that the distributions of slip and shear flow were similar. An obvious exception is a fixed-ended beam, where at the supports the shear force is a maximum and the slip is zero. Slip is also low at internal supports of continuous beams. A tedious calculation using the method of Ranzi et al. (2003), with data as above except that the ends are fixed, leads to the distributions of slip and slip strain shown in Figure A.2.

The maximum slip, near the points of contraflexure, is 0.17 mm, about two-thirds of the 0.244 mm found above, and its sudden return to zero at the ends of the fixed-ended beam requires high values of slip strain. The maximum value is 420×10^{-6}, which is about 25% of the yield strain of the steel beam, and far higher than in the simply-supported beam. In design it is usually assumed that the most highly loaded shear connectors are those near the supports. Partial-interaction elastic analysis of continuous beams would find higher longitudinal shears at the points of contraflexure.

In practice, bending moments in continuous beams are often found by elastic theory, ignoring slip and cracking of concrete. For moderate or high degrees of shear connection, it is found that slip causes negligible change in the bending moments.

For example, the distribution of total bending moment in the fixed-ended beam, shown in Figure A.3, has the identical points of contraflexure as in a beam without slip. However, at these points, the bending moment from axial force, $N_c d_c$, is not quite zero. It is 4.4 kN m, equal and opposite to the sum of the moments in the steel and concrete components. Similar conclusions follow from analysis of a propped cantilever.

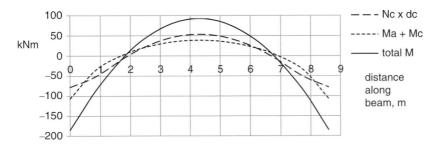

Figure A.3 Components of bending moment in a fixed-ended beam

References

Allison, R.W., Johnson R.P., and May, I.M. (1982). Tension field action in composite plate girders. *Proceedings of the Institution of Civil Engineers, Part 2*; 73: 255–276.

Anderson, D., Aribert J-M., Bode H., and Kronenburger H. (2000). Design rotation capacity of composite joints. *The Structural Engineer* 78 (6): 25–29.

ArcelorMittal (2016). *Slim Floor – An Innovative Concept for Floors*. Luxembourg: ArcelorMittal.

Basu, A.K. and Sommerville, W. (1969). Derivation of formulae for the design of rectangular composite columns. *Proceedings of the Institution of Civil Engineers, Supp.*: 233–280.

Beeby, A.W. and Narayanan, R.S. (2005). *Designers' Guide to Eurocode 2: Design of Concrete Structures*. London: Thomas Telford.

Bode, H., et al. (1996). Partial connection design of composite slabs. *Structural Engineering International* 6 (2): 53–56.

Brown, N. and Anderson, D. (2001). Structural properties of composite major-axis end plate connections. *Journal of Constructional Steel Research* 57: 327–349.

BSI (1990a). *BS 5950, Part 3, Section 3.1: Code of Practice for Design of Simple and Continuous Composite Beams*. London: British Standards Institution.

BSI (1990b). *BS EN 10025. Hot Rolled Products of Non-alloy Structural Steels and their Technical Delivery Conditions*. London: British Standards Institution.

BSI (2002a). *BS EN 1991: Eurocode 1: Actions on Structures – Part 1-1: General Actions*. London: British Standards Institution.

BSI (2002b). *BS EN 1991: Eurocode 1: Actions on Structures – Part 1-6: Actions during Execution*. London: British Standards Institution.

BSI (2002c). *BS EN 1990: Eurocode 0: Basis of Structural Design*. London: British Standards Institution.

BSI (2002d). *BS EN 1991: Eurocode 1: Actions on Structures – Part 1-2: Actions on Structures Exposed to Fire*. London: British Standards Institution.

BSI (2004). *BS EN 1994: Eurocode 4: Design of Composite Steel and Concrete Structures – Part 1-1: General Rules and Rules for Buildings*. London: British Standards Institution.

BSI (2005a). *BS EN 1993: Eurocode 3: Design of Steel Structures – Part 1-8: Design of Joints*. London: British Standards Institution.

BSI (2005b). *BS EN 10080: Steel for the Reinforcement of Concrete*. London: British Standards Institution.

Composite Structures of Steel and Concrete: Beams, Slabs, Columns and Frames for Buildings, Fourth Edition. Roger P. Johnson.
© 2019 John Wiley & Sons Ltd. Published 2019 by John Wiley & Sons Ltd.

BSI (2005c). *BS EN 1994: Eurocode 4: Design of Composite Steel and Concrete Structures – Part 2: General Rules and Rules for Bridges*. London: British Standards Institution.

BSI (2006a). *BS EN 1993: Eurocode 3: Design of Steel Structures – Part 1-3: Cold-formed Thin Gauge Members and Sheeting*. London: British Standards Institution.

BSI (2006b). *BS EN 1993: Eurocode 3: Design of Steel Structures – Part 1-5: Plated Structural Elements*. London: British Standards Institution.

BSI (2008). *BS 6472: Guide to Evaluation of Human Exposure to Vibration in Buildings – Part 1: Vibration Sources other than Blasting*. London: British Standards Institution.

BSI (2011). *BS EN 1090: Execution of steel and aluminium structures – Part 2:2008 + A1:2011: Technical requirements for steel structures*. London: British Standards Institution.

BSI (2014a). *BS EN 1992: Eurocode 2: Design of Concrete Structures – Part 1-1:2004 + A1:2014: General Rules and Rules for Buildings*. London: British Standards Institution.

BSI (2014b). *BS EN 1993: Eurocode 3: Design of Steel Structures – Part 1-1:2005 + A1:2014: General Rules and Rules for Buildings*. London: British Standards Institution.

BSI (2014c). *BS EN 1994: Eurocode 4: Design of Composite Steel and Concrete Structures – Part 1-1-2:2005 + A1:2014: General Rules – Structural Fire Design*. London: British Standards Institution.

BSI (2016). *BS EN 206:2013 + A1:2016: Concrete. Specification, Performance, and Conformity*. London: British Standards Institution.

Building Regulations 2010 (2013). *Approved Document B: Fire Safety, 2006 Edition Incorporating 2010 and 2013 Amendments*. London: HM Government.

Cassell, A., Chapman, J., and Sparkes, S. (1966). Observed behaviour of a building of composite steel and concrete construction. *Proceedings of the Institution of Civil Engineers* 33 (April): 637–658.

CEN (1992). *ENV 1994-1-1. Design of Composite Steel and Concrete Structures – Part 1-1: General Rules and Rules for Buildings*. Brussels: European Committee for Standardization.

Chung, K.F. and Lawson, R.M. (2001). Simplified design of composite beams with large web openings to Eurocode 4. *Journal of Constructional Steel Research* 57 (Feb.): 135–164.

Coates, R.C., Coutie, M.G., and Kong, F.K. (1988). *Structural Analysis*, 3rd ed. London: Chapman and Hall.

Costa-Neves, L., Vellasco, P.C.G.S., and Valente, I.B. (2016). *Perforated Shear Connectors in Composite Construction. Report to Technical Committee 11*. Brussels: European Convention for Constructional Steelwork.

Couchman, G. (2014). *Design of Composite Beams using Precast Concrete Slabs in Accordance with Eurocode 4*. Publication P401. Ascot: Steel Construction Institute.

Couchman, G. (2015). *Minimum Degree of Shear Connection Rules for UK Construction to Eurocode 4*. Publication P405. Ascot: Steel Construction Institute.

Couchman, G. and Way, A. (1998). *Joints in Steel Construction – Composite Connections*. Publication P213. Ascot: Steel Construction Institute.

ECCS (2016). Frame design: State-of-problem report prepared for the TC11 meeting, Delft, October. Brussels: European Convention for Constructional Steelwork.

Ellis, B.R. (2001). Serviceability evaluation of floor vibration induced by walking loads. *The Structural Engineer* 79 (6 November): 30–36.

Espinós, A., Albero, V., Romero, M.L., et al. (2017). Eurocode 4 based method for the fire design of concrete-filled steel tubular columns. *Proceedings of the 16th International Symposium on Tubular Structures*, Melbourne, Australia.

European Convention for Constructional Steelwork (1981). *Composite Structures*. London: The Construction Press.

Faber, O. (1956). More rational design of cased stanchions. *The Structural Engineer* 34: 88–109.

Feldmann, M., Kopp, M., and Pak, D. (2016). Composite dowels as shear connectors for composite beams – background to the German technical approval. *Steel Construction* 9 (2): 80–88.

Fisher, J. (1970) Design of composite beams with formed metal deck. *Engineering Journal of the American Institute of Steel Construction* 7 (July): 88–96.

Gardner, L. and Nethercot, D. (2004). *Designers' Guide to EN 1993-1-1 – Eurocode 3: Design of Steel Structures – Part 1.1: General Rules and Rules for Buildings*. London: Thomas Telford.

Goble, G. (1968). Shear strength of thin-flange composite specimens. *Engineering Journal of the American Institute of Steel Construction* 5: 62–65.

Grant, J.A., Fisher, J.W., and Slutter, R.G. (1977). Composite beams with formed metal deck. *Engineering Journal of the American Institute of Steel Construction* 1: 27–42.

Hendy, C.R. and Johnson, R.P. (2006). *Designers' Guide to EN 1994-2 – Eurocode 4: Design of Composite Steel and Concrete Structures – Part 2: General Rules and Rules for Bridges*. London: ICE Publishing.

Hicks, S. and Lawson, R.M. (2003). *Design of Composite Beams using Pre-cast Concrete Slabs*. Publication P287. Ascot: Steel Construction Institute.

Jaspart, J-P. and Weynand, K. (2016). *Design of Joints in Steel and Composite Construction*. Oxford: Wiley.

Johnson, R.P. (2006). Models for the longitudinal shear resistance of composite slabs, and the use of non-standard test data. In: *Composite Construction in Steel and Concrete V* (ed. R. Leon and J. Lange), 157–165. New York: American Society of Civil Engineers.

Johnson, R.P. (2008). Calibration of resistance of stud shear connectors in troughs of profiled sheeting. *Proceedings of the Institution of Civil Engineers, Structures and Buildings*, 161, SB3, 117–126.

Johnson, R.P. (2012). *Designers' Guide to Eurocode 4: Design of Composite Steel and Concrete Structures*, 2nd ed. London: ICE Publishing.

Johnson, R.P. and Allison, R. (1983). Cracking in concrete tension flanges of composite T-beams. *The Structural Engineer* 61B (March): 9–16.

Johnson, R.P. and Chen, S. (1991). Local buckling and moment redistribution in Class 2 composite beams. *Structural Engineering International* 1 (Nov.): 27–34.

Johnson, R.P., Finlinson, J. and Heyman, J. (1965). A plastic composite design. *Proceedings of the Institution of Civil Engineers* 32 (Oct.): 198–209.

Johnson, R.P., Greenwood, R. and van Dalen, K. (1969). Stud shear connectors in hogging moment regions of composite beams. *The Structural Engineer* 47 (Sept.): 345–350.

Johnson, R.P. and Molenstra, N. (1991). Partial shear connection in composite beams for buildings. *Proceedings of the Institution of Civil Engineers, Part 2*, 91 (Dec.): 679–704.

Kerensky, O. and Dallard, N. (1968). The four-level interchange between M4 and M5 motorways at Almondsbury. *Proceedings of the Institution of Civil Engineers* 40 (July): 295–322.

Kuhlmann, U. and Kürschner, K. (2004). Structural behaviour of horizontally lying shear studs. In: *Composite Construction in Steel and Concrete V* (ed. R. Leon and J. Lange), 534–543. New York: American Society of Civil Engineers.

Kuhlmann, U. and Raichle, J. (2008). Headed studs close to the concrete surface – fatigue behaviour and application. In: *Composite Construction in Steel and Concrete VI* (ed. R. Leon), 26–38. New York: American Society of Civil Engineers.

Lawson, R.M., Bode, H., Brekelmans, J., et al. (1999). 'Slimflor' and 'Slimdek' construction: European developments. *The Structural Engineer* 77 (20 April): 22–30.

Lawson, R.M. and Hicks, S. (2011). *Design of Composite Beams with Large Web Openings – in accordance with Eurocodes and the UK National Annexes.* Publication P355. Ascot: Steel Construction Institute.

Lawson, R.M., Lam, D., Aggelopoulos, E.S., and Nellinger, S. (2017). Serviceability performance of steel-concrete composite beams. *Proceedings of the Institution of Civil Engineers: Structures and Buildings* 170 (2): 98–114.

Lawson, R.M., Lim, J., Hicks, S. and Simms, W. (2006). Design of composite asymmetric beams and beams with large web openings. *Journal of Constructional Steel Research* 62 (6): 614–629.

Lawson, R.M., Mullett, D. and Rackham, J. (1997). *Design of Asymmetric Slimflor Beams using Deep Composite Decking.* Publication P175. Ascot: Steel Construction Institute.

Lawson, R.M. and Rackham, J.W. (1989). *Design of Haunched Composite Beams in Buildings.* Publication P060. Ascot: Steel Construction Institute.

Leskela, M.V. (2017). *Shear Connections in Composite Members of Steel and Concrete.* Report 138. Brussels: European Convention for Constructional Steelwork.

Liew, J. and Xiang, M. (2015). *Design Guide for Concrete-filled Tubular Members with High Strength Materials to Eurocode 4.* Singapore: Research Publishing Services.

Lorenc, W., Kosuch, M., and Rowinski, S. (2014). The behaviour of puzzle-shaped composite dowels. *Journal of Constructional Steel Research* 101: 482–499, 500–518.

Malik, A.S. (2009). *Joints in Steel Construction: Simple Connections.* Publication P212. Ascot: Steel Construction Institute.

Mottram, J.T. and Johnson, R.P. (1990). Push tests on studs welded through profiled steel sheeting. *The Structural Engineer* 68 (May): 187–193.

Mullett, D.L. and Lawson, R.M. (1999). *Design of Slimflor Fabricated Beams using Deep Composite Decking.* Publication P248. Ascot: Steel Construction Institute.

Newman, G.M., Robinson, J.T. and Bailey, C.G. (2006). *Fire Safe Design: A New Approach to Multi-Storey Steel-Framed Buildings*, 2nd ed. Publication P288. Ascot: Steel Construction Institute.

Newmark, N.M., Siess, C.P. and Viest, I.M. (1951). Tests and analysis of composite beams with incomplete shear connection. *Proceedings of the Society for Experimental Stress Analysis* 9 (1): 75–92.

Odenbreit, C. and Nellinger, S. (2017). Mechanical model to predict the resistance of the shear connection in composite beams with deep decking. *Steel Construction* 10 (3): 248–253.

Oehlers, D. and Bradford, M. (1995). *Composite Steel and Concrete Structural Members – Fundamental Behaviour.* New York: Elsevier Science Inc.

Patrick, M. and Bridge, R.Q. (1993). Design of composite slabs for vertical shear. In: *Composite Construction in Steel and Concrete II* (ed. W.S. Easterling and W.M.K. Roddis), 304–322. New York: American Society of Civil Engineers.

Patrick, M. and Bridge, R.Q. (1994). Partial shear connection design of composite slabs. *Engineering Structures* 16 (5): 348–362. London: Butterworth.

Porsch, M. and Hanswille, G. (2006). Load introduction in composite columns with concrete filled hollow sections. In: *Composite Construction in Steel and Concrete V* (ed. R.T. Leon and J. Lange). New York: American Society of Civil Engineers.

Rackham, J.W., Hicks, S.J., and Newman, G.M. (2006). *Design of Asymmetric Slimflor Beams with Precast Concrete Slabs*. Publication P342. Ascot: Steel Construction Institute.

Ranzi, G., Bradford, M., and Uy, B. (2003). A general method of analysis of composite beams with partial interaction. *Steel and Composite Structures* 3 (3): 169–194.

Roik, K. and Bergmann, R. (1992). Composite columns. In: *Constructional Steel Design – an International Guide* (ed. P.J. Dowling, J.L. Harding, and R. Bjorhovde), 443–469. London: Elsevier.

Shen, M.H. and Chung, K.F. (2017). Structural behaviour of stud shear connections with solid and composite slabs under co-existing shear and tension forces. *Structures* 9, 79–90.

Slutter, G. and Driscoll, G. (1965). Flexural strength of steel-concrete composite beams. *Proceedings of the American Society of Civil Engineers* 91 (ST2): 71–99.

Smith, A.L. and Couchman, G.H. (2010). Strength and ductility of headed stud shear connectors in profiled steel sheeting. *Journal of Constructional Steel Research* 66 (6): 748–754.

Smith, A.L., Hicks, S. and Devine, P. (2009). *Design of Floors for Vibration: A New Approach* (Revised). Publication P175. Ascot: Steel Construction Institute.

Spremic, M. and Markovic, Z. (2016). Recommendations for design of grouped shear studs in composite girders. *Report to Technical Committee 11*, European Convention for Constructional Steelwork, October.

Steel Construction Institute and British Constructional Steelwork Association (2014). *Joints in Steel Construction – Simple Joints to Eurocode 3*. Publication P358. Ascot: Steel Construction Institute.

Tata Steel UK (2016). *ComFlor Manual: Composite Floor Decking Design and Technical Information*. Shotton: Tata Steel.

Vasdravellis, G. and Uy, B. (2014). Shear strength and moment-shear interaction in steel-concrete composite beams. *Journal of Structural Engineering (American Society of Civil Engineers)* 140 (11): 1–11.

Vasdravellis, G., Uy, B, Tan, E.L., and Kirkland, B. (2013). Behaviour and design of composite beams subjected to bending and axial loading. In: *Composite Construction in Steel and Concrete VII* (American Society of Civil Engineers). http://ascelibrary.org/doi/abs/10.1061/9780784479735.018

Wang, Y.C. (2014). *Design of Reinforced, Concrete Filled, Hot Finished Structural Hollow Sections in Fire*. Tata Steel Europe. http://www.steelconstruction.info/Hollow sections in fire#Resources

Wang, Y.C., Burgess, I.W., Wald, F., and Gillie, M. (2012). *Performance Based Fire Engineering of Structures*. Abingdon: Taylor & Francis.

Way, A.G.J., Cosgrove, T.C., and Brettle, M.E. (2007). *Precast Concrete Floors in Steel Framed Buildings*. Publication P351. Ascot: Steel Construction Institute.

Wyatt, T. (1989). *Design Guide on the Vibration of Floors*. Publication P076. Ascot: Steel Construction Institute.

Index

Notes:

- references to 'beams' and to 'columns' are to composite members
- *see also* entries do not include resistance to fire; *see* 'fire…'

a

actions   xvii, 4–5
  accidental   4
  arrangements of   188–9
  characteristic   xvii, 5
  combination of   5–8
    accidental   4, 226
    characteristic   7, 69, 113, 148
    frequent   7, 148
    fundamental   7
    quasi-permanent   7, 148
  concentrated   11, 51, 60–2, 87, 155
  design   xvii, 5
  direct   4, 11–12
  effects of   xvii, 5–7, 131, 188
  favourable   5
  frequent   5
  horizontal   182, 187, 207–9
  impulsive   100, 117
  indirect   12
  leading   6, 186, 209
  permanent   4–5, 11, 51
  quasi-permanent   5
  repeated   33, 94
  representative   5
  variable   4–5, 11–12, 51
  see also loads; wind, effects of
analysis, finite-element   17, 142
analysis, global   8, 75, 97, 129–32

of continuous beams   149–56
  elastic   12, 129, 150–2, 185
  elasto-plastic   183
  first-order   172, 182–6
  of frames   129–31, 172–3, 184–6, 197, 205–6
  non-linear   17–18
  for profiled sheeting   169
  rigid-plastic   8, 75, 129, 149, 154–6, 186
    for composite slabs   169
  second-order   17, 182, 203, 208, 218
  uncracked   150
analysis, partial-interaction   41–3, 247–52
analysis of cross-sections   8–14, 96–7, 135–7
  see also beams, columns, etc.
anchorage, end   29, 58, 67
Annex, National *see* National Annex
axes   xviii

b

beams   1–2, 25, 73–99
  bending resistance of,
    with axial force   85–6
    hogging   133–49
    sagging   73–83, 106, 127
    see also webs, encased
  concrete-encased   2, 25
  continuous   16, 73, 129–68
  cross-sections of
    classification of   74–6, 133, 135, 157, 161
    critical   86–7, 138
    elastic analysis of   8–16, 75, 81–3, 96–8, 136–7

Composite Structures of Steel and Concrete: Beams, Slabs, Columns and Frames for Buildings, Fourth Edition. Roger P. Johnson.
© 2019 John Wiley & Sons Ltd. Published 2019 by John Wiley & Sons Ltd.